Peter Heering, Daniel Osewold (eds.)

Constructing Scientific Understanding through Contextual Teaching

Peter Heering, Daniel Osewold (eds.)

Constructing Scientific Understanding through Contextual Teaching

Frank & Timme
Verlag für wissenschaftliche Literatur

ISBN 978-3-86596-118-1

Herstellung durch das atelier eilenberger, Leipzig.
Printed in Germany.
Gedruckt auf säurefreiem, alterungsbeständigem Papier.

www.frank-timme.de

Contents

Preface

PETER HEERING, DANIEL OSEWOLD
Physics Education/History and Philosophy of Science, Carl-von-Ossietzky Universität, Oldenburg, Germany

Teaching is a question of context itself: as the Oldenburger Research Group Physics Education/History and Philosophy of Science accounts in their research for the historical context of science education, it appears to be consequential to focus on the role of the history of science in the teaching of science. Therefore, the goal of the *6th International Conference for the History of Science in Science Education* hosted in Oldenburg addressed the question of how historical context can promote the development of scientific understanding.

The richness of the 28 papers presented and the high quality of the discussions of the conference encouraged us to publish a volume, in which the majority of these stimulating papers is assembled. One reason why not all papers are to be found in this collection is due to the Editors tide timeline for the final paper submission. Some presenters had other obligations and could not submit their contributions in time. As editors, we considered it important to publish the obtained papers within a very short time as we feel that the topics presented and discussed are of major relevance to the future of our research field. This resulted in publication of nineteen conference papers in this volume.

We greatly appreciate the financial support of the German Science Foundation (Deutsche Forschungsgemeinschaft), the Ministry for Science and Culture of Lower Saxony, and the EWE Foundation (EWE-Stiftung), which made the realisation of this conference possible. We also thank the Carl-von-Ossietzky Universität Oldenburg for supporting the conference. Furthermore, our thanks go to the members of the Research Group Physics Education/ History and Philosophy of Science at the Carl von Ossietzky Universität that took the responsibility to make this conference a success. The publication of the conference proceedings was financially supported by the EWE foundation.

Peter Heering and Daniel Osewold
(The Editors)

Introduction

ROGER H. STUEWER
Program in the History of Science, Technology, and Medicine, University of Minnesota,
Minneapolis, MN 55455, USA

The papers that follow were presented at the Sixth International Conference for the History of Science in Science Education, which was held at the Carl-von-Ossietzky University of Oldenburg, Germany, July 10-15, 2006. They reflect the great diversity, vitality, and high quality of work on the many uses of the history of science in science teaching, to which the Oldenburg group has contributed so significantly for many years. Its leader, Falk Riess, is thus eminently qualified to sketch historically the use of historical experiments in physics teaching as presented below. The remaining papers then make clear that the use of history of science in science teaching has grown into a truly international endeavor: Science teachers from countries around the world are dedicating themselves to improving their courses at all levels of instruction by incorporating the history of science into them. The result is a remarkable diversity of means and approaches for improving the teaching of physics, astronomy, mathematics, and biology, as these papers demonstrate.

In physics, as part of an on-going research project aimed at replicating pioneering experiments to reveal their historical contexts and the difficulties in using them to produce reliable experimental data, Stephen Klassen shows how an understanding of the historical background and nature of Robert A. Millikan's oil-drop experiment can renew its pedagogical significance. Similarly, Arthur Stinner has connected the history of physics to classroom teaching through his large-concept approach as discussed in his new book. Other revealing endeavors are Shu-Chiu Liu's study of how grade-school students in Taiwan and Germany acquired their knowledge of models of the universe and of the nature of heat; Panagiotis Kokkotas and his colleagues' training program for primary-school teachers based upon the socioconstructivist and sociocultural approaches in studying the history of our understanding of freely falling bodies; and Barbara McMillan's fascinating teaching program for a Grade 4 curriculum on light that included historical stories and was aimed at addressing government-mandated learning outcomes in Manitoba.

In more specialized areas, Douglas Allchin challenges our understanding of the meaning of a physical law and the implications of that concept for science teaching. And Elizabeth Cavicchi presents a convincing case for the vital role that trust plays between teacher and undergraduate students, and among the students themselves, in carrying out historical experiments on light, mechanics, and electricity and in establishing a community of learners.

Then there is the crucial role of instruments. Gudrun Wolfschmidt shows how understanding the development and use of three instruments in magnetism – compasses, magnetometers, and radiotelescopes and satellites – provides an important context for teaching geophysics and astronomy. Pierre Lauginie emphasizes the pedagogical value of studying the attempts to determine the density of the Earth in the 18th century, the invention of the Cavendish torsion balance at the end of the 18th century, and its subsequent refinements, which yielded ever more precise values of the gravitational constant *G*. And Christian Sichau argues convincingly that the scientific instruments and apparatuses preserved in museums can be used as object stories to provide significant insights into the history and nature of science, and thus constitute the core strength of museums for science teaching.

Turning to astronomy, Jürgen Teichmann argues that three revolutions or evolutions in astronomy, in antiquity, in the 16th and 17th centuries, and in the 19th and 20th centuries, all of which were associated with much improved observational instruments, were closely connected to such fundamental philosophical questions as the uniqueness of life on Earth and man's place in the universe. And Don Metz presents a fascinating account of the enormous hardships that William Wales experienced in observing the transit of Venus of 1769 in northern Canada, which contributed to establishing the value of the astronomical unit.

In mathematics, Youjun Wang gives a captivating account of how ancient Chinese mathematicians solved mathematical problems involving the principle of congruency and the theory of limits by hands-on techniques. And Jeff Babb and James Currie sketch the history of the brachistochrone and its relationship to Snell's law, Fermat's principle of least time, and the calculus of variations, and show how this history motivated their students' understanding of the interplay between mathematics and physics.

In biology, David Rudge discusses a National Science Foundation program for middle school teachers that uses three episodes in the history of biology to promote their deeper understanding of the nature of science.

Finally, there are the generalists. Gábor Zemplén describes a provocative international baccalaureate theory-of-knowledge course based upon sociological and anthropological considerations that is aimed at showing students how scientists attempt to uncover sources of error and acquire reliable knowledge about the natural world. And Peter Heering argues that an examination of the great changes that occurred in the style of experimentation since the 18th-century Enlightenment can inform our science teaching today.

There is no Royal Road to effective science teaching, but the papers that follow provide a firm basis for confidence that the use of the history of science in science teaching, in one form or another, can contribute significantly to it on an international scale.

Teaching Science Lawlessly

DOUGLAS ALLCHIN
University of Minnesota, USA

Abstract. Boyle's law is the epitome of lawlike science in the classroom. Yet Boyle's law is not universal and invariant, as implied by the term 'law'. It does not hold at high pressures, low temperatures, or low volumes. It is based on constant temperature. It varies for different gases. Boyle's law—and other scientific "laws"—are not lawlike at all. An alternative is to teach a more complex view of nature, exhibiting only patches of regularity. In following scientists, teachers may focus on experiments as concrete models, such as Boyle's J-tube. Science education might well be ideally rooted in concrete apparatus and teaching experimental and analogical reasoning, not laws.

Introduction

Presented with a title like "Teaching Science Lawlessly," you may be envisioning science red in tooth and claw, chaotic nature careening out of control, or gangs of science teachers roaming through school hallways wielding huge Leyden jars ready to unleash electrical havoc on unruly students. That's not a bad image to start.

Let me couple that [Figure 1] with an appreciation: namely, for the opportunity to participate in this conference, where I might join in recognizing the achievements of the community here in Oldenburg under the Falk Riess' vision and leadership. I hope to elucidate how an image of lawlessness might serve this end: ultimately, as a tribute to the pioneering work here.

Towards that end, my central focus, and the occasion for analysis and commentary, is one of the most standard of benchmark concepts in introductory chemistry or physical science: Boyle's law. I focus on this law primarily because it is hard to find a science curriculum without it having a central place. It is central, I think, not because it informs students about how gases behave. Rather, it is a cultural icon that epitomizes the fundamental concept of a law. My aim is to profile an alternative that may better reflect science in real practice.

Most of you are no doubt familiar with the historical context of Boyle and his law. Boyle was interested in the "spring of air." Air exerts pressure when compressed. Borrowing from a device of Otto Guericke, Boyle enlisted Robert Hooke to build an "air pump". They did not generated a true vacuum, but the vessel did exhibit extremely low pressure. In 1660 Boyle published his findings on its many effects — on magnets, sound transmission, sealed bladders, burning candles, the life of small animals, and more. In response to criticism, Boyle further demonstrated the strength of the spring of air. He and Hooke compressed a small amount of air trapped in the end of a glass J-tube with increasing amounts of mercury. In a similar second set of experiments, they used the new air-pump to draw up the column of mercury to dilate the trapped air. In both cases, they recorded the volume of the trapped air and the corresponding height of the column of mercury (an indirect measure of its weight, or pressure). Students even today can graph his figures (Figures 2a and 2b) and see the inverse relationship of pressure and volume (Conant 1957).

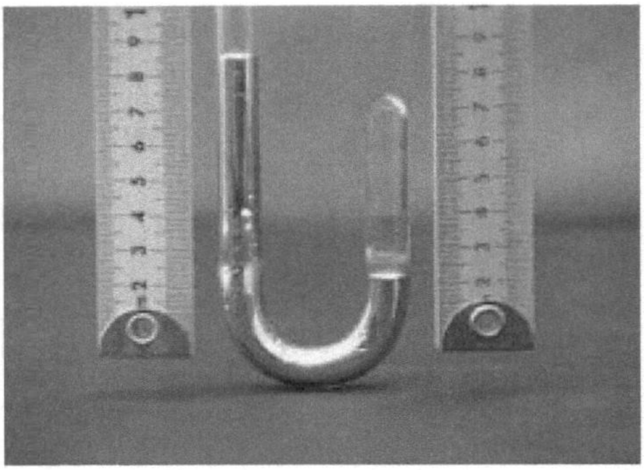

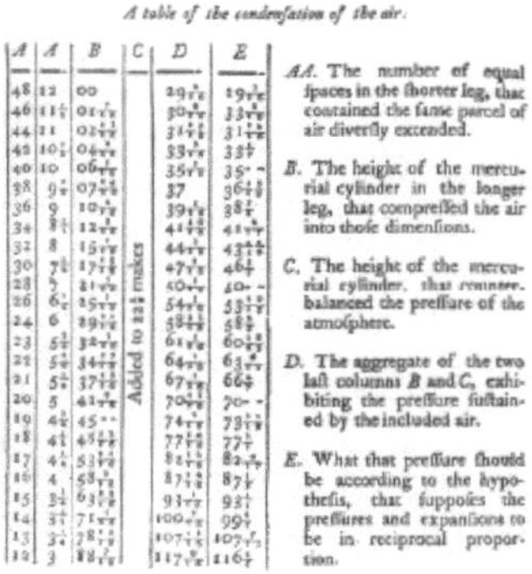

A table of the condensation of the air.

A	A	B	C	D	E
48	12	00	Added to 29⅛ makes	29 [illegible]	29 [illegible]
46	11½	01 [illegible]		30 [illegible]	33 [illegible]
44	11	02 [illegible]		31 [illegible]	31 [illegible]
42	10½	04 [illegible]		33 [illegible]	33 [illegible]
40	10	06 [illegible]		35 [illegible]	35
38	9½	07 [illegible]		37	36 [illegible]
36	9	10 [illegible]		39 [illegible]	38 [illegible]
34	8½	12 [illegible]		41 [illegible]	41 [illegible]
32	8	15 [illegible]		44 [illegible]	43 [illegible]
30	7½	17 [illegible]		47 [illegible]	46 [illegible]
28	7	21 [illegible]		50 [illegible]	50
26	6½	25 [illegible]		54 [illegible]	53 [illegible]
24	6	29 [illegible]		58 [illegible]	58 [illegible]
23	5¾	32 [illegible]		61 [illegible]	60 [illegible]
22	5½	34 [illegible]		64 [illegible]	63 [illegible]
21	5¼	37 [illegible]		67 [illegible]	66 [illegible]
20	5	41 [illegible]		70 [illegible]	70
19	4¾	45		74 [illegible]	73 [illegible]
18	4½	48 [illegible]		77 [illegible]	77 [illegible]
17	4¼	53 [illegible]		82 [illegible]	82 [illegible]
16	4	58 [illegible]		87 [illegible]	87 [illegible]
15	3¾	63 [illegible]		93 [illegible]	93 [illegible]
14	3½	71 [illegible]		100 [illegible]	99 [illegible]
13	3¼	78 [illegible]		107 [illegible]	107 [illegible]
12	3	88 [illegible]		117 [illegible]	116 [illegible]

AA. The number of equal ſpaces in the ſhorter leg, that contained the ſame parcel of air diverſly extended.

B. The height of the mercurial cylinder in the longer leg, that compreſſed the air into thoſe dimenſions.

C. The height of the mercurial cylinder, that counterbalanced the preſſure of the atmoſphere.

D. The aggregate of the two laſt columns *B* and *C*, exhibiting the preſſure ſuſtained by the included air.

E. What that preſſure ſhould be according to the hypotheſis, that ſuppoſes the preſſures and expanſions to be in reciprocal proportion.

Figure 2a. Boyle's data on the condensation of air, showing that volume decreases as pressure from a column of mercury increases.

There are many ways to express Boyle's law. My analysis will be based on a narrow, but explicit version:

$$\Delta P \propto 1/\Delta V$$

The change in pressure is inversely proportional to the change in volume. (As the volume increases, the pressure decreases, and vice versa.)

What's Not in Boyle's Law

The concept of scientific laws, such as Boyle's law, has a venerable tradition. Laws are empirically substantiated generalizations or regularities. For many, they are the basic units of scientific knowledge. Laws reflect a conception of nature as law-like or machine-like, as expressed in the mechanical philosophy, famously advocated by Boyle (Sargent 1995). Nature may thus be described by reducing phenomena to their parts and the laws that govern their interaction. Elucidating these laws is widely portrayed as the major goal of science. Familiar examples of laws might also include Snell's law of refraction, Galileo's law of the pendulum, Newton's laws of motion, Ohm's law of electrical resistance, and Mendel's law of independent assortment. As typically presented and interpreted, scientific laws are invariant and universal (Hempel 1966, 54, 58; Ziman 1978, 32; Kosso 1992, 52-60, 190; Woodward 2003, pp. 167, 236- 238, 265-266). Indeed, their very universality and invariance accounts for their value and authority as generalizations. Laws accordingly seem to earn central place in science education.

As Boyle's data show, Boyle's law certainly holds true in Boyle's J-tube. But is Boyle's law universal? Is it invariant? Is it truly lawlike? Does it hold in all cases? The answer is "no." It depends on context. *At high pressures*, the direct inverse relationship of pressure and volume breaks down (Figure 3). Henri Regnault (in 1852) and later Louis Cailletet noted this variation two centuries after Boyle's work. In modern terms, we might say that there is a limit to compression. Under very high pressures, a gas begins to behave more like a liquid than a gas. The scope of Boyle's law is limited. It only holds in the *restricted* domain of pressures up to approximately ten atmospheres. Never mind that these pressures are infrequently encountered in daily human life: the law is plainly not universal.

Perhaps the behavior of gases at high pressures is a rare, minor exception. Suppose one stipulates explicitly, then, that only certain pressures apply:

$$\forall P \not\gg 0 \Rightarrow \Delta P \propto 1/\Delta V$$

For all pressures not substantially greater than zero, the change in pressure is inversely proportional to the change in volume.

Figure 3. Cailletet's apparatus for examining gas under high pressure.

Is Boyle's law universal now? Well, no. At low temperatures, the inverse relationship breaks down, as well (Figure 4). Historically, this was noted by Thomas Andrews in the 1860s (Andrews 1869). Here, another context qualifies scope. But this case also introduces something unexpected: a new variable. The equation or formula does not even refer to temperature. An experimenter, for example, would need to be aware of temperature, if only to ensure that the temperature was not extremely low. The simple expression of Boyle's law hides this relevant variable entirely.

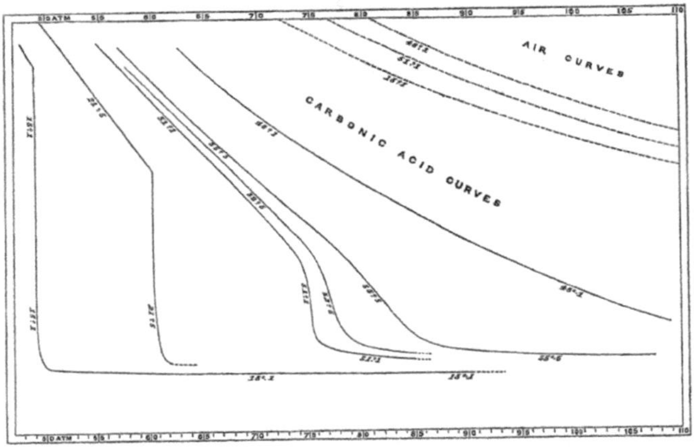

Figure 4. Effect of lower temperatures (successive curves to left) on pressure-volume relationship of carbon dioxide (Andrews 1869).

Well, then, let us add both these conditions, or provisos:

$$\forall T >> 0 \,\&\, \forall P \not>\not> 0 \Rightarrow \Delta P \propto 1/\Delta V$$

For all temperatures well above zero, and for all pressures not too high, the change in pressure is inversely proportional to the change in volume.

Is Boyle's law "fixed" now? For example, with the provisos of scope, is the law now invariant? Well, no. Temperature is indeed important. — Not just the range of the temperature, but also its holding constant. As Boyle himself (and others of his era) noted, gas pressure is sensitive to changes in temperature, as exemplified in several apparatus for measuring temperature (Figure 5).

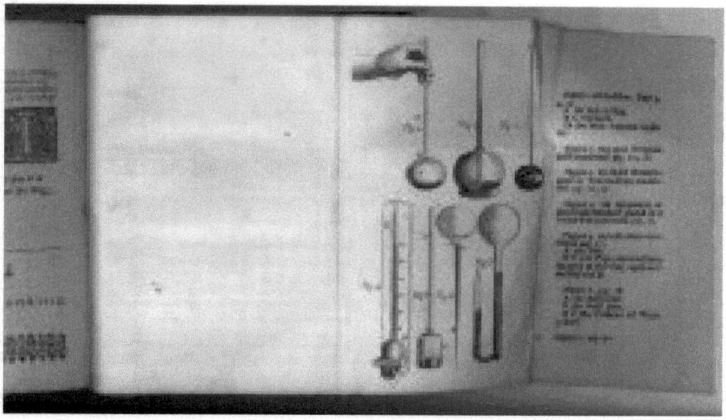

Figure 5. Instruments for indicating temperature change via changes in air volume.

A few decades later, of course, Jacques Charles (in 1787) and Joseph-Louis Gay Lussac (1802) formalized this in yet another law, which now bears (alternately) their names. Hence, while one may address the relationship of pressure and volume independently of temperature, temperature is nonetheless relevant. Constant temperature is a boundary condition. Sometimes it is expressed as a ceteris paribus clause: "all else being equal." But not all other things need to be equal. The amount of illumination, the relative motion of the system, or gravity exerted on it, have no effect on gas behaviour, so far as we know. No one need stipultate these as boundary conditions or imagine them in a *ceteris paribus* assumption. So specifying what must remain constant is important indeed if one expects the law to hold. The invariance of Boyle's law, ironically, depends on context.

Well, let us now add our additional boundary condition:

$$\forall T >> 0 \,\&\, \forall P \not\gg 0 \,\&\, \forall \Delta T = 0 \Rightarrow \Delta P \propto 1/\Delta V$$

For all temperatures well above zero, and for all pressures not too high, and at constant temperatures, the change in pressure is inversely proportional to the change in volume.

Is Boyle's law expressed fully now? No, still not yet(!). If the volume is very low (or the density very high), the volume of the gas molecules relative to the volume of the space between them becomes significant. Intermolecular interactions (London forces) become relevant. The behaviour of the gas changes noticeably. Moreover, because the size of the gas molecule matters, the variation is specific for each gas. Yet another factor qualifies Boyle's law. Even the qualifications are not regular. Moreover, any gas with strong polarities may also exhibit intermolecular forces, although of another sort. Such gases—carbon dioxide, for one—also vary from the simplified "ideal" (Figure 6). Ultimately, then, we must also take into account the nature or identity of the gas (or gases). Johannes van der Waals investigated these various dimensions of gas behavior in the 1870s and 80s, as well as the nature of some of the intermolecular forces, work that was recognized by a Nobel Prize in 1910. One can "correct" for the subtleties of molecular size and interactions, but the corrections (known as van der Waals constants) differ for each gas (Figure 7).

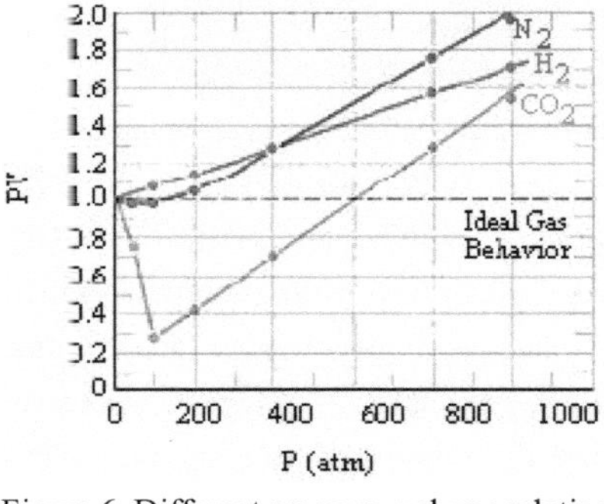

Figure 6. Different pressure-volume relationships for different gases.

Molecule	a (L^2-atm/mol^2)	b (L/mol)
He	0.03412	0.02370
Ne	0.2107	0.01709
H_2	0.2444	0.02661
Ar	1.345	0.03219
O_2	1.360	0.03803
N_2	1.390	0.03913
CO	1.485	0.03985
CH_4	2.253	0.04278
CO_2	3.592	0.04267
NH_3	4.170	0.03707

Figure 7. van der Waals contants for various Gases.

Thus, even his now well known generalized form of the gas law included variables specific for each gas:

$$\left(P + \frac{a}{V}\right)(V - b) = nRT$$

To be accurate, or realistic, Boyle's law must sacrifice universality. Boyle's law, like so many others, is only "lawlike" when qualified:

$\forall$ elastically compressible gases,

$\forall T >> 0 \,\&\, \forall P \not\gg 0 \,\&\, \forall V >> 0 \,\&\, \forall \Delta T = 0 \Rightarrow \Delta P \propto 1/\Delta V$

For all elastically compressible gases, for all temperatures well above zero, and for all pressures not too high, and for all volumes well above zero, and at constant temperatures, the change in pressure is inversely proportional to the change in volume.

A critical analysis of Boyle's law exposes and highlights this double irony: universality comes at the cost of limited scope; invariance, only with conditions. The lawlike world of "for all..." is inseparably coupled with a set of contingent "if-and-only-if"s. This is the lesson typically missing when teaching Boyle's law: what's *not* in Boyle's law — a world full of context and contingency, not simple rules.

Another irony lurks in Boyle's law. That is, Boyle himself never stated it. He certainly produced data commensurate with the law, yet he did not write an equation for the relationship. He surely understood the mathematics of "what the pressure should be according to the hypothesis" (Boyle 1662; Figure 2), yet he was generally disinclined to characterize nature mathematically (Shapin 1996, pp. 333-338).

Figure 8: Boyle

Boyle also embraced the concept of laws in science: he advocated mechanical philosophy as well as the theological framework of construing nature in terms of inherently divine laws (Boyle 1661). Yet Boyle did not regard his findings about the condensation and rarefaction of air as a universal law in the sense now accepted. Rather, he referred to the spring of air more modestly, as a 'habit of nature' or 'custom of nature': local, perhaps contingent, in nature (Boyle 1661; Sargent 1995; Shapin 1996, pp. 328-330, 338-350). He was well aware, for example, that temperature and "atmospheric tides" affected the Torricelli tube, today's mercury barometer (Boyle 1660, pp. 65, 123, 133; 1682, p. 50). Boyle's law is not strictly Boyle's. Teachers might thus profitably reflect on Boyle's own modest posture.

Rampant Lawlessness

One may well be tempted to imagine that there is no need for such caution, that Boyle's law is an exception and that most scientific laws are universal and invariant as promised. Not so. Consider Galileo's "law" of the period of the pendulum:

$$t = 2\pi\sqrt{\frac{l}{g}}$$

Figure 9: Pendulum

(where t= the time of the period, g=9.8 m/sec2; l is the length). This formula only works (and with limited precision, at that) for pendulums with small angles of swing (generally reported as less than 10°). The familiar equation is an approximation based on the broader characterization:

$$\frac{d\ x}{dt} + \frac{g}{l}\sin(x) = 0$$

(where x is the angle of swing). This may be universal, but it is rarely used. It cannot be solved analytically and requires iterative substitution even to approximate solutions. But even this expression assumes that the mass is concentrated on a single point and that no friction affects either the fulcrum or the interaction of the pendulum and its medium. These are not just boundary conditions, but unrealizable idealizations, as all the physicists here well know.

Ohm's Law of electrical resistance, too, has numerous exceptions: for example, at high current densities. Many common materials "violate" Ohm's Law: temperature-sensitive resistors (such as filaments in incandescent light bulbs, or sensors in digital thermostats); air (whose threshold resistance results in bolts of lightning); diodes (common electronics components), light-sensitive resistors, piezoelectrics (used in touch-sensitive switches), weak electrolyte solutions, varistors and high-vacuum electron tubes, as well as other more technical variants.

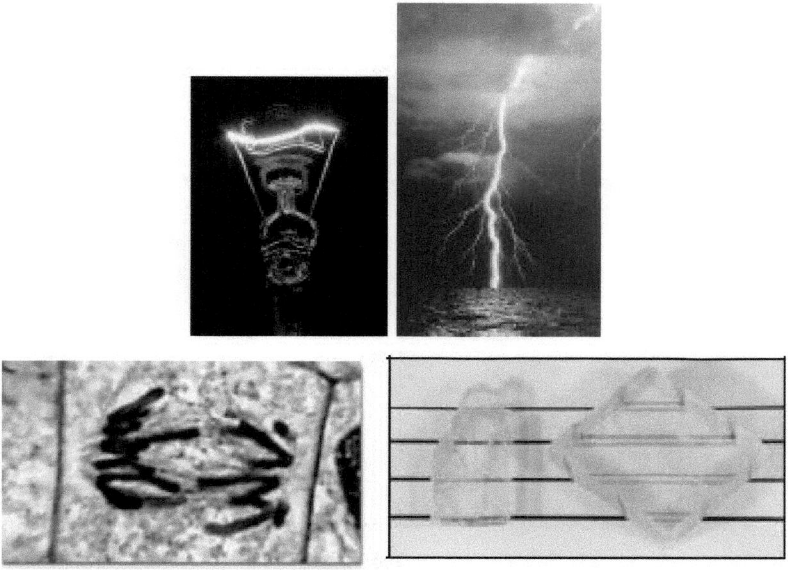

Figure 10: Light bulb, Lightning, chromosome and calcite

Newton's Laws of Motion do not apply at relativistic velocities (approaching the speed of light), and (like the pendulum law) exclude friction. Mendel's Law of Independent Assortment breaks down, as most students learn, for genes linked on the same chromosome. Snell's Law of Refraction does not apply for 'Icelandic spar', or calcite. In all these cases, laws seem very unlawlike. Scope circumscribes universality. Boundary conditions and exceptions limit invariance. That is, context matters.

One need not contend that these or other scientific laws are completely invalid. Nor need one discount their informativeness. The emphasis, instead, is on context. Few professional scientists or philosophy of science—nor likely anyone here—deny that laws are ultimately contextual. For example, Toulmin (1960, pp. 31, 63, 78-79, 87) underscores that laws have particular scope of application, or *domains*. He, along with Kuhn (1970), suggest that articulating this scope is a major function of scientific research. Still, Toulmin and others contend, the provisos are not part of the law. Of course, that is a philosophical sleight of hand. Only by erasing these conditions does a law acquire the illusion of universality. Laws, if not *false*, are illusory. They attain law-like status only by arbitrary specification of background conditions—a control which may be locally informative, but not universally valid.

Context depends on perspective. Interpretations of foreground and background rely on convention (Figure 11). So one may ask: why are laws central, and the exceptions "exceptions"? Why regard "laws" as primary? I am inviting a gestalt switch whereby cases now framed as "outside the scope" or "not within the boundary conditions" are framed instead as inside the system.

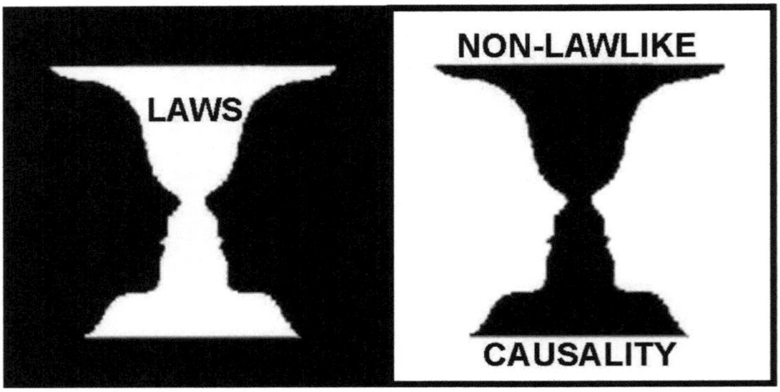

Figure 11. Gestalt

One denies laws privilege. Accordingly, the unlawlike earns parity. Boyle's law becomes a special case: no less true, but narrowly so, *contingent* on a particular context. It demarcates a patch of local regularity. The alternative perspective embraces the whole, of which Boyle's law is only a part. An unrepentantly staunch realist does not present an *idealized* gas law as even a simplification of all *real* cases. Laws are the *exception*, not the rule. They require special explanation, just as they need special experimental conditions to be observed in a lab. Laws are local, not universal. Generalities are contingent, not invariant.

Accordingly, one may be impressed by an apparatus where Boyle's law is observable, such as Boyle's J-tube. A change in volume is observed only when the system is closed. In an open system, such as the atmosphere, a gas under pressure does not change volume. It moves: hence, wind (Figure 12)! Indeed, the cases where Boyle's law holds are few: pistons (say, in internal combustion engines), industrial boilers, some ventilation systems.

Figure 12. Pressure in an open system creates wind, not a change in volume.

Boyle's law is not "basic" at all. It is an esoteric scientific fact. Few students will ever need to know or apply Boyle's law in either their personal or professional lives. Yet it is *de rigeur* for science education. The ultimate lesson of Boyle's law is not about how gases behave: it is about laws and characterizing nature as essentially lawlike — and that lesson is grossly misleading. We live in a world where weather happens. We need to teach equally about non-lawlike, or lawless, nature.

Lawless Causality

To accommodate lawless science, one needs to readjust the basic framework of causal thinking. But how can one think causally *without* a law? (Is it even possible?!) The mere image of lawless science, of course, bristles with potentially disruptive images of anarchy. It reeks of chaos and disorder. But the effect is purely linguistic. Consider the contradiction, for example, in claiming that a natural phenomenon violates or deviates from a law of nature. The *language* links law and order in the civil realm. We need to ensure that our thinking about science is not biased by meanings in cultural contexts. A lawless nature may seem to imply indeterminism. However, science without laws need not be any less deterministic.

Lawless science fits with an alternative conception of causality. Conventionally, laws reduce the world into component parts and causes. Laws are the basic units. Laws are like billiard balls (Figure 13). Each moving billiard ball expresses a mechanical vector: a cause. One ball hits another with an observable effect. The second ball may, in turn, hit another, and then perhaps another, leading to a causal cascade. Laws lead to thinking in terms of causal chains. Even where a cue ball hits a whole rack of fifteen balls, with scattering effects, all calculable (in principle) from the cue ball's original vector.

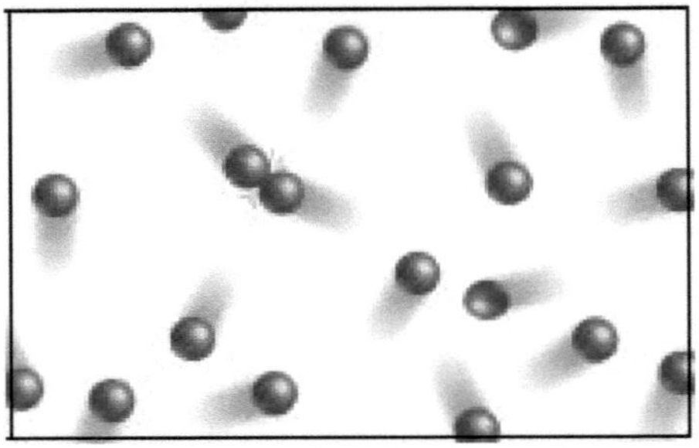

Figure 13. The kinetic theory of gases: molecules as metaphorical billiard balls.

The complex event is explained by *compounding* individual causal events, layering one cause upon another. This reductionistic view is nicely exemplified, of course, by the kinetic theory of gases, the modern explanation for Boyle's law. Gas molecules act like billiard balls colliding in three-dimensional space. The pressure is a collective effect of all the movements of the individual gas molecules, each with their own discrete contributing cause.

The alternative conception of causality is more holistic. It underscores multiple simultaneous causal factors. It resonates with a *state system* view of causality. That is, effects are not attributed to individual causes, or even to a composite of overlapping causes. Rather, effects are due to the state of the entire system. One identifies the whole as "the" cause, differentiating—but not privileging—individual elements. Pressure does not increase *merely* due to a decrease in volume, as suggested by Boyle's law. Rather, pressure increases due to the change in volume *and* extant pressure *and* temperature *and* overall volume *and* type of gas — *even when* the temperature does not change. The behavior of gases is always, not just sometimes, multivariate. Just so for pendulums, electrical resistance, inheritance, etc.

For further perspective on framing causality, consider the playful devices of Rube Goldberg (Figure 14). He delighted viewers by imagining how a modest action, such as emptying an ice cube tray or cooling hot soup, resulted from an elaborate and improbable series of events.

Figure 14. A Rube Goldberg device for making fresh orange juice. (Wolfe 1998)
Rube Goldberg is the ® and © of Rube Goldberg, Inc., www.RubeGoldberg.com

The humor arises chiefly from the convoluted causal pathway, underscoring the (otherwise humorless) association of simple causes with simple effects. Goldberg's elaborate mechanisms also highlight causal lineages, often labeled in an alphabetical sequence. At the same time, his devices are funny precisely by being incredibly contingent. If a rolling ball is not preset, a falling hammer not lifted, a spring not prewound, the causal cascade is interrupted. The design elicits a fun anticipation of several successive "only-if"s. The causal scenario is exceptionally fragile. Action depends on the whole structure at once. The arrangement and setup— the context—is critical. The causal lineage is merely an artifact of the causal structure—primed throughout, but not triggered. Goldberg's baroque and apparently silly devices, then, offer sophisticated commentary on reductionistic causal perspectives. Their whimsy helps celebrate a contingent view of causality.

Consider again the reductionist's billiard balls. Here, the status of the whole billiard table is important (Figure 15a). One cannot assume that balls are intially at rest: the system may already be complex and active, not passive in response. Nor can one assume the balls to be of the same material (like different gases perhaps?). They need not react in uniform ways: identical vectors underdetermine a collision outcome.

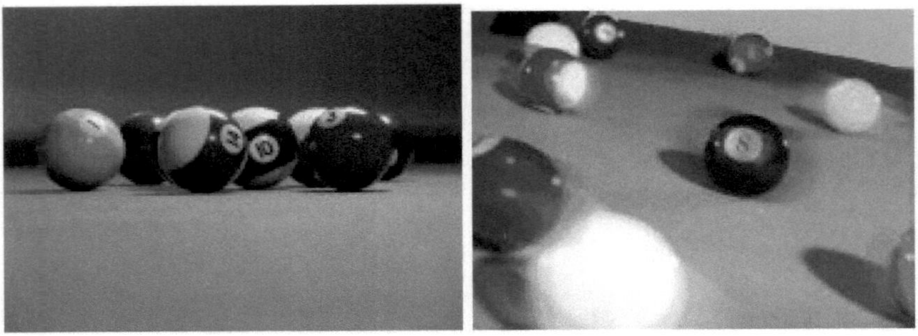

Figure 15. Billiard-ball causality, *not* assuming the table is flat or uniform.

The billiard table environment also matters: it may not be level, or remain level (anymore than the temperature of a gas). A tilted table affects all the components at once, even when the balls are not colliding (Figure 12b). Some causal factors may undulate the surface, generating different local conditions across the table, leading again to different responses for apparently similar collisions. Some balls may contain iron so changes in the magnetic field (context) may matter more than the interactions of individual balls (parts). With uniform balls on a level billiard table—which stays level—simple interactions may indeed be discernible. But that is a special case. The state system view looks well beyond the billiard balls—or individual gas molecules—as component causes. It emphasizes the whole and the contexts of the parts.

Let me note, at this juncture, that this presentation is derived and adapted from a larger project of mine, still in progress, on "The Gender of Boyle's Law." One could probe the history of laws in science and show that the approach is based on theology and a world view that values technology, control and mechanical philosophy (Steinle 2002). Or one could delve into the growing literature on models and mechanisms (Glennan 1996; Machamer, Darden and Craver 2000; Bechtel 2006), or on lawless nature (Cartwright 1999). Today, however, I want to profile further the concrete alternatives of lawless causality.

From Laws to Material Models

So: strip the lawlike status from Boyle's law and it may seem as if nothing remains. What remains, of course, is Boyle's J-tube (Figure 16) as an exemplar and concrete model, Kuhn's (1970) core sense of paradigm (see also Giere 1995, 1999).

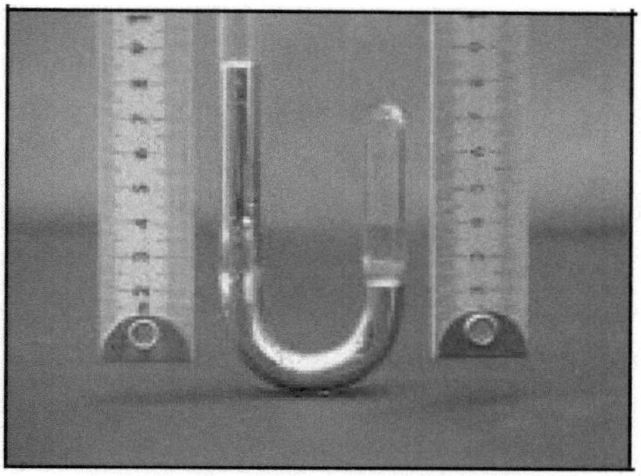

Figure 16. Boyle's J-tube (recreated).

A focus on models and particularity underscores the importance of material culture and experimental systems in investigation, as profiled recently by Hans-Jörg Rheinberger (1996). Here, one does not appeal to general laws. Rather, one compares other cases to Boyle's original. To the extent that the conditions are the same — including pressure, temperature, etc. — one expects to find similar behavior. The reasoning is primarily analogical, not deductive. It is also direct, not indirect. One does not detour through a general law, first by abstracting Boyle's J-tube, then by deciding whether a new case is an authentic example of the general law. Reasoning from case to case does require, however, clear attention to the variables and details of context. One quickly learns the basis and limits of similarity, an integral part of the exemplar, or paradigm. The lawless alternative to Boyle's law is Boyle's J-tube. It is eminently teachable.

Although lawless science challenges lawlike thinking, the legal system (on a cultural level) provides an instructive metaphor. In practicing law, especially in juridicial contexts, one distinguishes between statute (or code) law and common law. Statute law is based on rules, or codes: like the laws of nature. One assesses actions in reference to the law's general and explicit statements. One reasons deductively. Common law, by contrast, is case-based. One assesses actions based on precedent, or similar cases encountered in the past. Interpretation emphasizes the basis for similarity. One reasons chiefly analogically. Of course, numerous variables, or bases for similarity, are usually possible. The effectiveness of an analogy may be highly contingent. Context plays a major role. Yet the multitude of benchmarks can be beneficial, especially in interpreting complex cases. Under statute law, statutes may overlap and indicate conflicting interpretations. Case-based reasoning can often resolve this. Both

frameworks provide viable systems of law and interpreting justice. Imagine lawlike science resonating with statute law, "lawless" science with common law.
Accordingly, Boyle's J-tube is significant primarily as a precedent. To apply Boyle's law, for example, one depends on being able to match the particular conditions of Boyle's J-tube. It is no surprise, perhaps, that scientists who subsequently revised Boyle's law typically followed Boyle's model by using mercury in J-tubes. For example, with the construction of the Eiffel Tower (Figure 17), Louis-Paul Cailletet was able to build and support a column of mercury of unprecedented height. With it, he could examine the same system, but now with pressures up to a hundred atmospheres.

Figure 17. Eiffel Tower in 1889, where Louis Cailletet built a tall, high-pressure manometer.

He studied the limits to Boyle's law by probing variations in the original set-up. Indeed, scientists often focus primarily on *experimental systems,* rather than on specific concepts (Rheinberger 1996). The lineage of investigation is often shaped by the materials and opportunities at hand, not by a theoretical master-plan. While Boyle's J-tube may be local and contextual, it is no less valuable on that account. Indeed, its particularity functions to guide reasoning through analogy—and to help keep it in check. Historically, Boyle's J-tube —not the law— may be the more fundamental benchmark, and that may serve as the critical clue for what one teaches about the nature of science.
How, indeed, does one adapt teaching to accommodate the wider view of causality, to open student awareness of the context of laws and of lawless nature? The ultimate

import of my comments, here and at this time (under these *particular* circumstances), may finally be clear. If one were to assemble a science education curriculum founded on the importance of material systems and experimental exemplars, one could hardly do better than follow the model set by this institution. Here is a paradigmatic environment for teaching science without laws. In teaching with historical perspectives and replicas of original apparatus (Figure 15), the role of precedent in science is made clear.

Students equally learn the role of concrete experimental models. While there are other aims at work here, the instructional framework supports appreciation of scientific reasoning and practice beyond laws.

Let me highlight (perhaps celebrate) just two examples of what one may learn from this approach. Each underscores the importance of understanding the materiality, not just the concepts of experimental practice. Recreating the work of Joule on the mechanical equivalent of heat (Figure 18) would seem simple enough: after all, it's been done before.

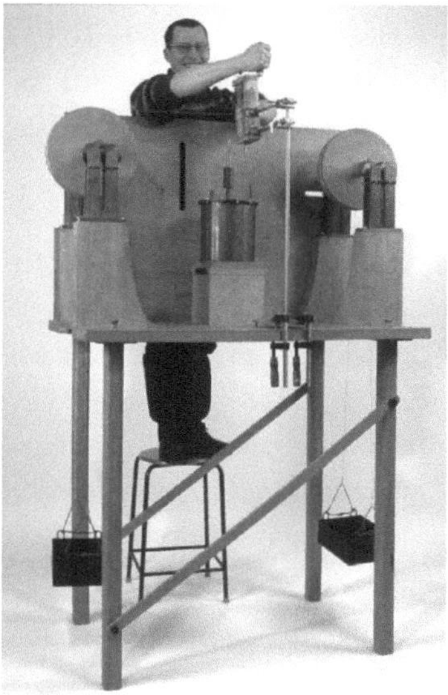

Figure 18. Recreation of Joule's apparatus for investigating the mechanical equivalent of heat.

However, efforts here led to important discoveries. One factor they traced was particular dimensions of the paddle that rotated through the water chamber of the calorimeter (Figure 19), an apparently modest variable of unsuspected significance (Heering 1992).

Figure 19. Water chamber and paddle in Joule's experiment.

Joule's findings were context-dependent indeed. In other work, they tried to replicate Coulomb's work on electrostatic attraction. Work with the apparatus was so delicate that they ultimately questioned whether Coulomb could have produced the results he reported (Heering 1991). Both cases echo the lesson that the materiality of experiment is more complex than the simple equations and laws that educators often present as the foundation of science.

One need not assume that the Oldenburg group aims explicitly or even intentionally to foster understanding of lawless science. Indeed, as these two examples illustrate, a major goal seems to be modeling how the classic laws developed historically. The laws are still prominent landmarks of science. But this is not inconsistent with Boyle's modest posture, or an appreciation of the limited context of laws. I want to highlight how the Oldenburg model implicitly guides students through experimental thinking and thus promotes the type of understanding I am advocating. I am hoping that others may learn from the experience here and, with instruments in hand, fruitfully pursue the prospects of teaching science lawlessly.

References

Andrews, T. 1869. "On the continuity of the gaseous and liquid states of matter." *Proceedings of the Royal Society* 18:42-45.

Bechtel, W. 2006. *Discovering Cell Mechanisms*. New York: Cambridge University Press.

Boyle, R. 1660. *New Experiments Physico-Mechanical, Touching the Spring of Air*. London.

Boyle, R. 1661. *Certain Physiological Essays*. London.

Boyle, R. 1662. *New Experiments Physico-Mechanical, Touching the Spring of Air*, 2nd edition. London.

Boyle, R. 1682. *A Defence of the Doctrine Touching the Spring and Weight of the Air*. London: Miles Flesher.

Cartwright, N. 1999. *The Dappled World*. Cambridge, UK: Cambridge University Press.

Conant, J.B. 1957. *Case Histories in Experimental Science*. Cambridge, MA: Harvard University Press.

Gay-Lussac, J.L. 1802. "The Expansion of Gases by Heat." *Annales de Chimie* 43:137-.

Giere, R.N. 1995. "The skeptical perspective: Science without laws of nature." In *Laws of Nature: Essays on the Philosophical, Scientific and Historical Dimensions*, F. Weinert, ed., De Gruyter (Berlin), pp. 120-138.

Giere, R.N. 1999. *Science Without Laws*. Chicago, IL: University of Chicago Press.

Glennan, S. 1996. "Mechanisms and the nature of causation." *Erkenntnis* 44:50-71.

Heering, P. 1991. "On Coulomb's inverse square law." *American Journal of Physics* 60:988-996.

Heering, P. 1992. "On J.P. Joule's determination of the mechanical equivalent of heat." In *The History and Philosophy of Science in Science Education*, S. Hills (ed.), Queens University (Kingston, ON), 1:495-505.

Hempel, C. 1966. *Philosophy of Natural Science*. Englewood Cliffs, NJ: Prentice-Hall.

Kosso, P. 1992. *Reading the Book of Nature*.

Kuhn, T.S. 1970. *The Structure of Scientific Revolutions*, 2d. ed. Chicago, IL: University of Chicago Press.

Machamer, P., L. Darden and C. Craver. 2000. "Thinking about mechanisms." *Philosophy of Science* 67:1-25.

Regnault, H. V. 1847. "Relations des expériences entreprises par ordre de Monsieur le Ministre des Travaux Publics, et sur la proposition de la Commission Centrale des Machines a Vapeur, pour déterminer les principales lois et les données numériques qui entrent dans le calcul des machines a vapeur." *Mémoires de l'Académie Royale des Sciences de l'Institut de France* 21:1-748..

Rheinberger, H.J. 1996. *Toward a History of Epistemic Things*. Stanford, CA: Stanford University Press.

Sargent, R. 1995. *The Diffident Naturalist*. Chicago, IL: University of Chicago Press.

Shapin, S. 1996. *A Social History of Truth*. Chicago, IL: University of Chicago Press.

Steinle, F. 2002. "Negotiating experiment, reason and theology: The concept of laws of nature in the early Royal Society." In *Ideals and Cultures of Knowledge in Early Modern Europe*, H. Detel and C. Zittel, eds., Akademie Verlag (Berlin), pp. 197-212.

Toulmin, Stephen. 1960. *The Philosophy of Science*. New York: Harper and Row.

Waals, J. van der. 1910. "The equation of state for gases and liquids." Nobel Lectures.

Woodward, J. 2003. *Making Things Happen*. Oxford: Oxford University Press.

Ziman, J. 1978. *Reliable Knowledge*. Cambridge, UK: Cambridge University Press.

The Brachistochrone and Related Curves: Implications for Teaching the History of Calculus

JEFF BABB; JAMES CURRIE

Department of Mathematics and Statistics, University of Winnipeg, Canada

Abstract: In 1696 Johann Bernoulli obtained the brachistochrone, the curve of quickest descent, and posed it as a problem to the mathematicians of Europe. This paper examines the fascinating history of the brachistochrone and explores its relation to Snell's law of refraction, Fermat's principle of least time and the calculus of variations. A computerized demonstration for comparing various selected curves of descent is discussed. Pedagogical aspects of the brachistochrone problem are considered in the context of an undergraduate course in the history of calculus.

Introduction

Each year the University of Winnipeg offer several sections of an undergraduate mathematics course entitled MATH-32.2901/3 History of Calculus (Babb, 2005). The course examines the main ideas of calculus and surveys the historical development of these ideas and related concepts from ancient to modern times. The student audience is typically quite mixed in terms of mathematical aptitude and interest. In teaching the history of calculus (H of C), an effort is made to ensure that students obtain an appreciation for mathematics in an historical context and also learn about several aspects of mathematics and calculus, including proof concepts. The problem of determining the brachistochrone is a very appropriate topic for the course, as it has a rich history involving the leading mathematicians and physicists of the seventeenth and eighteenth centuries, offers opportunities to present relatively straight-forward proofs for a number of relevant theorems, and provides an introduction to the calculus of variations.

Three Curves

According to Kline (1972), three curves of major interest to the mathematicians of the seventeenth century were the cycloid, the isochrone and the brachistochrone. The following definitions of these curves are paraphrased from the Penguin Dictionary of Mathematics, 2nd Edition (1996): A *cycloid* is the curve generated by a specific point on the circumference of a circle as the circle rolls (without slipping) along a straight line. An *isochrone*, or *tautochrone*, is a curve along which a free-sliding particle will descend to the lowest point of the curve in a fixed time, irrespective of its starting point on the curve. A *brachistochrone* is a curve along which a particle will slide in the shortest time from an initial point to another lower point, not directly beneath the first.

Brief History of the Cycloid and Isochrone

The cycloid was discovered in the early sixteenth century by the mathematician Charles Bouvelles (Cooke, 1997). In the 1590s, Galileo conjectured and empirically demonstrated that the area under one arch of the cycloid is approximately three times the area of the generating circle. That the area is exactly three times that of the generating circle was proven by Roberval in 1634 and Torricelli in 1644 (Boyer and Merzbach, 1991). Roberval constructed the tangent to the cycloid in 1634 (Struik, 1969). According to Cooke (1997), constructions of the tangent to the cycloid were independently discovered circa 1638 by Descartes, Fermat and, and slightly later by Torricelli. In 1659, Christopher Wren also determined the length of a cycloidal arch, showing it to be exactly four times the diameter of the generating circle (Stillwell, 1989).

Most of these discoveries concerning the cycloid are inaccessible to students with no calculus background. Remarkably, Roberval's historical construction (Struik, 1969) of a tangent to the cycloid is quite accessible to students, as it uses only the parallelogram law for vector addition. The proof is presented in H of C:

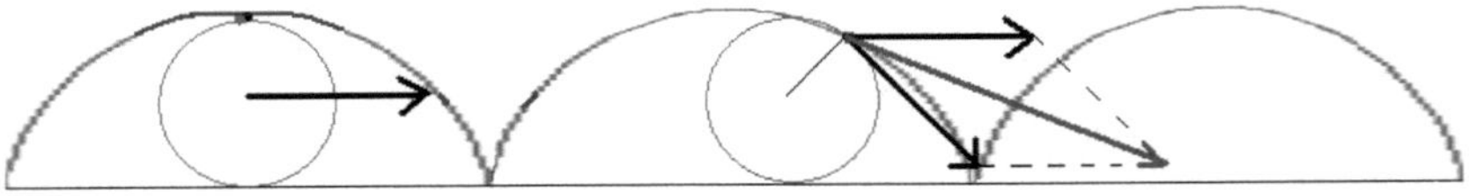

Figure 1. Finding a tangent to the cycloid

The proof can almost be presented "without words". (See Figure 1.) Follow the path of the "tracing" point on the generating circle of a cycloid. The motion of this point at a given instant has a *transverse* component, corresponding to the transverse motion of the center of the circle; it also has a component *normal to a radius of the circle*, since the circle rolls. These components are of equal magnitude, since the circle rolls without slipping. The resultant of the motions is found by the parallelogram law, and is the tangent to the cycloid.

In 1659, Huygens discovered that the cycloid is a solution to the isochrone or tautochrone problem; he showed that a particle sliding on a cycloid will exhibit simple harmonic motion with period independent of the starting point (Stillwell, 1989). Huygens published his discovery of the cycloidal pendulum in *Horologium oscillatorium* in 1673 (Boyer and Merzbach, 1991). The cycloidal pendulum also features in Newton's Principia (Gauld, 2005). In 1696, Johann Bernoulli demonstrated that a brachistochrone is a cycloid (Erlichson, 1999).

Brief History of the Brachistochrone Problem

In the section entitled *Naturally Accelerated Motion* from the *Third Day* of his 1638 masterpiece, *Dialogues Concerning Two New Sciences*, Galileo Galilei compared rates of descent along selected curves.

With his Proposition VI Galileo established the law of chords:

> 'If from the highest or lowest point in a vertical circle there be drawn any inclined planes meeting the circumference, the times of descent along these chords are each equal to the other'. (Galilei 1638/1952, p.212)

Galileo applied Euclidean geometry in his proof. In the H of C course, the law of chords is demonstrated by deriving expressions for the velocities and descent time, and noting that the expression for descent time is independent of the upper point of the chord along the circular arc. The expression for descent time may be obtained by considering the component of the force of gravity along the inclined chord. Alternatively, Nahin (2004) derives the descent time by applying the principle of the conservation of mechanical energy.

Lattery (2001) discusses an interesting approach suggested by Matthews in 1994 for leading students towards a derivation of the law of chords. Students are asked to consider the following thought experiment:

> 'Suppose a ball is released at some point A on the perimeter of a vertical circle and rolls down a ramp to point B, the lowest point on the circle ... The ramp may be rotated about point B. For what angle will the time of descent along chord AB be the least? ' (Lattery, 2001, p. 485)

This way of posing the problem may help students to greater appreciate the surprising result. It also leaves the option of discussing the distinction between sliding and rolling motion.

With his Proposition XXXVI, Galileo proved that the descent time from an upper point along the lower quadrant of a vertical circle to the bottom point of the circle is quicker along two consecutive chords than along a direct chord:

> 'If from the lowest point of a vertical circle, a chord is drawn subtending an arc not greater than a quadrant, and if from the two ends of this chord two other chords be drawn to any point on the arc, the time of descent along the two latter

chords will be shorter than along the first, and shorter also, by the same amount, than along the lower of these two latter chords.' (Galilei 1638/1952, p.234)

Galileo began his proof by a clever application of the law of chords, but then completed it by a fairly involved geometric argument. In the H of C course, the proof is completed using a much shorter method, proposed by Erlichson (1998), based on the principle of the conservation of mechanical energy.

Galileo may also have been the first to consider the problem of finding the path of quickest descent, as suggested by the initial statement of his Scholium to Proposition XXXVI:

> 'From the preceding it is possible to infer that the path of quickest descent from one point to another is not the shortest path, namely, a straight line, but the arc of a circle'. (Galilei 1638/2002, p.583)

Many researchers, such as Stillman Drake and Herman Goldstine, have concluded that Galileo incorrectly claimed that a circular arc is the general curve of quickest descent (Erlichson 1998 and Erlichson 1999). However, Erlichson (1998), argues that Galileo restricted himself to descent paths that used points along a circle:

> 'In an important Scholium to Proposition 36, Galileo attempted to establish that the swiftest descent, given the constraint of using points on the circular arc, was via the circular arc itself'. (Erlichson 1998, p.344)

Galileo's claim is based on an argument that the descent time for a particle along a twice-broken path would be less than that along a once-broken path, and that the descent time along a multiply-broken path would be even less. According to Nahin (2004), Galileo's claim is correct but his reasoning was flawed; Erlichson (1998) noted that Galileo's method of proving Proposition XXXVI holds for descent from rest, but fails to generalize to situations in which the particle is initially moving.

In early 1696, Johann Bernoulli solved the problem of finding the curve of quickest descent; he showed that the brachistochrone was a cycloid. Later, in June of that year, he posed the problem in the journal *Acta Eruditorum* (Erlichson, 1999). Johann Bernoulli's problem appeared as follows:

> ' PROBLEMA NOVUM, ad cujus Solutione Mathematici invitantur.
>
> " Datis in plano verticali duobus punctis *A* et *B*, assignare mobili *M* viam *AMB*, per quam gravitate sua descendens, et moveri incipiens a puncto *A*, brevissimo tempore perveniat ad alterum punctum *B*."' (Woodhouse 1810, pp.2 – 3)

An English translation, by Dr. Lincoln La Paz, is:

'If two points A and B are given in a vertical plane, to assign to a mobile particle M the path AMB along which, descending under its own weight, it passes from the point A to the point B in the briefest time.' (Smith, D.E. 1929, p.644)

The problem was also solved by Jakob Bernoulli, Leibniz, L'Hôpital and Newton. Newton's solution was published anonymously in the *Philosophical Transactions of the Royal Society* in January, 1697. Solutions by Johann Bernoulli, Jakob Bernoulli, Leibniz and Newton were published in *Acta Eruditorum* in May, 1697. According to Stillwell (1989), the most profound was Jakob Bernoulli's solution, which represents a key step in the development of the *calculus of variations*. Jakob Bernoulli recognized that the *brachistochrone problem* involved minimization *over a variable curve.*

Law of Reflection

The *law of reflection* states that if a ray of light strikes a mirror, then the angle of incidence equals the angle of reflection (Serway and Jewett, 2004). This law was first noted by Euclid in the fourth century BCE (Ronchi, 1957), and was explained using a principle of least distance by Heron of Alexandria in the first century CE (Cooke, 1997). In the H of C course, the law of reflection is demonstrated using Heron's *principle of least distance*. Fermat showed, in the mid-seventeenth century, that the law may also be derived from a *principle of least time* (Nahin, 2004).

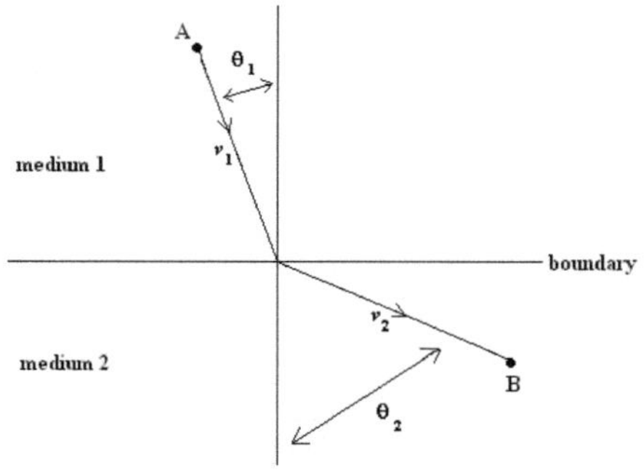

Figure 2: Snell's law of refraction

Law of Refraction (Snell's Law)

The *law of refraction* states that when a ray of light crosses the boundary between transparent media, it experiences a change in direction characterized by the relation

$\sin \theta_i / \sin \theta_r = k$

in which θ_i is the angle of incidence, θ_r is the angle of refraction and k is a constant dependent on the nature of the two media. (See Figure 2.) This law was discovered experimentally by the Dutch physicist Willebrord Snel circa 1621; in English it is known as *Snell's law* (Nahin, 2004). Snel noticed that, for a ray of light passing from medium 1 into medium 2, if medium 1 is less dense than medium 2, then $k > 1$; that is, upon entering medium 2, the light ray bends towards the normal to the boundary (Nahin, 2004).

Derivation of Snell's Law Using Fermat's Principle of Least Time

In the mid-seventeenth century, Fermat demonstrated that Snell's law of refraction may be derived from a principle of least time (Nahin, 2004). In the H of C course, the law of refraction is demonstrated using Fermat's principle of least time and an abbreviation of the approach outlined by Nahin (2004):

Consider a ray of light crossing the boundary between two transparent media, medium 1 (upper) and medium 2 (lower). Let v_i denote the speed of light in medium i, for $i = 1, 2$. With reference to Figure 3, let T denote the transit time for a light ray travelling from point A in medium 1, through point B at the boundary, to point C in medium 2. Then

$$\begin{aligned} T &= (\text{length of } AB) / v_1 + (\text{length of } BC) / v_2 \\ &= [h_1^2 + x^2]^{1/2} / v_1 + [h_2^2 + (d-x)^2]^{1/2} / v_2 \end{aligned}$$

To obtain the path of least time, it is necessary to determine x so that the transit time T is minimized.

$$\begin{aligned} dT / dx &= x / (v_1 [h_1^2 + x^2]^{1/2}) - (d-x) / (v_2 [h_2^2 + (d-x)^2]^{1/2}) \\ &= \sin \theta_1 / v_1 - \sin \theta_2 / v_2 \end{aligned}$$

The necessary condition for a minimum, namely that $dT / dx = 0$, yields the requirement that

$$\sin \theta_1 / \sin \theta_2 = v_1 / v_2$$

Thus, Snell's media-dependent constant is $k = v_1 / v_2$.

It should be noted that Fermat achieved the minimization using his *method of adequality*, which is comparable to differentiation; however, he had to introduce some approximations, as he did not know how to apply his adequality method directly to expressions involving square roots. Nahin (2004) gives a detailed presentation of Fermat's solution.

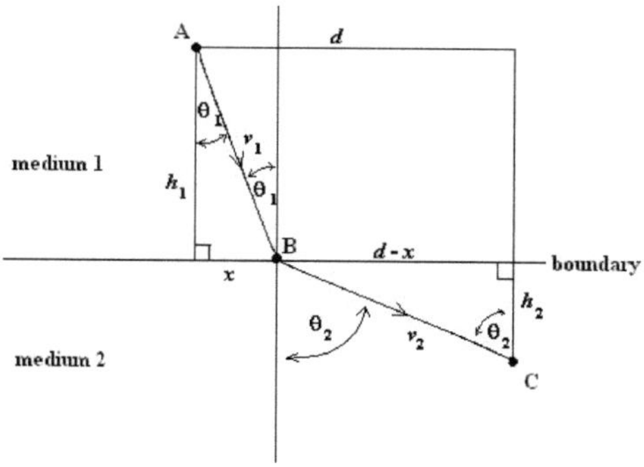

Figure 3: Derivation of Snell's law

Johann Bernoulli's Solution to the Brachistochrone Problem

Consider a light ray travelling through two transparent media, from point A in medium 1 (upper) to point B in medium 2 (lower). Let θ_1 denote the angle of incidence and θ_2 the angle of refraction. Let v_1 and v_2 denote the speed of light in the respective media. Suppose that medium 2 is less dense than medium 1, so that $v_2 > v_1$. Then, since

$$\sin \theta_1 / \sin \theta_2 = v_1 / v_2 < 1$$

$\theta_2 > \theta_1$, and the light ray bends away from the normal to the boundary. (See Figure 3.) Note, also, that

$$\sin \theta_1 / v_1 = \sin \theta_2 / v_2 = \text{constant}$$

Now, consider a similar situation with a light ray traveling downward through many layered transparent media, with each medium less dense than the layer above it. The speed of light increases in the successive media as it progresses through deeper layers and the ray of light bends further away from each successive normal to the boundary at point of contact. (See Figure 4.)

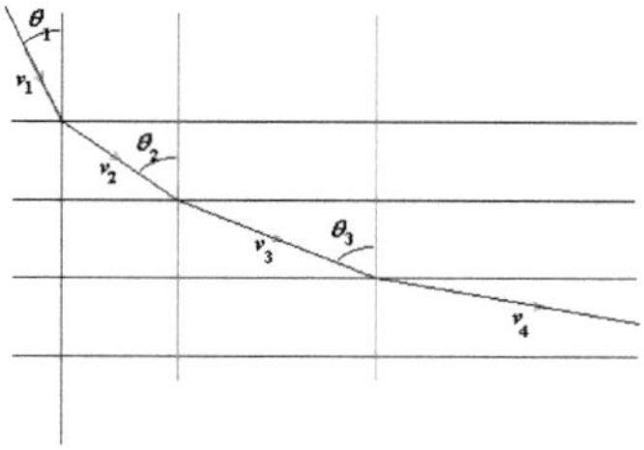

Figure 4: Refraction through multiple layers

Note also, that:

$$\sin \theta_1 / v_1 = \sin \theta_2 / v_2 = \sin \theta_3 / v_3 = \ldots = \text{constant}$$

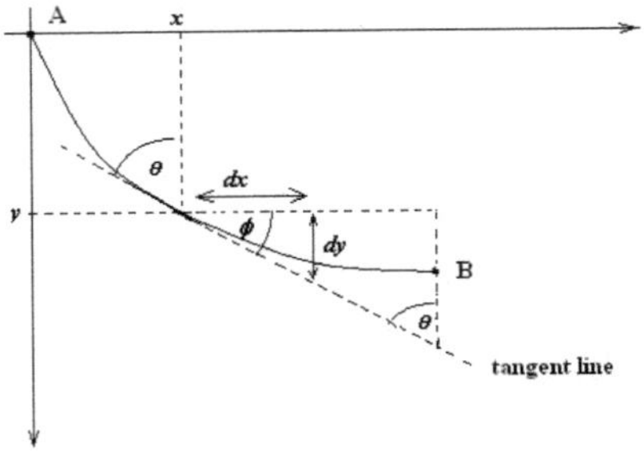

Figure 5: Johann Bernoulli's proof

Letting the number of layers increase without bound and the thickness of each layer decrease towards zero, the path of the light ray becomes a smooth curve. (See Figure 5.) At each point along the curve

$$\sin \theta / v = c \qquad (1)$$

where c is some constant. Johann Bernoulli realized that equation (1) could be used as a starting point to derive the curve of quickest descent.

The principal of conservation of mechanical energy implies that a particle of mass m starting at rest and falling through a vertical height of y will acquire a speed

$$v = (2gy)^{1/2} \qquad (2)$$

Bernoulli realized that a particle falling along the curve of quickest descent from A to B must satisfy both equations (1) and (2). From the highlighted triangle in Figure 5

$$\sin \theta = \cos \phi = 1 / \sec \phi = 1 / [1 + \tan^2 \phi]^{1/2} = 1 / [1 + (dy/dx)^2]^{1/2}$$

Thus

$$\sin \theta / v = \text{constant} = 1 / (v\,[1 + (dy / dx)^2]^{1/2})$$

where $v = (2gy)^{1/2}$. This yields the following nonlinear differential equation:

$$y\,[1 + (dy / dx)^2] = k \qquad (3)$$

where k is some constant. Algebraic manipulation of the differential quantities, dx and dy, yields the equation

$$dx = dy\,[y / (k - y)]^{1/2} \qquad (4)$$

which Bernoulli recognized as a differential equation describing a cycloid. In the translated words of Johann Bernoulli:

> 'from which I conclude that the Brachystochrone is the ordinary Cycloid'
> (Struik 1969, p .394 translation of Johann Bernoulli 1697)

To avoid confusion when examining Bernoulli's equation that corresponds to equation (4) above, please note that in his May 1697 paper in *Acta Eruditorum*, the usual labelling of x and y coordinates is reversed so that his paper also reverses the roles of dx and dy.

Equation (4) may be further manipulated to obtain parametric equations for x and y; for details, see the excellent accounts by Simmons (1972), Erlichson (1999) and Nahin (2004).

The Calculus of Variations

The *calculus of variations* deals with the problem of finding extreme values of integrals. In the simplest case, the objective is to determine a functional relationship

$$y = f(x)$$

such that an integral

$$\int_a^b g(x, y)\, dx$$

is a maximum or a minimum. Special cases of the calculus of variations include isoperimetric problems and the *brachistochrone problem.*

Computerized Demonstration of the Brachistochrone

It may be difficult for students to grasp the nature of the minimization problem involved in finding the brachistochrone. We are not finding the tangent or area of some particular *given* curve, as is usually the case in calculus. Instead, we must search among *all possible hypothetical curves* to find that which allows least time descent. The nature of the problem also precludes physical demonstration; we may build an apparatus to demonstrate sliding descent on a particular curve, or we may "race" physical beads on two or three particular curves; however, it is hard to conceive how one would physically demonstrate beads descending on enough curves to allow students to conceptualize descent on an *arbitrary* curve. Here the computer comes to the rescue:

Using the symbolic programming language *MAPLE*, we simulate the descent under gravity of a particle along a curve as follows:

1. Curves are given parametrically: $x = x(s)$, $y = y(s)$, $a \leq s \leq b$.
2. A given curve is systematically sampled at $n + 1$ points $s_i = a + i\,(b - a)/n$, $i = 0, 1, \ldots, n$.
3. The curve is modeled by straight-line segments, from (x_i, y_i) to (x_{i+1}, y_{i+1}), $i = 0, 1, \ldots, n - 1$.

4. The motion of a particle under gravity down the digitized curve is modeled in a straight-forward way: If the particle enters the straight-line segment from (x_i , y_i) to (x_{i+1} , y_{i+1}) moving at speed v_0, its acceleration under gravity will be $a = g \sin\theta$, where g is the acceleration (downward) due to gravity, and $\sin\theta = (y_i - y_{i+1})/\sqrt{(x_i - x_{i+1})^2 + (y_i - y_{i+1})^2}$. The length of the segment is $D = \sqrt{(x_i - x_{i+1})^2 + (y_i - y_{i+1})^2}$. At time t after entering the segment, the particle has moved distance $d = v_0 t + ½\, at^2$ along the segment. The time T spent traversing the segment is then the solution of $D = v_0 T + ½\, aT^2$, so that the particle enters the next segment with a speed of $v_0 + aT$.
5. The positions of particles on several digitized curves can thus be worked out as parametric functions of time. The *MAPLE* command *animate*() is then capable of presenting trajectories of two or more of these particles simultaneously, as they each move along their underlying curves. Our code as currently implemented contrasts the motion of particles along two desired curves.

We have found that $n = 11$ already gives very smooth-looking approximations. Using our software, we "race", for example, a particle on a cycloid arc against a particle on a straight-line ramp (see Figure 6), or a particle on a circular arc. We can in fact race along any (parametric) curve suggested by students. The same software will illustrate Galileo's law of chords.

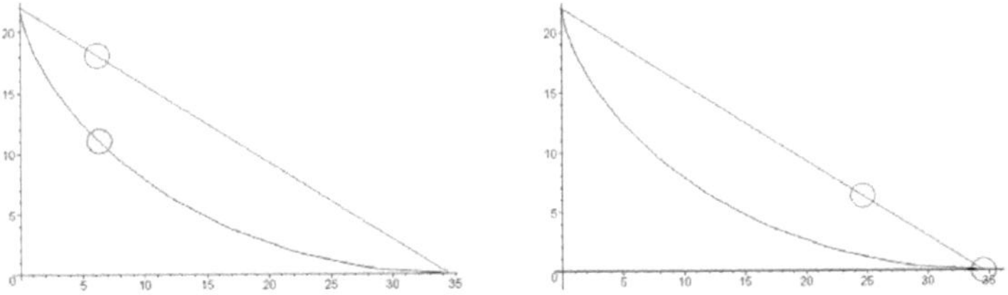

Figure 6: Screen snapshots from *MAPLE* race: cycloid vs. straight-line ramp

Jakob Bernoulli's Solution to the Brachistochrone Problem

Although Johann Bernoulli's solution to the brachistochrone problem impresses us with its elegance, it is tailored to a very specific application. Jakob Bernoulli's more methodical approach generalizes, and in fact became the basis of the calculus of variations. In fact, neither of the Bernoulli's solutions uses calculus explicitly in the foreground. Each of the brothers, by a different method, sets up a differential equation, and having found this equation, declares the problem solved. (Struik, 1969). Their quickness at this early date to recognize a differential equation of the cycloid is striking!

Since the calculus is in the background, Jakob Bernoulli's solution to the cycloid may be outlined to H of C students. We give our abbreviated presentation of his proof below. It relies on the use of similar triangles, some (typical) hand-waving regarding infinitesimals, and the concept of *stationary points* of functions.

A *stationary point* of a function is one at which the function's rate of change is zero. In the ordinary calculus, we recognize that local extrema of a function occur at stationary points. Jakob Bernoulli extended this idea to the brachistochrone problem. Since the brachistochrone minimizes descent time, the rate of change of descent time must be zero with respect to infinitesimal variation of the brachistochrone path. Consider Figure 7.

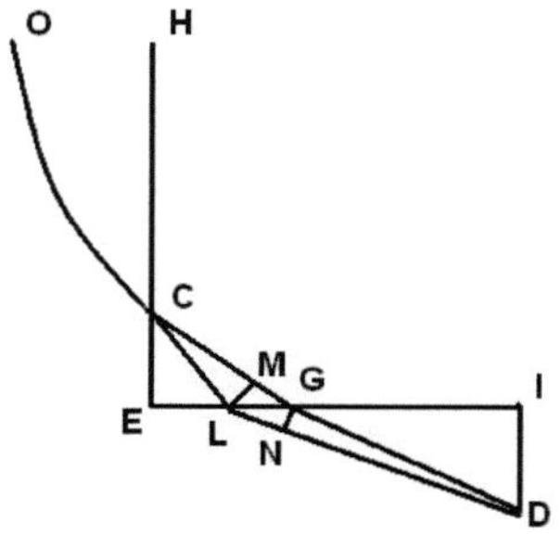

Figure 7: Jakob Bernoulli's solution of the brachistochrone problem

Let curve *OCGD* be a small section of the brachistochrone. Letting y measure vertical drop from O, a particle moves along the brachistochrone with instantaneous speed $\sqrt{y}$ at any point, by considering potential versus kinetic energy. We consider *CG* to be so short that a particle moves along *CG* with constant speed $\sqrt{|HC|}$ where $|HC|$ denotes the length of *HC*. Similarly, we assume a constant speed $\sqrt{|HE|}$ on *GD*.

We vary the path by moving G an infinitesimal distance horizontally, to L. As the brachistochrone is stationary, the descent time along *OCLD* must also be minimal.

Add construction lines *ML* and *NG* such that triangles ΔCML and ΔDNG are isosceles. The descent times along *CM* and *CL* are thus equal, as are descent times along *GD* and *ND*. As the total descent times along *OCGD* and *OCLD* are to be equal, the descent time along *MG* must equal that along *LN*, and

$$|MG|/\sqrt{|HC|} = |LN|/\sqrt{|HE|}. \qquad (1)$$

As we are dealing with infinitesimal distances, we may consider ML to be an arc of a circle centered at C, and LMG to be a right angle. By similar triangles, then, $|MG|$ / $|LG| = |EG|$ / $|CG|$. If we let x measure horizontal distance, and s arc length along the brachistochrone, this can be rewritten as

$$|MG| / |LG| = dx / ds \text{ on segment } CG.$$

By an analogous argument,

$$|LN| / |LG| = dx / ds \text{ on segment } GD.$$

Dividing by $\sqrt{y}$ on each segment, and applying equation (1), we find that

$$\sqrt{y}dx / ds = k \qquad (2)$$

with the same constant k on both segment CG and segment GD. We conclude that equation (2) holds everywhere on the brachistochrone. Jakob Bernoulli recognized this as a differential equation of the cycloid.

Note that various mysteries involving infinitesimals take place; segment CG is only short, while LG is infinitesimal. Also, isosceles triangle ΔCML contains two right angles, and this is essential to the argument with similar triangles. Again, Bernoulli interchanged x and y, which can be confusing to a modern reader.

Conclusion

We have traced the thread of quickest descent problems and the brachistochrone from Galileo, through Fermat and Roberval, to the Bernoullis and the dawn of the calculus of variations. We have spelled out in detail a selection of mathematical results which we have presented to our H of C students. These results include mathematical content ranging in difficulty from geometry, through vectors, to differential equations. A themed unit on the cycloid could show students at different levels the strong interplay between mathematics and physics: Geometry informs optics (Fermat), optics informs kinematics (Johann Bernoulli), and kinematics informs geometry (Roberval).

Of particular interest is Johann Bernoulli's beautiful *Gedankenexperiment*, whereby a falling particle becomes a ray of light, moving through media arranged without regard for the possibility of actual physical construction. Again, the purely mathematical and hypothetical nature of the frictionless bead in the brachistochrone problem motivates our software demonstrations to students. The digital sampling of curves in our *MAPLE* code echoes the (finite) layering of media by Johann Bernoulli. A careful examination of the arguments of the Bernoullis introduces students to the interesting philosophical and technical issues related to infinitesimals/differentials in mathematics and physics.

References

Babb, J.: 2005, 'Mathematical Concepts and Proofs from Nicole Oresme: Using the History of Calculus to Teach Mathematics', *Science and Education* 14(3-5), 443-456.

Boyer, C.B. & Merzbach, U.C.: 1991, *A History of Mathematics*, 2nd edition, John Wiley and Sons, New York.

Cooke, R.: 1997, *The History of Mathematics: A Brief Course*, John Wiley and Sons, New York.

Erlichson, H.: 1998, 'Galileo's Work on Swiftest Descent from a Circle and How He Almost Proved the Circle Itself Was the Minimum Time Path', *American Mathematical Monthly* 105, 338-347.

Erlichson, H.: 1999, 'Johann Bernoulli's brachistochrone solution using Fermat's principle of least time', *European Journal of Physics* 20, 299-344.

Galilei, G.: 1638/1952, *Dialogues Concerning Two New Sciences,* translated by H. Crew and A. de Salvio, Encyclopaedia Britannica, Chicago.

Galilei, G.: 1638/1974, *Dialogues Concerning Two New Sciences,* translated by S. Drake, University of Wisconsin Press, Madison.

Gauld, C.: 2005, 'The Treatment of Cycloidal Pendulum Motion in Newton's *Principia*' in *The Pendulum: Scientific, Historical, Philosophical and Educational Perspectives,* Springer, The Netherlands, pp.115-125.

Kline, M.: 1972, *Mathematical Thought from Ancient to Modern Times*, Oxford University Press, New York.

Lattery, M.J.: 2001, 'Thought Experiments in Physics Education: A Simple and Practical Example', *Science and Education* 10: 485-492.

Nahin, P.J.: 2004, *When Least is Best: how mathematicians discovered many clever ways to make things as small (or as large) as possible*, Princeton University Press, Princeton, NJ.

David Nelson, ed.: 1998, *The Penguin Dictionary of Mathematics*, 2nd Edition Penguin, University of Manchester, Manchester.

Ronchi, V.: 1957, *Optics: The Science of Vision* (reprinted in 1991), Dover Publications, New York.

Serway, R.A. & Jewett, J.W.Jr.: 2004, *Physics for Scientists and Engineers with Modern Physics,* 6th Edition, Thomson Brooks/Cole, Toronto.

Simmons, G.F.: 1972, *Differential Equations with Applications and Historical Notes*, McGraw- Hill Book Company, New York.

Smith, D.E.: 1929, *A Source Book of Mathematics, Volume Two* (reprinted in 1959), Dover Publications, New York.

Stillwell, J.: 1989, *Mathematics and Its History*, Springer-Verlag, New York.

Stinner, A. & Metz, D.: 2003, 'Pursuing the Ubiquitous Pendulum', *The Physics Teacher* 41, pp.38-43.

Struik, D.J. (editor): 1969, *A Source Book in Mathematics 1200-1800*, Harvard University Press, Cambridge.

Woodhouse, R.: 1810, *A History of the Calculus of Variations in the Eighteenth Century*, Chelsea Publishing Company, Bronx, New York (reprint of *A Treatise on Isoperimetrical Problems and the Calculus of Variations*, originally published 1810).

Mirrors, swinging weights, light bulbs...: Simple experiments and history help a class become a scientific community

ELIZABETH CAVICCHI

Dibner Institute for the History of Science and Technology, MIT, Cambridge MA, USA

Abstract: Nine "honors" undergraduates at a public university found unexpected surprises when asked by their teacher to look in a mirror at an angle, swing weights on a fishing line, and light a bulb with a battery. Euclid, Galileo, Volta, and others long ago explored these 'simple' experiments seriously, and the students – exposed to historical accounts and apparatus, and encouraged to observe carefully – began to see why. Old queries were repeated with modern instruments: timing a pendulum with a cell phone, dissecting an "Energy Saver" bulb, beaming lasers into jello, and congealing ice cream with liquid nitrogen! Respect deepened for historical predecessors, and for each other's inventions and thoughtful conjectures. When encouraged to develop a shared understanding of both history and phenomena, the students recast themselves as thoughtful and reflective explorers of the physical world. They became a community of "scientists" and gained firsthand insight into how shared knowledge is created.

Introduction

A class has many eyes, hearts and minds, each having potential for engaging with others and the subject. That diverse potential can emerge in unexpected ways where relationships matter and the class becomes a community. In the process they create shared knowledge, in which everyone has a part. This paper describes such a class, where so-called simple experiments and science history were the context for explorations and discussion among nine undergraduates.

The early twentieth century American philosopher John Dewey envisioned the organic interplay of life and learning as fostering a "miniature community ... in close interaction with ... experience beyond school walls" (Dewey 1916, 360). David Hawkins, leader in the Elementary Science Study of the 1960s, saw teacher and learners and subject as making up "a community of subject-matter and engagement which extends *beyond* the circle of their intimacy" (Hawkins 2000, p. 49). For both Dewey and Hawkins, the vitality of teaching and learning relationships enables participants to reach out to materials, each other, and to resources and people not present in the classroom.

Some recent science education literature depicts classrooms where relationships contribute to learning (Gallas 1995, Hammer 1997, Osborne 2001, Zembylas 2004;

Zion 2004). For example, by reflecting on distrust in urban schools – and realizing that trust is essential for doing science investigations -- an African-American chemistry teacher sought to develop mutual trust with those in his class (Sconiers 2000). This teacher's finding about trust correlates with philosophical and historical analyses of science, where trust is foundational to conducting science as a community (Hardwig 1991, Rolin 2002, Shapin 1994; Grasswick 2004), enabling shared evidence, critiques and dissent, to the extent that relationships among individuals and communities are inextricable from the new knowledge being generated.

These relational qualities of knowing and learning underlie 'critical exploration', the method of research and pedagogy developed by Eleanor Duckworth (2005, 2001, 1999, 1991) from the clinical interviewing of Piaget (1926/1983, 1936/1954) and Inhelder (1974). To set conditions for critical exploration, a teacher seeks to create a classroom environment where students feel safe to explore authentic materials and express tentative understandings as these evolve. Simultaneous with *students'* explorations of the subject, the *teacher* explores how they engage with it and their evolving activities and inferences. The teacher does this research so as to respond to students in ways that may keep their investigation going and widen possibilities by which they may engage the subject and apprehend its behaviors or patterns. Relationships among teacher, learner and subject, along with the history of what goes on in critical explorations, are inseparable from this work and from the new knowledge, the "wonderful ideas" (Duckworth 1973) that students generate through exploring.

In communicating critical exploration as research, narratives are derived from the experience with supporting documents (such as photos, journals, and assignments) and accompanied by the reflective writing of teacher and participants. Such narratives of learning together with the accompanying confusions, uncertain understandings and relationships become, in turn, a resource for further exploration. In these narratives, community becomes conspicuous, as Duckworth wrote in opening her account of the floating and sinking investigations of a class of adults: "This is a story about the collective creation of knowledge: its multiple beginnings; its movement forward, backward, sideward; its intertwining pathways" (Duckworth 1986).

Setting

A new science course offered through in the honors program at the public university was the setting for the critical explorations presented here. The students in the course were beyond their first year of college and all but one had chosen non-science majors.

Many expected that taking a science course meant formulas and cookbook labs. But instead, together we explored pendulums, mirrors, batteries and bulbs, magnets, prisms and tuning forks.
The design of this course was an extension of my earlier teaching research with students' science experimenting in relation to history (Cavicchi 2007, 2006, 2005, 1999). Beginning with in-class explorations of physical science materials and outside-class assigned activities using materials and readings, by mid-term I asked the students to pursue an exploratory project of their own (or with a partner) outside of class. Class sessions for the first half of the semester included exploratory activities and discussions. We read scientific and reflective writing by scientists such as Euclid (translated in Kheirandish 1999, Smith 1999), Gilbert (1600), Galileo (1632, 1638), Payne-Gaposchkin (1984), Sacks (2001), Vermeij (1997) and essays about historical experimenting (Abbott 1987, Bonnet 1994, Corn 1996, Dibner 1947, 1964, Klein 1971, Needham & Ling 1962, Settle 1996, Simms 1977, Smith 1982). In the middle of the term, I arranged field trips to: the Burndy rare book library and Westinghouse-Hibben Historical Lamp Collection; the MIT Museum for its nineteenth century surveying tools; and a 1911 power substation remaining from Boston's early electric railway. During the final meetings, each student or pair involved their classmates in exploratory experiences with the science materials of their project. Projects included: papermaking, the sounds of guitar strings, optical illusions, dissecting electrical appliances, making visible the paths of light in materials.
The three following sections of this paper excerpt from the class' critical explorations with: swinging weights; mirrors; and light bulbs. In each of these three cases, exploratory work that originated with assigned exercises involving the whole class, was continued in the journal assignments, projects, and shared activities that were initiated by individual students. Across these stories, the class moves from doing groupwork to becoming a fluid evolving experimental community. Historical science readings and experiments are an inspiration for this development. As the students become explorers of their own, the questions they explore have to be their own. They become more like Galileo and other historical experimenters, the more they are working from their own curiosity and observations.

Swinging Weights

After two weeks of class experimenting with pendulums, and reading excerpts from Galileo, we began one session by reading a children's story about an African boy, Chima, and his time-keeping rock-on-a-string (Savage 1967). One student, Annamarie compared swinging Chima's boulder with swinging a pebble. She said Galileo found weight did not "affect the length of time it took to go back to rest." Then another student, Samantha, described a check up on Galileo she had done on her own at home. She had timed the successive swings of a pendulum, and found these times all about the same. Her careful findings attracted no interest, since the class was becoming focused on the weight coming to rest. Samantha next invited us to do an experiment: "Is there … a certain length of something that ensures that it will have an arc that you could time?" (Transcript, 2005)[1]

Out of the class' evolving discussion came a specific challenge: to make a pendulum that would go for one second. The first of many such whole-class experiments, what surprised me most was the class' desire to work together, not in pairs. Experimenting together, their community gelled.

Soon Samantha and her partner Andrew took the stage, tying a weighted string to a coat-rack. Annamarie said, "ready, set, go" (Figure 1 left). The half-swing was over so quickly that Annamarie missed reading my watch, which had no stop. Everyone laughed. Annamarie asked Andrew: "could you be the eyes ….say stop?" A classmate, Devin, interrupted; her cellphone had a stopwatch. Timing 9/10 second with it, Devin exclaimed "Galileo didn't have a cellphone!"

They lengthened the string (Figure 1, right). Devin's next reading went the wrong way, at 8/10 second. This was perplexing. Lucienne, who had often restated Galileo's finding that shorter strings gave shorter times, now questioned whether our string should shorten. Devin disagreed saying "Maybe we are just better at timing". When the next reading came in low, Samantha *lengthened* the string again. No one dissented. The first swing time at this new length, .97 second, gave the appearance of closing in on the second. Devin's outlook evolved. She urged: "we could do it multiple times … you have to get one second more than one time in order to know that the right time is there" (Transcript, 2005).

In the playfulness of doing many trials came more thoughtfulness about experimenting. If Andrew forgot to say stop; laughter cathartically related the experimenters with the onlookers. When her times varied, Devin asked me: "there's a lot of human error involved, right?" Andrew instantly concurred, saying "a lot!" Spotting Devin, he said, "I'm not saying on your part!" Amid this teasing was the link making Devin and Andrew part of the same time instrumentation.

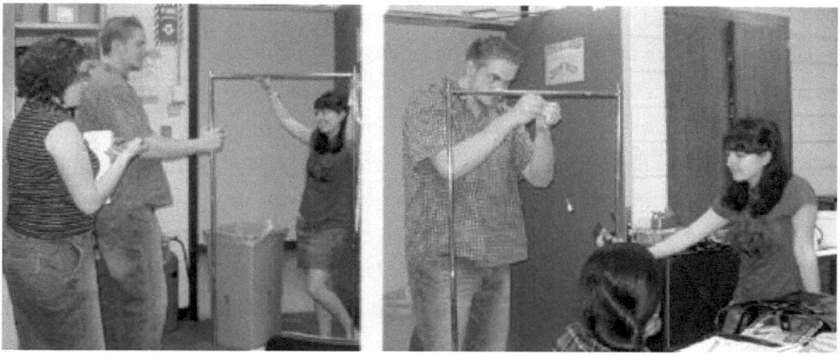

Figure 1. Left: One student (right) lets go of the weight from a string tied to a coat-rack support, while others hold the rack and time the swing. Right: readjusting the length of the pendulum string.

One reading was way off. On their own, the students stopped to critique their experiment. Anna supposed the pendulum's release height might affect its time but Lucienne disagreed: Galileo found only length mattered. Then Lucienne suggested: "do you guys want to make an average of all these times?" After experimenting resumed, Annamarie, in her role of taking data, noticed that "the last two [*times*] were really close" to each other. To her, the closeness in these numbers showed that there was more consistency in *doing* the experiment. Once she collected ten runs since the break, Annemarie read off the data. Devin computed the average at .94 seconds. She concluded the string had to be longer yet. Reflecting on how the experiment evolved, Annamarie asked me: "is it really ok that we just kind of ignored the first ones, because … there were some things that we needed to straighten out?" Responding that experimenting is about revising and redoing, I observed that they would go on finding more to change.

Andrew brought up the children's story; Chima never spoke of the rock stopping, but we saw it does. As Annamarie drew on the previous week's discussion to reconstruct the relationship between a pendulum's diminished arc and slower speed, she gained clarity about Andrew's question of its stopping, and said: "clearly at the end point, that's a change in time, right, and it is no longer moving…" Annamarie's highlighting of the endpoint helped Lucienne to speak what was troubling her, and Lucienne said: "if you have the same length of string … I thought Galileo meant that it took the same time for it to come to a stop." Devin read aloud from Galileo: "when suspended from the string pulled from 90 degrees or one half of a degree, it will employ the same time in passing through the least as through the largest of these arcs" (Galileo, 1638/1914,

p. 97, 141). To Devin, it was clear that Galileo meant that the time was the same for big and for small arcs but Lucienne still wondered: did he meant the time to come to a stop, or to go through one arc? As Devin reread Galileo's sentence aloud, Annamarie's insight about the endpoint helped her to articulate this subtle distinction in Galileo's text to Lucienne and the others. Devin said: "he says, "it would employ the same time in passing through the least as through the largest of these arcs." So if it is going to stop, it wouldn't be in an arc so it would have to be still moving [for the time to apply]." Now as Lucienne finally concurred, Devin remarked that that Galileo had made this great discovery but did not boast about it: "he's a humble man".

It was the class' discovery too, woven from many sources and observations in discussions, historical reading, and experimenting. Much remained for us to refine, as shown in the disparity between the half-meter length of the class' experimental seconds pendulum, and the full meter length of a true "seconds" pendulum. What the class created was more than an answer, it was awareness of consistencies in nature and of their own investigative *and ever-evolving* process. This process encompassed observant concerns about the slowing motion; imagining a boulder and a pebble; improvisation with string, mounts and timers; and confusion in relating Galileo's text to what they saw. Trust was growing through the class' interactions, from trust in each other's role in timing that became apparent as more consistent readings, to trust in the integrity of each other's ideas. For example, through Devin's respect for Lucienne's interpretation of time-to-stop and Annamarie's analysis of stopping, Galileo's text became more accessible. Galileo was part of this community too, moving from cryptic words to being at the crux of the students' observations, to exemplifying a scientist's humility.

Periodicity again engaged us on the day two musicians in our class, Peter and Lucienne, shared their project work with the guitar. Both were curious to see the vibrating string's waveform. Peter had once observed it while playing his guitar in the flickering light of his computer screen, and both had read about it in a historical account (Figure 2, left; Sauveur 1701). I brought out a strobe light. Awhile passed before anyone saw the faint effect. Andrew wanted the room lights off. An emergency light would not go off so Aaron and Andrew climbed on the table, duct-taping black cloth over it.

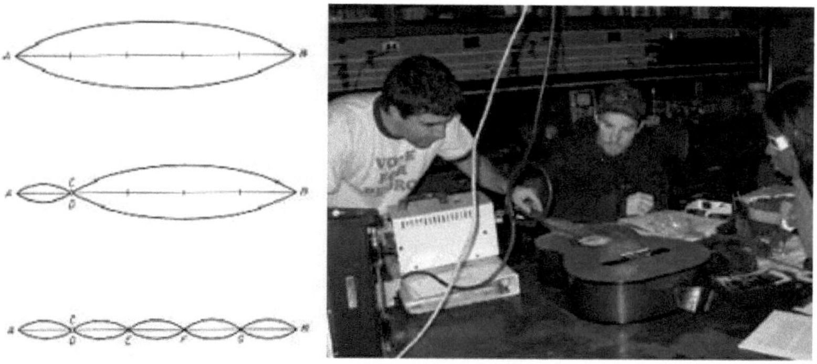

Figure 2. Left: Historical diagram of waves on a string (Sauveur 1701). Right: Dialing a strobe light (white box on left) through different frequencies to match the vibration of a plucked string on the guitar.

The added darkness made a huge difference. Everyone was enthralled. Waves on the string stood out, warbling, wobbling, and becoming still where frequencies of string and strobe nearly matched. On playing chords, Lucienne exclaimed "oh look at that!" Peter detected; "the smallest dissonance… awesome." More explorations followed. Peter invented an experiment (Figure 2, right): to select a guitar string whose frequency he knew as 660, and have Aaron dial the strobe meter to see where they matched. He said: "I'm going to start playing a note…. We'll see if it comes out to roughly what it should be." The perfect match at 660 on Aaron's meter, astonished Peter: "it's crazy that it came that close." This experiment with its matched frequencies was inextricable from the class community and their fluidly evolving experiences of plucking the guitar string, darkening the room, adjusting the strobe, wondering at the wave.

'Simple Experiments' with Mirrors, Light Rays, and Liquid Nitrogen Ice Cream

The next part to my story is from the class' work with mirrors, and shows how a plane mirror came to be far more complex than anyone anticipated. When the course started, the students initially considered such activities too "simple", obvious, beneath them. Only in observing these things themselves did they find there was much going on they did not understand. Peter later reflected:

> There were some [in class] who thought they'd known exactly what the outcomes would be, and I myself had thought… that I knew…. Needless to say …(probably all) of us, myself included, had been dazzled… when experiments… had not gone as we expected. [Some] experiments were complex, but it was the

seemingly simple ones that seemed to stun everyone the most when their basic assumptions had proven false. (Tusi 2005)

The mirror activity we did the fourth week was such a seemingly "simple" exercise having a surprising outcome. My idea for it originated in the class' curiosity about mirrors of long ago. A reading on ancient mirrors reported that wealthy Romans had metal mirrors large enough for someone to see their entire body (Melchior-Bonnet 1994, p. 11). I asked the class to consider the size that such a mirror would have to be, by exploring as a test case the size on the mirror occupied by their whole head's image when viewed at different distances (for another treatment see Kipnis 1993). Lucienne could not picture how there could be any specific mirror size, and said: "Cause if you have a tiny mirror … it seems like if you back away a lot you could see yourself … eventually you are going to get … where … you can see your feet." Samantha had backed six feet away from a mirror at home, and said her image appeared nine feet away. Andrew queried "you don't think it's double?" Jenniemae, who had used a reference gauge, said: "When I did it, I thought my reflection was the same distance away [i.e. inside the mirror], because … I could see the ground floorboard … I had a better sense of perspective."

I suggested doing this together. Anna held our oval mirror while Lucienne and Jenniemae took turns backing away while watching their head's image. At first, Anna had the mirror at waist-level but her viewers kept prodding, "Higher Higher" until it was finally at face level. It took practice, walking and holding, walking back and forth, to get where someone actually saw their face throughout the whole course of backing up. Noticing something odd, Lucienne said "I don't think it [*image*] changes in size at all."

When Anna took her turn (Figure 3), Lucienne put tape onto the mirror around the outlines of Anna's imaged face.

Figure 3. Viewing the mirror image of your face while backing away from the mirror.

Anna backed up, in her words "believing that my face would obviously shrink in the mirror." (Tsui 2005) Surprise deepened with some realization that things seen in mirrors are not quite what one might expect, in what she spoke while moving: "... my face ... if I keep like ... It's weird. It still fits between the lines!" An effort to measure both head and its taped image was inconclusive; not quite a ratio of 2/1. Peter tried it too. He interpreted the outcome along with the shift from viewer's perspective to that of the mirror plane which had eluded the class and left them with flawed expectations. Peter said: "[At] a different distance, it [*the image*] looks like a different size because you are further away. So you look smaller *to you*, but in the mirror, you are still the same... size."

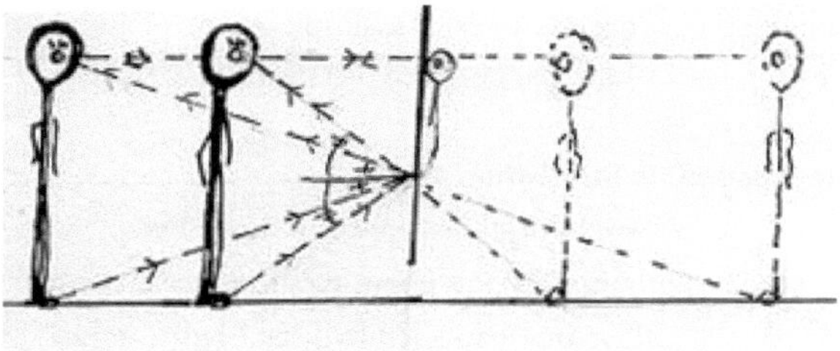

Figure 4. Ray diagram for a person viewing their full body in a mirror at two different distances from the mirror. A ray passes from the person's foot, to the mirror, where it reflects at equal angles to their eye; another ray, normal to the mirror, passes from eye, to mirror, back to eye. The height occupied on the mirror, by the image, is the same for both positions of the person.

At the board, I sketched the ray diagram (Figure 4) for someone viewing their own reflection at two positions, relating it to what the class previously worked out about equal-angle reflection. But I soon became aware that my synthesis was not in contact with theirs. For them, the heart of this exploration was a shift in view of an image seen as a whole, as Peter said, not its composition from linear paths. Anna's end-of-term project, however, to make visible the ray of light moved the class more into observing these light paths. Anna prepared a fishtank of dirty water dosed with cream. Her classmates shone flashlights and laser pointers into it. So much creamer went in that soon the flashlight's glow, and even the laser's, was only a diffuse scatter, but Anna's jello better showed the laser's straight path.

Anna wanted to see "how light reacts to smoke and vapors" (Tsui 2005). To do this, we poured the lab's liquid nitrogen into a Styrofoam cup in another clear fishtank. Here the laser's linear red path appeared, as did the vaprous glow of the forming cloud. Anna incited a new challenge at low temperature: mixing up the café's creamer and sugar with liquid nitrogen. Would it make ice cream?! That phase transition was tricky to reach, but Aaron persisted in stirring and readjusting milk and coolant proportions, Peter in checking the confection, and Andrew in supplying more "liquid N". We all had some – even the reluctant teacher – helped by Aaron's open encouragement and Andrew's repeated assurance, "it's good, it's good."

Wonderful questions arose – such as were we eating "liquid N" itself? -- spontaneous and genuinely evoked by what went on in the fishtank. The class as a community created new optical effects, such substances as ice cream, and understandings of light's paths and their own development as experimenters. Anna reflected that original historical experiences: "of awe and puzzlement are similar to our own…[S]imple observations are crucial in the understanding of something new…[O]ur seemingly insignificant comments … in class over… a simple beam of light are important…in the exploration of science" (Tsui 2005).

Light Bulbs and Roslindale Substation

The last case from the class explorations concerns light bulbs; here, the light bulb, which we use everyday without thinking about them, aroused confusion, delight, and understanding. One week, after they had used two-lead bulbs in class, I sent everyone home with a flashlight bulb, battery and wire,. No one reliably lit this bulb. So I posed the materials again to the whole class. Soon, bulbs glowed, tentatively for some, with increasing brilliance for Peter and Aaron who teamed up batteries, adding in more until they blew it out.

At the board, the class tried to diagram the bulb (Figure 5, left). Two prongs went down from the fine wire that glows but no one could see where they terminated in its base. Peter asked to break one open; I said yes! He cracked it so carefully the filament stayed intact. They hooked up a battery; the filament glowed, smoked and went out. Peter and Aaron dismantled the tiny bulb's base, but without identifying the electrical path. Next I provided small bulbs that had no base to obscure the two filament prongs' exit. No one could get them lit! Smoke appeared at the battery tops where a wire coming from the battery base was twisted into *both* leads of the small bulb. Devin however, clamped a wire to each lead, connected these across a battery, but saw no light! We rechecked all the connections, a tiny glow appeared, barely discernible in the lit room. The battery line-up of Aaron and Peter was soon applied to boost the light (of this higher-voltage bulb).

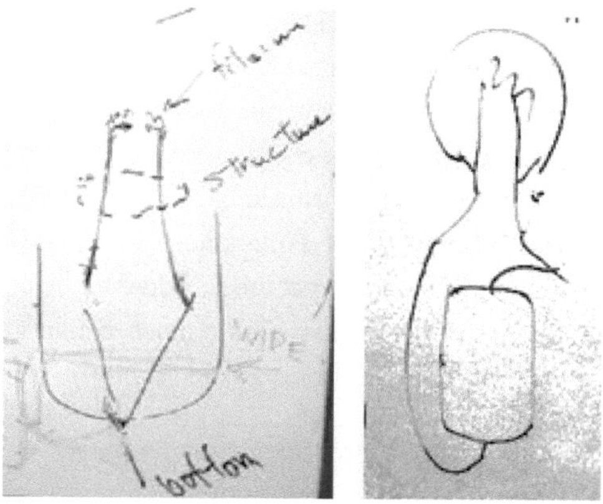

Figure 5. Left: Class blackboard diagram of the filament inside U-shaped bulb casing, showing class uncertainty about how the filament leads exit the bulb. Right: Blackboard diagram showing that filament leads exit the bulb and contact with wires that end on opposite terminals of the battery (rectangular shape).

For the class, Devin sketched what she'd done: battery contacts meeting the two wires of her globe bulb (Figure 5, right). Jenniemae said both battery ends had to be involved. Taking off from that, Anna completed the circuit reversibly in her mind, saying: "like the battery has two ends, one [filament] prong has got a positive and one … has … a negative? … And wires are connected to the positive … and negative side of the battery. It completes like the circle." Devin brought up the enigmatic flashlight bulb; how could Anna's idea work where it had, as she thought, "two wires coming out of the bottom?" I asked how else wires might be involved. By a thought experiment, Peter pictured attaching a wire to each battery end and then using these wires to light a flashlight bulb, he asked: "So does one wire go to each?"

By reflection and question integrating the class' experiences, Peter inferred how the circuit got through the obscuring casing. Yet this circuital analysis was unstable. Later on, Peter and Aaron produced so many shorts, they claimed "a cloud of smoke always ends up … out of everything we do." Only after redoing connections in one "failed" experiment, did Peter "take a step back" to recover again in a new context, his circuital "discovery".

When we visited the Burndy Library's Historical Bulbs, Samantha made a visible analogy between our class' explorations and historical ones; she wrote: "…experimentation, as we have learned, is obviously try everything that occurs ...; Galileo and Ptolemy clearly thought up every variance…the common theme is perseverance and imagination; observing each variance, in every possible way" [Pitchel 2005].

In contrast with the delicate bulbs, electrical history took on an immersive scale during the class visit to a 1911 power substation of Boston's transit system (Figure 6, left). Peter described it in these words: "When entering the building, I had no idea what was going on … In the center there were three enormous converters … Behind the control board … the multitude of wires, fuses … amazing. … [Tusi 2005]" Andrew said "the basement offered a whole other world of exploration. [Lix 2005]" There was also a human story to be told, hinted at by the closet toilet, ladder to the ceiling, "DANGER" signs (Figure 6, right), and stray work gloves. As much as the grand space and its electrical installation had the power to transport us bodily to another time, it also raised the question: was it safe to work here?

Figure 6. Left: Inside the 1911 power substation, with control panel on the left, and rotary converters on the right. Right: Danger signs in the work area.

While the historical light bulb display accentuated creativity and diversity, the message of what the class called "Roslindale Substation" bore more on community – both the historical network of streetcar lines, passengers, conductors and electrical trainees, and our class experience. How deeply history and community pervaded the class showed in the final session, when Aaron pulled out a "Roslindale Substation" specimen light bulb! He ceremoniously dissected it alongside modern bulbs including an "Energy Saver" fluorescent, a long straight florescent tube, and an ordinary incandescent.

The broken Roslindale bulb passed, from hand to hand, carried wonder and intense curiosity (Figure 7). And the reflections that arose along with it bespeak a community having the organic life that Dewey and Hawkins advocated. We see this from Lucienne's personal awareness in writing

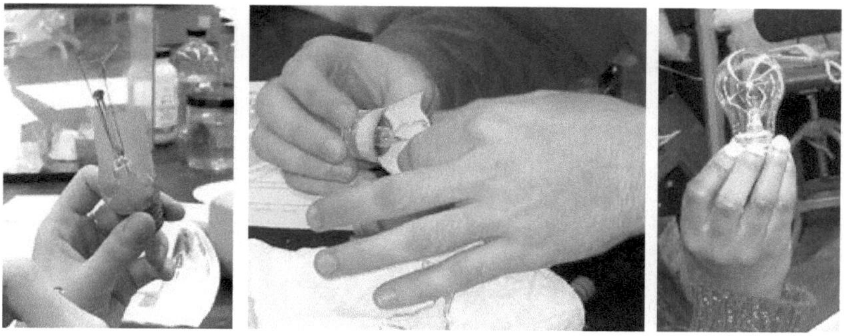

Figure 7. Historical dissect bulb (left) and modern bulbs (dissected, middle; intact, right) in students' hands.

> As much as I am developing an adventurous spirit, I am also learning about how I learn and how I approach unknowns" (Pierre 2005)

to Samantha's sensitivity to the evolving whole:

> It was during class when we bounced ideas and observations off each other, that I began to see the web of thought processes that could stem from a simple activity" (Pitchel 2005),

to Andrew's inclusion of the class with the flow of science history:

> Who knows, one of us might be the modern day Galileo, Aristotle, Plato, or Bern Dibner (Lix 2005).

Community, History, and Experimenting

The historical record inspired members of this class, both for what they noticed about the natural world, and for how they went about learning and exploring more. The students developed a sense of early historical research, not just through word and example, but also within their own experience in empowering ways. They were not "recreating history", but instead creating experiments, *being* Galileo by treading that ground anew for themselves. Lucienne realizes herself as an "adventurous spirit"; Andrew wonders about being the next Galileo.

At the start of the term, doing science experimentally was unfamiliar. As the class became involved in developing experiments and community of their own, the experiences of science history gained relevance in questions and issues that mattered to them. For example, when Devin reread Galileo's insight about pendulum swing times, the class's understanding of pendulum stopping became clearer, while Aaron's dissection of historical and modern bulbs provoked comparisons in materials and design. An aspiration for teaching science emerges from these narratives and examples: that *our students* may become the explorers and adventurers in their own lives and for the future.

Acknowledgment
For their creative and collaborative experimenting, I am happy to acknowledge and thank Devin Bramhill, Jenniemae Cronan, Annamarie Fallon, Andrew Lix, Lucienne Pierre, Samantha Pitchel, Aaron Shaw, Peter Tusi, and Anna Tsui. For support in teaching this course, I thank Dick Cluster, Joyce Morrissey, and Rajini Srikanth of the University of Massachusetts Honors Program. Lab support and materials were facilitated by: Yvonne Vaillencourt, Paul Foster, Steve Arnason at University of Massachusetts; Wolfgang Rueckner of Harvard; James Bales and Edward Moriarty of the MIT Edgerton Center. Class field trips were hosted by Deborah Douglas at the MIT Museum; Philip Cronenwett, Elaheh Kheirandish, and Anne Battis at the Burndy Library; Frances Boyle, Michael O'Malley and MBTA staff at the Roslindale substation, along with interpretation by Gil Cooke. The Dibner Institute for the History of Science and Technology and the Tufts Department of Computer Science provided historical resources, internet access and other support. For encouragement, resources, and example trials for historical and experimental teaching, I thank: Peter Heering, Fallk Riess, Giuliano Pancaldi , Paula Labbe, David Pantalony, Phylis Morrison, Sam Schweber, Roy Veatch, Violet, Elise, and Julia Cavicchi; Yvonne Pappenheim, Laura Allen, Jamie Bell, David Williams, Kristin Buchholz, Rachel Riemer, Kathleen Stevens, Richard Whitney, Brandon Pezely. Ben Marsden, Matthew Harpster and James Bales discussed a version of this paper. For her mentoring in explorative teaching, I thank Eleanor Duckworth; for thoughtfulness in teaching examples, I thank Fiona McDonnell, Alythea McKinney, Bonnie Tai, Lisa Schneier, Petra Lucht, Peter Heering, Claryce Evans, Wolfgang Rueckner, Ryan Tweney. Alva Couch, Alanna Connors, Phil and Roy Veatch are co-experimenters. I thank the organizers Peter Heering and Daniel Osewold of the 6th International Conference for the History of Science in Science Education for the opportunity to present this paper on July 13, 2006 in Oldenburg Germany. This work is dedicated to the memory of Philip Morrison.

Notes

[1] All further quotations from class discussion are excerpted from the class transcripts of fall 2005.

References

Abbott, B.: 1987, 'The Beauty of Physics: Berenice Abbott', NY Academy of Sciences, NY.

Cavicchi, E.: 2007, 'Historical Experiments in Students' Hands: Unfragmenting Science through Action and History', *Science and Education*, in press.

Cavicchi, E.: 2006, 'Science Experimenting with Old and New Things,' in Cavicchi, E., Chiu, S., McDonnell, F., 'Development through Creativity and "Wonderful Ideas": Student Explorations of Art and Science', Jean Piaget Society, Baltimore MD, June 2, 2006.

Cavicchi, E.: 2005, 'Exploring Water: Art and Physics in Teaching and Learning with Water', in France, R. (ed.), *Facilitating Watershed Management: Practical Approaches and Case Studies,* Rowman & Littlefield, Lanham MD.

Cavicchi, E.: 1999, 'Experimenting with Wires, Batteries, Bulbs and the Induction Coil: Narratives of Teaching and Learning Physics in the Electrical Investigations of Laura, David, Jamie, Myself and the Nineteenth Century Experimenters – Our Developments and Instruments', Dissertation, Harvard University, Cambridge MA.

Corn, J.: 1996, 'Object Lessons/Object Myths? What Historians of Technology Learn from Things', in Kingery, W.D., *Learning from Things: Method and Theory of Material Culture Studies*, Smithsonian, Washington DC.

Dewey, J.: 1916 (1944), *Democracy and Education*, Free Press NY.

Dewey, J.: 1902 (1956), 'The Child and the Curriculum', in *The Child and the Curriculum and The School and Society*, University of Chicago Press, Chicago IL.

Dibner, B.: 1947, *Doctor William Gilbert*, Burndy Library, Norwalk CT.

Dibner, B.: 1964, *Alessandro Volta and the electric battery*, Franklin Watts, NY.

Duckworth, E.: 2005, 'Critical Exploration in the Classroom', *The New Educator*, 1(4), 257-272.

Duckworth, E.: 2001, 'Teaching/Learning Research' in Duckworth, E. (ed.), *"Tell Me More": Listening to Learners Explain*, Teachers College Press, NY.

Duckworth, E.: 1999, 'Address', unpublished transcript, Institute on Teaching and Learning (with Eleanor Duckworth), Miquon School, Conshohocken PA, January 22-23, 1999.

Duckworth, E.: 1991, 'Twenty-Four, Forty-Two and I Love You: Keeping it Complex', in Duckworth, E., *"The Having of Wonderful Ideas" and Other Essays on Teaching and Learning*, Teachers College Press, NY 1987/1996.

Duckworth, E.: 1986, 'Inventing Density', in Duckworth, E. (ed.), *"Tell Me More": Listening to Learners Explain*, Teachers College Press, NY, 2001.

Duckworth, E.: 1973, 'The Having of Wonderful Ideas', , in Duckworth, E., *"The Having of Wonderful Ideas" and Other Essays on Teaching and Learning*, Teachers College Press, NY 1987/1996.

Galileo, G.: 1632, *Dialogue Concerning the Two Chief World Systems – Ptolemaic & Copernican*, Drake, S., trans., University of CA Press (1967).

Galileo, G.: 1638, Two New Sciences, Crew, H & de Salvio, A., trans., Dover NY (1914).

Gallas, K.: 1995, *Talking their way into science: Hearing Children's Questions and Theories, Responding with Curricula*, Teacher's College Press, NY.

Gilbert, W.: 1600, *De Magnete*, Motteley, trans, Dover NY (1958).

Grasswick, H.: 2004, 'Individuals-in-Communities: The Search for a Feminist Model of Epistemic Subjects', *Hypatia*, 19, 85-120.

Hadzigeorgiou, Y.: 1999, 'On problem situations and science learning', *School Science Review*, 81, 43-53.

Hammer, D.: 1997, 'Discovery Learning and Discovery Teaching', *Cognition and Instruction*, 15, 484-529.

Hardwig, J.: 1991, 'The Role of Trust in Knowledge', *Journal of Philosophy*, 88, 693-708.

Hawkins, D.: 2000, 'Reconstruction in Education: A Perspective on Dewey's Development', in Hawkins, D., *Roots of Literacy*, University of Colorado Press, Boulder CO 2000.

Inhelder, B., Sinclair, H., and Bovet, M.: 1974, *Learning and the Development of Cognition*, S. Wedgwood, trans., Harvard University Press, Cambridge MA.

Kheirandish, E.: 1999, *The Arabic version of Euclid's optics, Kitab Uqlidis fi ikhtilaf al-manazir / edited and translated with historical introduction and commentary* Springer NY.

Klein, A.: 1971, *The Hidden Contributors: Black Scientists and Inventors in America*, Doubleday, NY.

Kipnis, N.: 1993, *Rediscovering Optics*, BENA Press, Minneapolis MN.

Lix, A.: 2005, Final Paper, Honors 290F.

Matthews, M.: 2000, *Time for Science Education*, Kluwer NY.

Melchior-Bonnet, S.: 1994, *The Mirror: A History*, K H Jewett, trans., Routledge NY.

Needham, J., and Ling, W.: 1962, *Science and Civilization in China*, v 4, Cambridge University Press, Cambridge UK.

Osborne, M. D., and Brady, D. J.: 2001, 'The Magical and the Real in Science and in Teaching: Joy and the Paradox of Control', in Barton, A. C., and Osborne, M. D., *Teaching Science in Diverse Settings: Marginalized Discourses and Classroom Practice*, Peter Lang, NY.

Payne-Gaposchkin, C.: 1984, *Cecilia Payne-Gaposchkin: an autobiography and other recollections*, Haramundanis, K., ed., Cambridge U Press, Cambridge UK.

Peck, M. S.: 1987, *The Different Drum: Community Making and Peace*, Simon & Schuster, NY

Piaget, J.: 1926 (1983), *The Child's Conception of the World*, J. and A. Tomlinson, trans., Rowman & Allanheld, Totowa NJ.

Piaget, J.: 1936 (1954), *The Construction of Reality in the Child*, M. Cook, trans, Basic Books, NY.

Pierre, L.: 2005, Final Paper, Honors 290F.

Pigott, V.: 1992, 'Ancient Metal-Mirror Making in South India: Analyzing a Mysterious Alloy', *Journal of Metallurgy (JOM)*, 38-40.

Pitchel, S.: 2005, Final Paper, Honors 290F.

Rolin K.: 2002, 'Gender and Trust in Science', *Hypatia*, 17, 95-118.

Sacks, O.: 2001, *Uncle Tungsten: Memories of a Chemical Boyhood*, Knopf NY.

Sauveur, J.: 1701, "General System of Sound Intervals and Its Application ot Sounds of All Systems and All Musical Instruments", Lindsay, R., trans., in *Acoustics: Historical and Philosophical Development*, Lindsay, R., ed., Dowden, Hutchinson & Ross, Pennsylvania, 1973.

Savage, M.: 1967, *Chima Makes a Clock*, African Primary Science, Nairobi.

Sconiers, A. D. and Rosiek, J. L.: 2000, 'Historical Perspective as an Important Element of Teacher's Knowledge: A Sonata-Form Case Study of Equity Issues in a Chemistry Classroom', *Harvard Educational Review*, 70, 370-404.

Settle, T.: 1996, "Galileo's Experimental Research", Max-Planck Institute for History of Science, Berlin.

Shapin, S.: 1994, *A Social History of Truth*, University of Chicago Press, Chicago IL.

Simms, D.L.: 1977, 'Archimedes and the Burning Mirrors of Syracuse', *Technology and Culture*,18, 1-24.

Smith, A. M.: 1999, *Ptolemy and the Foundations of Ancient Mathematical Optics: A Source Based Guided Study*, American Philosophical Society, Philadelphia PA.

Smith, A. M.: 1982, "Ptolemy's Search for a Law of Refraction: A Case-Study in the Classical Methodology "Saving the Appearances" and its Limitations", *Archive for the History of the Exact Sciences*, 26, 221-240.

Transcripts, fall 2005.

Tsui, A.: 2005, Final Paper, Honors 290F.

Tusi, P.: 2005, Final Paper, Honors 290F.

Traweek, S.: 1988, *Beamtimes and Lifetimes: The world of high energy physicists*, Harvard University Press, Cambridge MA.

Verkaik A. and J.: 'Skychasing', Skyart, Elmwood Ontario.

Vermeij, G.: 1997, *Privileged Hands: a scientific Life*, Freeman, NY.

Zembylas, M.: 2004, 'Emotion Metaphors and Emotional Labor in Science Teaching', *Science Education*, 88, 301-324.

Zion, M., Sleak, M., Shapra, D., Link, E., Bashan, E., Brumer, M., Orian, T., Nussinowitz, R., Court, D., Agrest, B., Mendelvici, R., Valanides, N.: 2004, 'Dynamic, Open Inquiry in Biology Learning,' *Science Education*, 88, 728-753.

Educating and Entertaining: Using Enlightenment Experiments for Teacher Training

PETER HEERING

Physics Education/History and Philosophy of Science, Carl-von-Ossietzky Universität, Oldenburg, Germany

Abstract. One aspect of scientific understanding is the notion of science as a cultural activity. In this respect, offering students the opportunity to redo historical experiments can play a central role, as their unlikeness to modern experiments is obvious. However, this is not only a question of materials and procedures, but also of a changing notion of how scientific knowledge is to be constructed and what purpose science has to serve. Thus, being confronted with historical standards and notions can enable students to realize their own beliefs of how science is to be done. At the same time, historical experiments can qualify students to develop or strengthen a conceptual scientific understanding.

In this respect, it appears to be particularly remarkable that experiments from the Enlightenment have not yet received very much attention. This is even more astonishing as these experiments were – according to historians of science – particularly designed in order to be performed by (or in front of) a group of lay persons. Moreover, in this period, entertainment goes hand in hand with natural philosophy. This stands in contrast to experiments that are designed later and are based on the exclusion of a general audience from the process of producing scientific knowledge. In this paper, I will describe at first the characteristics of the experimental practice that can be related to the Enlightenment. This practice will be compared to one that can be taken as the basis of the subsequent standard. In the second part of this paper, I will discuss how these experiments with their respective characteristics can be used in attempting to obtain different aims in educational situations.

Introduction

The implementation of historical experiments in teaching situations is one of the topics that have been promoted by the Oldenburg group for almost two decades. In doing so, various experiments have been used either with teacher students or with students at secondary schools. Even though we lack up to now 'hard empirical evidence', it can be argued from a series of individual experiences that this approach could be used as a successful addition in various teaching situations. Yet, in my view, a weakness has to be attested in these accounts (and this does include my own work): The historical experiments have been treated as just one group of experiments. Consequently, basically no differentiation between various historical experiments has been made in the published accounts. This deficit results at least in part from the focus of these papers. They either described the experiences that were made in specific educational situations (Rieß 1995, Heering 2000, Sichau 2000, Rieß 2001, Heering/Müller 2002) in which historical experiments were used, or they discussed specific experiments and their potential for teaching situations (Heering 2001, Heering 2003).

To develop a more differentiated account of potential use of historical experiments, I will at first sketch different experiments with respect to the style of experimentation that can be attributed to these experiments[1]. In a second step, this characterisation will serve as a criterion for a discussion of the potential implementation in teaching situations. However, I am not intending to indicate that specific experiments are to be used in teaching situations. Rather, I aim to show that particularly experiments from the Enlightenment could be useful in a specific manner in science teaching.

Experiments from the Enlightenment – The example of the solar microscope

The experimental practice of the Enlightenment can basically be connected with devices such as the electrostatic generator and the air pump; experiments with both devices became extremely popular in that period. However, other (smaller and cheaper) apparatus were also very popular and can be considered characteristic with respect to the style of experimentation. Among the latter is the solar microscope, a projection microscope that was promoted from the 1740s onwards and became extremely popular in the second half of the eighteenth century[2].

Simplified, a solar microscope can be described as a combination of a mirror, a condensing lens and a microscope, in many cases a single lens microscope. The instrument is used in a darkened chamber; it is placed in the shutter of the window in a manner that the mirror is outside the room and the microscope inside. With the help of the mirror, sunlight is reflected onto the condensing lens which focuses the light. The (transparent) specimen is placed shortly before the focal point, consequently it is completely illuminated. With a microscope, an image of the specimen is projected onto a screen that is placed in the room.

I had the opportunity to use two solar microscopes at the Deutsches Museum Munich. One of them was made by the Dollond workshop; this instrument can be regarded as the best quality available at that time. The other solar microscope was made by Junker and was meant to be used in schools. In using the solar microscopes, certain characteristics of the style of experimentation became obvious[3]. First of all, the Dollond instrument could be used without any formal training[4]. Moreover, the resulting projections are (seemingly)[5] visible to everyone. This is a significant difference from common microscopes where the image is a virtual image that is created in the eye of the observer. Consequently, one has to learn how to look at the specimen. And, in demonstrating the projected images to an audience, it turned out that the projections create a situation that encourages a discourse of the audience.

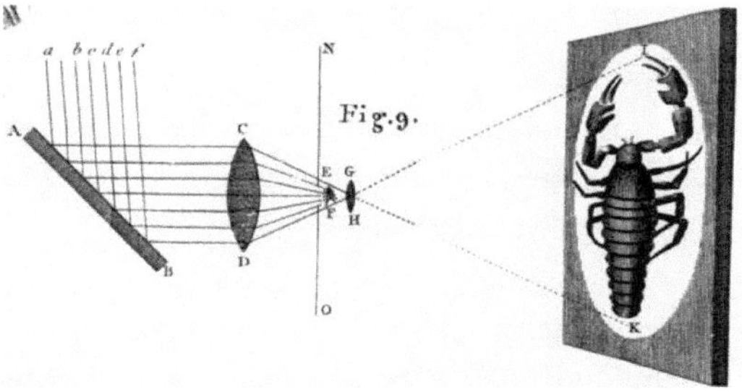

Figure 1. Working principle of the solar microscope. From Adams 1794, with permission of the Bakken Library and Museum, Minneapolis

Above all, the latter aspect was already discussed by historians of science with respect to the cultural meaning of the solar microscope. In doing so, this device is clearly distinguished from other microscopes which are described as being considered inadequate with respect to specific social standards:

> "The most challenging obstacle to overcome in the effort to associate science with politeness was dispelling contemporary caricatures of those interested in natural philosophy as solitary, pedantic, asexual, melancholic bores, more comfortable with the microscopic creatures found in pond water than with the polite ladies and gentlemen of society. To counter this image, polite science explicitly associated the study of natural philosophy with conversation, the fundamental activity of polite society" (Walters 1997, p. 126).

As Walters had also remarked, particularly these aspects can be identified as being irrelevant to the practice with the solar microscope. Moreover, this device offers actually an acceptable alternative:

> "Indeed, solar microscopes strongly encouraged common observation ... The popularity of these instruments in the eighteenth century undoubtedly derived in large part from their ability to illuminate the religiously and aesthetically inspirational microscopic world in polite social settings" (Walters 1997, p. 141, a comparable statement can be found in Stafford 1991, p. 360).

Walkers' description seems to correspond nicely to a characterisation from the eighteenth century. With respect to the advantages of the solar microscopes (which he also sold) the London instrument maker G. Adams pointed out:

> "There are many Conveniences in this, which no other microscope has, … Besides this particular Property it hath, that the Numbers of People may view an Object at the same Time, and may point to different Parts thereof, and by discoursing on what they see, may understand each other better, and more probably find out the Truth, than when they are obliged to look one after the other" (Adams 1747, p. 11)[6].

However, there is one aspect which I consider to be highly important, as it gives this phrase a different focus. This gets obvious when another aspect is taken into consideration that has been made explicit by Schickore:

> "(t)he observer's mental activities were understood to be one major factor conditioning the outcome of microscopical investigations. … If the observers exerted strict control over their mental operations, they could prevent erroneous results" (Schickore 2001, p. 128).

Taking Schickore's observation into consideration, a shift of the focus gets accessible with respect to Adams' statement quoted above: It is not only the opportunity to look at microscopic images in a polite social setting that made the solar microscope attractive. It is actually the opportunity to develop jointly (explicit) knowledge about the microcosm as well as (implicit) competences with respect to microscopic seeing (which means to identify what is actually looked at on the screen). In this respect the attempt to develop or improve literacy with respect to the microcosm seems to be an issue that had relevance in these discussions and that therefore played an important role for the design of experiments.

The style of experimentation that has been identified with respect to the practice with the solar microscope is not limited to this device. Other styles can be identified that show comparable characteristics. Among them are several apparatus' that were used in electrical demonstrations that also combined educational as well as entertainment features; among them are, for example, Franklin's bells or discharge plates[7].

However, as I discussed elsewhere (Heering 2006), towards the end of the eighteenth century a new style of experimentation was developed which significantly differs from the one discussed up to now. This style is of particular interest in teaching situations, as, in retrospect, it can be taken as a relevant step of the development of modern scientific practice.

An exclusive approach to scientific practice – the example of Coulomb

One of the instruments that can be taken as being a materialisation of this style of experimentation is the torsion balance that had been developed by the French military engi-

neer Charles Augustin Coulomb. This device, first introduced in 1785, was transformed into an icon of electricity by the French Academy of Science. Moreover, for a certain period of time, it was made to be the device that could decide whether or not a phenomenon was electric[8]. However, the practice with this instrument differed significantly from the established one. This becomes clear from the analysis of the practice with the torsion balance, but also from the reaction of German scientists to this device (Heering 1994).
As the experiments with the torsion balance and the characteristics of the style of experimentation connected with this device were discussed elsewhere (Heering 1998, Heering 2001), I am just mentioning some of the results very briefly. A relevant aspect in this style lies in the purpose of the experiments with the torsion balance; they were designed to come to a mathematical description of the behaviour of the electrical fluid. Consequently, the experimental results were communicated in a mathematical form. These results were determined through few measurements. In this respect, it was the singular event that was considered as being relevant; consequently, no statistics were employed. The torsion balance was – as well as other instruments that can be related to this style of experimentation – highly sensitive. Accordingly, the measurements were highly error sensitive. Moreover, in contrast to the experiments described in the preceding part, the experiments were carried out without witnesses.
As in the case of the solar microscope, this style of experimentation is not limited to this instrument[9]. Another example is the ice-calorimeter introduced by Lavoisier and

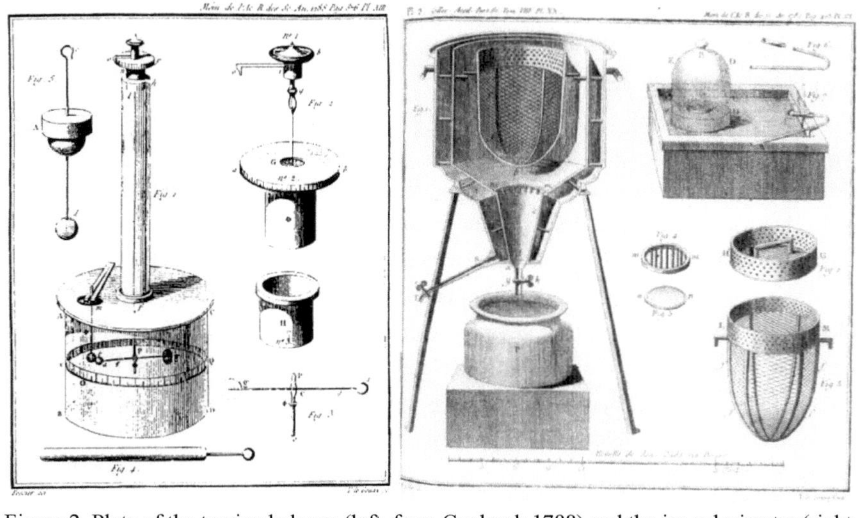

Figure 2. Plate of the torsion balance (left, from Coulomb 1788) and the ice calorimeter (right, from Lavoisier/Laplace 1787), with permission of the Landesbibliothek Oldenburg.

Laplace almost simultaneously. One aspect that appears to be remarkable to me is the status that both the torsion balance and the ice-calorimeter have achieved until today. Both can be regarded as being icons of specific fields, which can be inferred from the observation that copper plates of both instruments are still reproduced in various science text books, even if they have no specific historical approach. This can be taken as an indication that these experiments (or at least the apparatus) are seemingly valuable with respect to the formation of modern physicists.

The use of historical experiments in science education – the example of Coulomb

If one looks slightly closer at how the experiments with the torsion balance and with the ice-calorimeter are discussed in physics textbooks, it becomes obvious that they are only discussed on a principle level. Neither their historical context nor the actual experimental practice is a topic in the teaching. Instead, the experiments are discussed as being straightforward and easy to perform. Even though the actual setups are normally not available at schools or universities, there exists, at least in case of Coulomb's law, a didactical device that can be used (according to the claim that is made by the manufacturer) to determine Coulomb's law[10].

However, attempting to use a reconstructed torsion balance can produce a lot of difficulties. Actually, an accurate measurement with this device is not possible at school and even hardly at university level. Yet, the device can be used in both situations (Heering 2000, Heering 2001). What can be accomplished in using the device, is to make students aware of the difficulties that are connected with the experiment, to discuss issues such as the 'rhetoric of numbers' that can be identified in Coulomb's account as well as the question of what is to be considered as a 'proper experimental proof'.

Moreover, by using this experiment after having carried out other electrical experiments that show the characteristics of the style of experimentation discussed at the beginning of this paper, the students can understand that scientific practice undergoes changes and that standards are not time-independent. Or, using another approach, the rejection of Coulomb's work by German scientists of the 19th century can be implemented in this course, thus showing that scientific laws are not simply 'found' and instantaneously accepted, but that the criteria for the acceptance of an experiment and the results are more complex than this oversimplified image seems to suggest.

Two important aspects for the argument of this paper should have become obvious from these sketched approaches to how to teach Coulomb's law in a historical manner. First of all, experiments with instruments such as the torsion balance require a thorough embedding in a frame that has been constructed under educational aspects. This frame has a twofold meaning: It is relevant with respect to the motivation of the students – the torsion balance itself is not that easy to understand and thus does not appear to be very motivating in the beginning. But what seems to be more important: The experiments need to be introduced by some description that gives the students an idea of what the device does – and the reason why. Thus, students do not understand the experiment and the result, unless they have acquired sufficient scientific background knowledge.

Moreover, the frame is also relevant to the outcome of the course: It enables the students to discuss issues that go beyond the scientific aspects of the inverse square law. As a result of such an approach, the students are not only enabled to develop an understanding of the physical content (Coulomb's law)[11], but they can also gain insights into epistemological questions and develop an understanding of the nature of science.

The use of historical experiments in science education – the Enlightenment experiments

Things are somewhat different with respect to experiments that show the characteristics I have discussed for the solar microscope. As one example of how these experiments can be used in educational settings, I would like to discuss electrical experiments from the eighteenth century[12]. Among these experiments are the electrical kiss, the ignition of alcohol (Fig. 3) or the amplification effect of the Leyden jar (Fig. 4) and its discharge through a human chain.

Figure 3. Electric ignition of alcohol (from Desaguliers 1751), with permission of the Landesbibliothek Oldenburg.

Of course, one could argue that the major difference between these experiments lies in their character: Whilst Coulomb's experiment is a precision measurement, the others are demonstrations of phenomena and either semi-quantitative or not quantitative at all. However, in my view this argument misses one important point. As I pointed out in the discussion of the styles of experimentation, Coulomb's experiment was designed to be exclusive with respect to a lay audience, whilst the experiments I am discussing in this part were designed to be performed in front of or even by lay persons in order to educate them. And in my view, the major distinction between these experiments lies in this aspect of the style of experimentation.

This distinction makes it possible to introduce the experiments to the students in different manners. Whilst a text is necessary describing the concept and the physics background of the experiment in case of the torsion balance experiment, this is not the case with the discharging experiments. They were designed in a manner that the actual experimental procedures were easy to understand. Consequently, in the part of the reader which deals with these experiments, we include for some of them just the historical copper plates as an indication of what could be done (see Fig. 3). In other examples (such as the Leyden jar), we also use historical accounts of the experiment to give the students a starting point for their own experimentation[13].

It is important to give the students a narrative of the introduction of the Leyden jar, as the image (Fig. 4) is not self-explanatory. Moreover, it also appears to be important from a motivational point of view to describe the difficulties the historical actors experienced, because the device (a bottle with water and a nail) does not seem to be too motivating, yet, combining it with the historical narrative creates a context in which the device has a different meaning. And from this meaning, the students are interested in working with the device and in developing an understanding of its behaviour. This necessity arises also, as the tutorial for these experiments does not contain the actual concept of physics by which this experiment is explained.

Actually, this is not necessary to carry out these experiments. As a result, there is no inherent conceptual frame that is necessary to develop an understanding of the experiment. However, the students still have to develop an understanding what is happening in the experiment. If one observes the students in working on these experiments, it is striking that they are very motivated; these experiments obviously appear attractive to students – at least they develop such an attitude during the course of experimentation[14].

Figure 4. Discharge of a Leyden jar (from Nollet 1753), with permission of the Bakken Library and Museum, Minneapolis.

Having just an image (and no pre-set conceptual frame) is of consequence to the experimenting: the students are able to carry out the experiments according to the image, however, some of them also start to modify some of them, partly in order to improve the initial version, partly to develop a conceptual understanding. Thus, when they realize that they have problems in understanding the behaviour of the experimental set-up, they still feel competent enough to achieve an understanding by doing some more experiments. One aspect that seems to be very important in this respect lies in the initial conception of the experiments:

Figure 5: Students experimenting with condensing electroscopes and Leyden jars (Foto Th. Biedermann)

They were designed to be understood by lay persons. Even though they are not free of theory, they are transparent in a sense that they do not appear as black boxes; on the contrary. Working with a phial filled with water in which a nail is placed does appear to be a very simple piece of equipment. Yet, when students are starting to charge and discharge the water, they quickly realize that the behaviour of this device is not that easy to understand or predict. However, to them it is not a complicated device that is to be blamed, but they feel that they should be able to find out what this bottle does – and the reason why[15]. As a result, they start to modify their experiments and start to develop an inquiry into what this device does[16]. In some cases (particularly when experimenting with the Leyden jar and with Volta's condenser electroscope, see Fig. 5), this may lead to a practice which can be described as being comparable to what Steinle has called 'exploratory experimentation' in an epistemological context (Steinle 2005) but which also shows remarkable parallels to what Reinhold called 'open experimenting' (Reinhold 2000).

In working with the experimental set-ups, the students develop their individual understanding – even if it is not coherent to the modern scientific interpretation of the experimental outcome – from their prior conceptions. However, as they are not doing these experiments on their own but in groups, they have to discuss their sense of the experiment (which does include the interpretation) with their fellow students. In doing so, they develop a sort of common interpretation in their group. Moreover, as the concept of the lab course includes frequent exchange with the teachers, they also get reflections on their interpretations from another source. As a result, they are developing a thorough interpretation of the experimental situation. As these students have already studied physics for some three to four years, this may sound slightly irritating in case of these seemingly simple experiments. However, experiences have shown that in particular the individual development of an interpretation of these experiments requires a conceptual scientific basis that a significant number of students have not yet reached – despite their formal training and despite their ability to perform complex calculations in the field of theoretical electrodynamics. In this respect, it is particularly striking that a significant number of students still find it difficult to understand the Leyden jar[17], but also effects such as electrostatic induction as it appears for example in the electrophorus or in the condenser electroscope. These deficits are overcome during the course, which gets clear in a very explicit manner, when those students who have already worked with the condenser electroscope are confronted with the electrophorus. Now, they are very quick in giving an insightful description and explanation of the behaviour of the device. Thus, from confronting students in such an open manner with experiments from the eighteenth century, they are able to develop a profound conceptual understanding in the field of electricity by working with the apparatus[18].

Conclusion

In this paper, I have discussed the educational use of experiments from two different styles of experimentation. The use of experiments from the style (that can be related to experiments such as Coulomb's demonstration of the inverse square law in electrostatic) can particularly serve to develop an understanding of epistemological aspects. In this respect, these experiments are excellent examples for an observation Lind pointed out very explicitly: "Among didacts [of physics, PH] seems to exist a vast consensus that the most important application of the history of physics in teaching lies in the epistemological consolidation of physics" (Lind 1992, p. 41). Even though I am far from rejecting this claim, yet, an addition seems to be necessary. As I have tried to show, experiments such as Coulomb's require a basic scientific understanding. Thus, they can only be carried out by students who have already acquired an appropriate body of knowledge. This is different with the experiments that are related to the style of experimentation connected to the Enlightenment. Here, the electrical experiments[19] do not require a preliminarily conceptual understanding of the instruments' behaviour. However, they are appropriate for the students to develop this conceptual knowledge in working on these experiments.

Up to now, I have discussed the educational use of experiments that originated from two different styles of experimentation individually[20]. However, such a discussion could be misleading, if it is taken as a competitive comparison. Experiments from both styles of experimentation develop an even stronger potential in science teaching when they are combined adequately.
One of the major deficits that empirical studies on the conceptions of teachers as well as teacher students have identified lies in the idea of the existence of the scientific method (see also the contribution of Kokkotas et al. in this volume). In my paper, it should have become explicit that experimental practice is not stable but that the style of experimentation changes (and together with the standard, the criteria of what counts as a scientific experiment change as well[21]). By experiencing experiments which have an incompatible style of experimentation, the students get the opportunity to realise that there is no such thing as 'the experimental method'. Consequently, teaching Coulomb's law with the torsion balance after the students have made experiences with electrical experiments from the Enlightenment, has turned out to be particularly beneficial, as the students are able to experience the dissimilarities of experiments that were initially made in roughly the same period and in the same field.

Notes

[1] 'Styles of experimentation' is an epistemological concept I am using with reference to Ludwik Fleck's concept of style of thought and thought collective (see Cohen/Schnelle 1986, Heering 2006).

[2] Moreover, in the nineteenth century the solar microscope was characterised as being a toy instead of a scientific instrument, whilst the status of devices such as the electrostatic generator or the pump was not changed.

[3] The experiences made in using the solar microscope are discussed and analysed in Heering (2006, 2007).

[4]This was slightly different with the Junker device – however, this is understandable as the instrument was meant to be used in schools. Therefore, the teacher could be expected to develop the skills necessary for operating the instrument.

[5]This qualification is necessary, as it is not possible to identify the specimen from the projected image on the screen without prior experiences. However, different from other microscopes, it is not necessary to learn how to look at the image (for this distinction see Fleck 1947).

[6]An adequate discussion of Adams' notion of 'truth' would go beyond the scope of this paper. To deal with this term in a pragmatic manner, it could be interpreted as 'commonly accepted knowledge'.

[7] I do not include in this list the electric generator or the Leyden jar, even though they could also be used in experiments that can be seen in this context. However, these devices can also be (and have been) used in experiments that are related to a different style of experimentation.

[8] On the torsion balance as the device for French scientists to decide whether the voltaic pile produces electricity or another fluid see Brown (1969) and Sutton (1981).

[9] These characteristics of the style of experimentation are not only identified from the analysis of Coulomb's torsion balance experiments, but are also comparable to the findings of the analysis of Coulomb's work on the diurnal variation of the magnetic field of the earth (Heering/Osewold 2005) and the torsion of metal wires (Heering 2006b)

[10] This device is a Schürholz balance which has a bifilar suspension and is less sensitive than Coulomb's balance. These didactical devices have conducting spheres and are connected to a power supply to compensate charge leakage (see for example Lauginie 2003, http://www.hanmacco.co.kr/products/ed_pdf/page093.pdf, Oct. 15th, 2006).

[11] At university level, the students are of course already familiar with Coulomb's law when they work with the torsion balance. However, at secondary school level, the electrostatic inverse square law can be introduced by using the torsion balance.

[12] These experiments have been used for teaching at university level as well as at school level. As I have not used the solar microscope in teaching situations, I consider them to be more appropriate for a discussion of what educational aspects can be achieved with experiments that are developed in this style of experimentation. The standard discussion of electricity and electrical experiments in this period is still Heilbron (1979), however, see also Hochadel (2003) whose study has a specific focus on the non-academic experimenters and thus provides an excellent addition to Heilbron's work.

[13] In case of the Leyden jar, the students get texts initially published in Winkler (1746) and Priestley (1772). Descriptions of how to build the apparatus that can be used for such experiments are found in Habben/Mehrle (1994) and Bakken (1995), images can be found for example in Teichmann (1998) and Schiffer (2003).

[14]Some students change their attitude towards the experiments where they are using their own body for discharging. Some of them are afraid of electrical shocks initially, but this attitude changes during the course, when they realize that these shocks are not harmful but entertaining (Heering 2000, Olberding 2006).

[15] In this respect the historical context of the Leyden phial being an important innovation in eighteenth century electricity research is meaningful. At least partly due to this attribution, some of the students consider the experiment a scientific one which they feel they have to understand.

[16] A comparable behaviour of students has been described by Cavicchi in her account of students experimenting on electromagnetic experiments from the nineteenth century (Cavicchi 2003).

[17] Remarkably enough, quite a lot of them identifies the Leyden jar as a condenser – yet, when asked to give an interpretation of the behaviour of the condenser in the experiments

they have carried out, they fail to do so. A more thorough analysis of the conceptual deficits and the adequateness to overcome these deficits through the discussion of some of the electrical experiments from the Enlightenment is planned for the next year. Apart from that, we are also attempting to analyse the motivational potential of these experiments in more detail. Here, a first pre-study (Olberding 2006) has given results which indicate that the experiences sketched in this paper can be validated in a more quantitative study.

[18] At the same time, the students also understand that these experiments were scientific according to the standards of the mid-eighteenth century. Thus, they also realise that scientific practice is changing and is time-dependent.

[19] This potential is not limited to the electrical experiments. A comparable potential seems to exist, for example, in case of didactic demonstrations in the field of mechanics, that were developed in Leyden by Musschenbroek and 'sGravesande in the 1720s (see Riess et al. 2005, Degendorfer 2006).

[20] It should be clear that there exist more than just the two styles of experimentation I am discussing in this paper.

[21] As already mentioned, for French researchers the torsion balance became the device for identifying the electric fluid. This can be taken as an example where the criteria of what could be considered 'scientific' were significantly changed in one community. Remarkably, this device was not immediately adopted by researchers outside France; in Germany, it was explicitly rejected at the beginning of the nineteenth century (see Heering 2001).

References

Adams, G.: 1747, *Micrographia Illustrata, or, The Knowledge of the Microscope Explain'd: Together with an Account of a New Invented Universal, Single or Double, Microscope, Either of which is capable of bein applied to an Improv'd Solar Apparatus*, 2nd Ed., Printed for the Author, London

Adams, G.: 1794, *Lectures on Natural and Experimental Philosophy, considered in it's present state of Improvement: Describing, in a familiar and easy manner, the principal phenomena of Nature; and shewing, that they all co-operate in displaying the Goodness, Wisdom, and Power of God*, R. Hindmarsh, London, 5 Vols.

Bakken Library and Museum: 1995, *Sparks and Shocks: Experiments from the Golden Age of Static Electricity*, Dubuque (Iowa), Kendall/Hunt.

Brown, T.M.: 1969, 'The Electric Current in Early Nineteenth-Century French Physics', *Historical Studies in the Physical Sciences* **1**, 61–103

Cavicchi, E.: 2003, 'Experiences with the magnetism of conducting loops: Historical instruments, experimental replications, and productive confusions', *American Journal of Physics* **71**, 156-167

Cohen R.S.; Schnelle, T.: 1986, *Cognition and Fact: Materials on Ludwik Fleck*, Reidel, Dordrecht.

Coulomb, C.A.: 1788, 'Sur l'électricité et le magnétisme, premier mémoir, ...', *Mémoires de l'Academie Royale des Sciences pour l'Année 1785*, 569–577.

Degendorfer, K.: 2006, *Ausgewählte Experimente Galileis zur Analyse des freien Falls und deren Umsetzung als Demonstrationsversuche durch Musschenbroek und s'Gravesande. Experimenteller Nachvollzug und physikalische Analyse*, final thesis, Augsburg.

Desaguliers, J.T.: 1751, *De Natuurkunde uit Ondervindingen*, 2 vols., Isaak Tirion, Amsterdam.

Fleck, L.: 1947 (1986), 'To look, to see, to know', in R.S. Cohen and T. Schnelle (eds.), *Cognition and Fact: Materials on Ludwik Fleck*, trans., Dordrecht, 1986, 129–151.

Habben, D.; Mehrle, U.: 1994, *Vom Bernstein zur Volta-Säule: Geschichte der Elektrizität im Unterricht*, Redaktions- Gemeinschaft Soznat, Marburg.

Heering, P.: 2007, 'The Enlightened Microscope – Re-enacting and Analysing Projections with 18th Century Solar Microscopes', *British Journal for History of Science*, under consideration.

Heering, P.: 2006b, 'Regular Twists: 'Replicating Coulomb's Wire-Torsion Experiments', *Physics in Perspective* **8**, 52–63.

Heering, P.: 2006, *Experimentierstile in der zweiten Hälfte des 18. Jahrhunderts*, habilitation thesis, Hamburg.

Heering, P.: 2003, 'Rejected Historical Experiments and their Use for Science Teacher Training', in D. Metz (ed.), *Proceedings of the 7th International History, Philosophy and Science Teaching Group meeting*, Winnipeg, 366-376.

Heering, P.: 2001, 'Using reconstructions of historical devices for teaching purposes: The example of Coulomb's law', in R. Mason (ed.), *The Uses of History in Science Education: Proceedings of the the Third International Seminar for the History of Science and Science Education* (Winnipeg: University of Manitoba).

Heering, P.: 2000, 'Getting Shocks: Teaching Electrostatics with Historical Experiments at Secondary School Level', *Science and Education* **9**, 363–373.

Heering, P.: 1998, *Das Grundgesetz der Elektrostatik. Experimentelle Replikation und wissenschaftshistorische Analyse*, DUV, Wiesbaden

Heering, P.: 1994, 'The replication of the torsion balance experiment: The inverse square Law and its refutation by early 19th-century German physicists', C. Blondel and M. Dörries (eds.), *Restaging Coulomb: Usages, Controverses et réplications autour de la balance de torsion*, Leo S. Olschki, Florence, 47–66.

Heering, P.; Müller, F.: 2002, 'Cultures of Experimental Practice – An Approach in a Museum', *Science and Education* **11**, 203-214.

Heering, P.; Osewold, D.: 2005, 'Ein Problem, zwei Wissenschaftler, drei Instrumente', *Centaurus* **47**,115–139.

Heilbron, J.L.: 1979, *Electricity in the 17th and 18th Centuries*, University of California Press, Berkeley.

Hochadel, O.: *Öffentliche Wissenschaft: Elektrizität in der deutschen Aufklärung*, Wallstein, Göttingen.

Lauginie, P.: 2003, 'Cavendish and Coulomb Torsion Balances: Only one Technique, only one Law; Two Balances, two Aims, two Styles, two Destinies', D. Metz (ed.), *Proceedings of the 7th International History, Philosophy and Science Teaching Group meeting*, Winnipeg 504–517.

Lavoisier, A.L.; Laplace, P.S.: 1787, 'Mémoire sur le Chaleur', *Mémoires de l'Academie Royale des Sciences pour l'Année 1787*, 355-408.

Lind, G.: 1992, *Physik im Lehrbuch 1700 - 1850: Zur Geschichte der Physik und ihrer Didaktik in Deutschland*, Springer, Berlin.

Nollet. J.A.: 1753, *Essai sur l'Électricité des Corps*, 3rd edn., Paris, Freres Guerin.

Olberding, I.: 2006, Elektrostatische Unterhaltungsexperimente des 18. Jahrhunderts: Experimentelle Analyse und didaktische Betrachtung ausgewählter Beispiele, final thesis, Oldenburg.

Priestley, J.: 1772, *Geschichte und gegenwärtiger Zustand der Elektricität*, Gottlieb August Lange, Berlin & Stralsund (Reprint Th. Schäfer, Hannover 1983).

Reinhold, P.: 2000, 'Open Experimenting – An Approach to Structure Science Teaching and Learning', S. Hopmann; K. Riquarts; I. Westbury (eds.) *Teaching as a Reflective Practice. The German Didaktik Tradition*, Erlbaum, Mahwah, New Jersey, 295-318.

Rieß, F.: 2001, 'Historical Experiments – A Modern Approach to Physics Teaching?', J. Turlo (ed.), *Science and Mathematics Teaching for the Information Society*, Top Kurier, Torun, 41 – 52.

Rieß, F.: 1995, 'Teaching Science and the History of Science by Redoing Historical Experiments', F. Finley; D. Allchin; D. Rhees; S. Fifield (eds.), *Proceedings of the Third International History, Philosophy, and Science Teaching Conference*, Vol. II, University of Minnesota, Minneapolis, 958–966.

Rieß, F.; Heering, P.; Nawrath, D.: 2005, 'Reconstructing Galileo's Inclined Plane Experiments for Teaching Purposes', *Online Proceedings of the Eighth International History, Philosophy, Sociology & Science Teaching Conference* (http://www.ihpst2005.leeds.ac.uk/papers/Riess_Heering_Nawrath.pdf)

Schickore, J.: 2001, 'Ever-Present Impediments: Exploring Instruments and Methods of Microscopy', *Perspectives on Science* **9/2**, 126-146.

Schiffer, B. M.: 2003, *Draw the lightning down: Benjamin Franklin and Electrical Technology in the Age of Enlightenment*, University of California Press, Berkeley.

Sichau, C.: 2000, 'Practising Helps: Thermodynamics, History, and Experiment', *Science and Education* **9**, 2000, 389-398.

Stafford, B. M.: 1991, *Body criticism: imaging the unseen in Enlightenment art and medicine*, Cambridge (Mass.).

Steinle, F.: 2005, *Explorative Experimente: Ampère, Faraday und die Ursprünge der Elektrodynamik*, Steiner, Stuttgart.

Sutton, G.: 1981, 'The politics of science in early Napoleonic France: The case of the voltaic pile', *Historical Studies in the Physical Sciences* **11/2**, 329–366

Walters, A. N.: 1997, 'Conversation pieces: Science and Politeness in Eighteenth-Century England', *History of Science* **35**, 123-154.

Winkler, J.H.: 1746, *Die Stärke der Electrischen Kraft des Wassers in gläsernen Gefäßen*, Bernhard Christoph Breitkopf, Leipzig (Reprint Hüthig, Heidelberg 1983).

Pedagogical Renewal of the Millikan Oil Drop Experiment

STEPHEN KLASSEN

University of Winnipeg, Winnipeg, Manitoba, Canada, R3B 2E9

Abstract: The Millikan oil drop experiment has been characterized as one of the 'most beautiful' physics experiments of all time and, certainly, as one of the most frustrating of all the exercises in the undergraduate physics laboratory. In this paper the historical background of the oil drop experiment is outlined, emphasizing those details humanizing the protagonists Millikan and Fletcher. The difficulty in doing speculative research without depending on presuppositions is stressed, based on the historical account and on the insights of university physics students exposed both to the historical account and the student experiment. Difficulties current students have in performing the experiment are discussed from the perspective of Hodson's (1993) framework and student observations. Lastly, prospective historical materials are outlined that may be used to encourage student insight into the fundamental nature of electricity. It is proposed that these four aspects are essential as a basis for the pedagogical renewal of the Millikan oil drop experiment.

Introduction

In a 2002 article in *Physics World* (Crease, 2002), Robert Crease reported on a poll that had been conducted by him to determine which physics experiments the journal's readers considered the 'most beautiful' of all time. Among the top ten was Millikan's oil drop experiment. In the words of one reader, 'the experiment can leave no–one in doubt that electrical charge is quantized and that modern physics is real, observable, and true' (Crease 2002, p. 19). Crease observed that, disturbingly, many of his respondents seemed to think that the experiments they were proposing had been conceived, or could be carried out and understood, with considerable ease. However, Crease notes, 'it hardly needs saying that the various experiments demonstrating quantum–mechanical principles were not born simply, and have not made the world simpler' (Crease 2002, p. 20). Due to the pressure in science education towards the decontextualization and distillation of scientific knowledge this insight of historians and philosophers of science is not promoted. Rather, contemporary textbooks often produce a caricature of experiments such as Millikan's in textbooks (Niaz and Rodríguez 2005; Rodríguez and Niaz 2004).

A similar kind of dynamic seems to be at work where the Millikan oil drop experiment is being used in student labs. Modern adaptations of the original Millikan apparatus have been designed so as to simplify the performance of the experiment as much as possible, tending to give the false impression that Millikan's measurements were as effortless as they were 'beautiful'. The notion that the Millikan experiment is uncomplicated may also be promoted by laboratory instructions that implicitly or explicitly convey the impression that the experiment is 'quite simple' (College of St. Catherine 2006). Most undergraduate physics students will, at some time in their laboratory work, perform a 'simplified' re–construction of the oil drop experiment using commercially available apparatus. Although students may consider the effort to determine the value of 'e' worthwhile in cases where the procedure is carefully designed and there is sufficient instructor assistance, it is well known that the exercise is not straightforward. Kruglak concluded that 'the experiment remains perhaps the most frustrating of all the exercises in the undergraduate laboratory' (1972 p. 769).

There is a gap, then, between the legendary greatness of the original Millikan experiment and the practical difficulties in the student laboratory. That is to say, students cannot anticipate the challenges that the oil drop experiment represents and hardly identify with the experiment's originators. The historical aspects of the original experiment will, then, need to be developed with the view of assisting students to approach the experiment with a better knowledge of what was involved and the motivations of and challenges faced by the originators. This aspect can then be applied to the re–design of the pedagogical underpinnings of the lab.

Historical Background[1]

Millikan's interest in the elementary electrical charge may have begun in 1895, when he spent a year in Europe. In October he travelled to Berlin where Max Planck worked. Millikan attended a series of lectures on theoretical physics given by Planck. The subject on every physicists mind in Berlin at that time was the nature of cathode rays. When Millikan returned from Europe, the next spring, he took up a research assistantship with Michelson at the University of Chicago. The very next year, in 1897, J. J. Thomson determined that cathode rays are composed of electric corpuscles. Millikan was obviously inspired by the major developments in physics at the time. However, like other great physicists before him, for example, Ohm and Kelvin, his motivation was also financial. By 1906 he had been at the University of Chicago for ten years and had only become an assistant professor, with no major scientific publications. That year he began to build a large house for his growing family which turned out not to be affordable on his salary. In order to turn his fortunes as a scientist, Millikan directed all his energies outside of teaching to determining the value of the elementary electrical charge, at first re–doing the investigations of H. A. Wilson, using clouds of water droplets. This initial work produced results that were somewhat more accurate that those of Wilson, since Millikan had used a 1000–Volt bank of batteries that allowed him to separate out a single water droplet. Millikan took the opportunity to travel to Winnipeg during the summer of 1909 to attend the meeting of the British Association in late August. The mathematical and physical science section meeting, under the presidency of Lord Rutherford, took place at what is now the University of Winnipeg in the Wesley Hall building (Duckworth 1993). Rutherford, in his presidential address, spent a considerable time arguing for the existence of atoms but noted that 'it has not yet been possible to detect a single electron by its electrical or optical effect, as in the case of alpha particles' (in Fletcher 1982, p. 45). Sitting in the audience was Robert Millikan. He would have the opportunity to contradict the statement of Rutherford a few days later when he would announce that he had been able to isolate tiny water droplets containing only a single elementary electrical charge and measure a value for that charge. However, the experimental method was severely limited and the accuracy of the result was not good.

Millikan's scientific career might have floundered from here on, had it not been for the arrival of a bright young graduate student into his life. Millikan had, several times, counselled this physics student, Harvey Fletcher, who was hoping to gain admission into the Ph.D. program. It seems that the two had developed an immediate liking for each other. In early December of 1909, Fletcher finally succeeded in obtaining an appointment with the busy Millikan, who had invited Fletcher to the research lab for a tour and a chat about a possible Ph.D. thesis project. Millikan's Ph.D. student at the time, Louis Begeman, was also present. The three talked about the research to measure the charge on the electron. Millikan let Fletcher look into the microscope as water droplets formed in the chamber. Fletcher recalls that 'the water forming the droplet evaporated so fast that it would only stay in view for about 2 seconds, so it was difficult to get more than a rough estimate of the charge' (Fletcher 1982, p. 44). The three began to discuss alternatives to water droplets, for example, mercury and oil. At the end of the brief discussion, Millikan said to Fletcher, 'There is your thesis; go try one of these substances which will not evaporate' (Fletcher 1982, p. 44).

It would be an understatement to call Fletcher enthusiastic at that instant. He wasted no time in going directly to a drug store to buy a perfume atomizer, returning to the lab to put together a crude apparatus out of parts and components he was able to find. By that evening he was able to spray oil across the parallel plates and look through the telescope. In his words, 'The field was full of little starlets, having all colours of the rainbow. The larger drops soon fell to the bottom, but the smaller ones seemed to hang in the air for nearly a minute. They executed the most fascinating dance. I had never

seen Brownian movement before. The tiny droplets were being pushed first that way and then this way by the actual molecules in the air surrounding them. I could hardly wait until I could try an electrical field upon them to see if they were charged.' (Fletcher 1982, p. 45). It was not until the next day that Fletcher finished connecting a 1000–Volt battery bank, the required switch, and a Voltmeter. Again he sprayed the oil and observed the tiny drops falling and moving about with Brownian motion. He recalled, that 'As soon as I turned on the switch some of them went slowly up and some went faster down. I was about to scream as I know then that some were charged negatively and others positively.' (Fletcher 1982, p. 46). That day Fletcher played with his crude oil drop apparatus and was able to determine a 'fairly decent' value for 'e'.
It was not until a few days later that Fletcher was able to get the busy Millikan to come to the lab to see his setup. Millikan was greatly excited by what he saw, concluding that it should be possible to get an accurate value of 'e' by this method. From that day forward, Millikan stopped working with his previous Ph.D. student, Begeman, and worked nearly every afternoon for two years with Fletcher. Begeman who, in his view, never received proper credit from Millikan for his key role in the initial experiments was understandably hurt (Hanson 1997). Millikan and Fletcher's work was written up in the series of famous papers that ultimately resulted in the Nobel Prize for Millikan. Fletcher, on the other hand, received his Ph.D. based on a paper Millikan agreed he could publish as single author about Brownian motion.
From 1916 and on, Millikan was nominated regularly for the Nobel Prize until he finally received it in 1923 (Holton 1988). However, the work on the oil drop was not without its problems and continued to be done on its own merits as late as 1941 (Binnie 2003). The central issue dogging the experiment related to the viscosity of air. Since the value of 'e' was determined by other methods, a discrepancy became evident and led to the re–determination of the viscosity of air, which was recognized to be the likely cause of the disagreement of values of 'e' determined by the oil drop method and other methods.
The difficulty in achieving a highly accurate value for 'e' due to not knowing a good value for the viscosity of air was only one of the challenges facing Millikan. In Germany, Felix Ehrenhaft also undertook to measure the charge on small droplets or other microscopic bodies. Millikan and Ehrenhaft, however, had completely different philosophies of experimental method. Ehrenhaft scrupulously recorded and used all his measurements. Although he did not say so publicly, Millikan checked his calculations for 'e' on each drop and discarded data he did not like for one reason or another. Millikan achieved a clear atomicity in his charge values, whereas Ehrenhaft obtained a

complete distribution of charge values lower than those of Millikan. The result was a vigorous war of words over whose methods and conclusions were correct. The controversy between Millikan and Ehrenhaft has been exhaustively studied, analysed, and debated in the literature (Holton 1988; D. Goodstein 2001; Franklin 1981, Niaz 2000) and a recapitulation of the details here is not necessary.

During Fletcher's lifetime, persistent rumours circulated that Millikan had cheated him out of the credit for inventing the oil drop experiment. Fletcher, although disappointed at not having co–authored the first paper on the oil drop with Millikan, considered Millikan a true friend and never lost his gratitude for the opportunities that Millikan gave him. For this reason, Fletcher never allowed his autobiography to be published and his role in the oil drop experiment was only exposed posthumously when the editor of *Physics Today* learned of the existence of Fletcher's autobiography from the writer of Fletcher's obituary published in *Physics Today* in October of 1981.

Fletcher established his own fame, but not with the experiments he had begun at the University of Chicago. He subsequently accepted a position with Bell labs (at the time Western Electric) to research sound and acoustics in 1918. There he took on the most challenging problem of quantifying and modelling how we hear and understand speech. He was the co–inventor, with Leopold Stokowski, of stereophonic recording. Whenever we listen to a particularly good recording of our favourite musical selection, we should remember that we have Harvey Fletcher to thank. Engineer and acoustical researcher J. B. Allen writes that

> [t]he problems that Fletcher and his colleagues studied were so complicated, and took so many years, that it has been difficult to appreciate the magnitude of their accomplishments. ... Everyone who has ever used the telephone has reaped the benefit provided by this man and his genius. ... Bell invented the telephone, and Edison made it into a practical device. Harvey Fletcher may not be as well known as these men today, but his scientific contributions to the fields of telephony, hearing, and human communication are absolutely unsurpassed. ... I would describe Harvey Fletcher as the singular intellectual force in the development of present–day communication acoustics and telephony. (Allen 1999, p. 1825)

Fletcher was a great physicist in his own right and, furthermore, there is ample reason not to discount his important role in the development of the oil drop experiment. The story of Fletcher, his key role in inventing the oil drop method and measuring 'e', and his friendship with Millikan have not been widely known up to this point. I find that students are very receptive to the story and that they identify with Fletcher.

THE POINT OF THE STORY

The story of Millikan and Fletcher comprises a fruitful background for attempts to improve the situation vis–à–vis student frustrations in performing the current simplified versions of the oil drop experiment. There are a number of aspects of the story that may be singled out and, perhaps, even amplified. The human aspect—the twists and turns that brought about the successful completion of the oil drop experiment and the resulting effects, personally on Millikan and Fletcher is always of interest to students. In my experience, students will be curious about what motivated the protagonists to take particular courses of action, for example, Millikan's decision to publish the first paper as a single author. The conflict between Millikan and Ehrenhaft presents a unique opportunity to highlight the complex nature of scientific methodology and provides students with a basis upon which to begin thinking about the nature of science. Students will naturally ponder whether they should be guided by the traditional 'scientific method' or their presuppositions. Students will also begin to recognize the scientific, technological, and methodological complexity of the oil drop experiment which made (and still makes) it difficult to obtain consistent and accurate results. Lastly, students should recognize that the real objective in designing the oil drop experiment was the physics of fundamental particles rather than the physics of the experimental design. These issues will provide a basis, both for analysing existing instructional materials and for altering the instructional approach.

Student Difficulties with the Millikan Experiment

A typical set of laboratory instructions for performing the oil drop experiment will have as its stated objective 'to measure the charge of the electron, and show that it is quantized with smallest value of 1.6×10^{-19} *C*' (Izmir Institute of Technology 2006). The 'theory' section of the instructions provided for the student will typically detail the physics of a falling oil drop in equilibrium with the Stokes drag forces. However, it does not encourage students to think about what evidence is provided by their measurements on oil droplets for the existence of an elementary electrical charge and what this charge has to say about the nature of the electron.

Students often find laboratory exercises, as exemplified by the oil drop experiment, to be, at the best, challenging, and, at the worst, confusing. They see experimental work from a completely different standpoint than the teacher or researcher. From students' perspectives the goals are primarily to follow, sometimes, meaningless instructions and to get the 'right' answers (Hodson 1993; Lunetta 1998; Petrosino 1998). The laboratory presents a daunting set of tasks for the student, the purposes for which are not at all clear in the student's mind. According to Hodson's influential review of student practical work, in the typical laboratory the student must (a) understand the nature of the problem, (b) understand the procedure, (c) develop a theoretical perspective, (d) read, comprehend, and follow directions, (e) insure that they are getting along with their partner, (f) operate the apparatus and collect data, and (g) interpret results and write a report (Hodson 1993). Hodson's categorization scheme of student difficulties provides a useful framework from which to analyse student difficulties. The first three issues will be considered here.

Understanding the Nature of the Problem. The simplistic formulation of the objective for the typical Millikan instructions misrepresents the complex underlying historical issues that had to be resolved before the electronic charge could be quoted with unqualified confidence. In his 1913 paper, Millikan listed several of these complicating issues. He pointed out that the drag force on an oil droplet of one micron diameter does not follow the Stokes resistive force because of the presence of Brownian motion. The droplets literally experience pockets of empty space between molecules where the resistive force is zero. The correction to the Stokes resistive force was then made by Millikan through a combination of theoretical considerations and empirical factors. Another assumption was that the surface tension of the droplet does not affect its density, i.e., that the density of a droplet is the same as the density of the oil in bulk at that temperature. Millikan was able to show that the elementary charge did not vary with droplet size, implying that the smallest droplet had the same density as at the limit of infinite size (i.e., the oil in bulk).
These underlying complications or assumptions, however, relate to the experimental method and not to the electron, *per se*. Both negative and positive charges are observed in the experiment. When positively–charged droplets are measured, the charge involved is not that of the electron, but of the atomic nucleus (something that was not entirely clear to Millikan at the outset, since atomic models were just then being formulated). It would, then, be more correct to characterize the measurement as one of the 'elementary electrical charge'.

Understanding the procedure. The procedure of the Millikan experiment involves the simultaneous measurement and monitoring of a number of factors. These include timing of the falling droplet without an electric field applied, that for a droplet rising under the influence of an electric field, determining the temperature inside the droplet chamber, and measuring the barometric pressure. When an observed droplet is very small, then Brownian motion is significant and the timing data will vary much more than the experimental timing uncertainty would predict. As was discussed above, the first investigator to see oil droplets and measure their electrical charge was not Millikan, but his graduate student, Harvey Fletcher. In a filmed interview, Fletcher recalled in significant detail his first attempts at measuring the charge on oil droplets and what he saw at the time (Fletcher 1963). Students would benefit from seeing this first–hand narrative description of the original investigation since it brings the event to life and relates their current re–capitulation of the experiment to the real experiences of another student. The entire movie is available for downloading on–line (see Fletcher 1963).

It would be misleading, however, to give students the impression that experimental results from the oil drop experiment are completely straightforward—the results and investigations of Felix Ehrenhaft, would indicate otherwise (Niaz 2000). As we have seen, Ehrenhaft measured the elementary electrical charge at the same time as Millikan, and he not only came to the conclusion that the elementary charge can be less than that that of the electron, but he hotly disputed Millikan's findings.

Developing a Theoretical Perspective. The usual approach to the student Millikan experiment confuses the theory underlying the method with theory relating to the electron, *per se*. The theory of the *method* is, for the most part, a simple application of first–year dynamics. The only complication relates to the resistive force produced by the air as the drop moves through it. However, the point of the experiment is (a) to determine whether there is a fundamental electrical charge that is indivisible and (b) the value of that charge. An ancillary question might be: Is the fundamental electrical charge identical with the electron? As for a theory of the origin of the electron itself, Lorentz, Zeeman, Thompson, and Bohr began to shed some light on the matter and this is more properly the subject of a separate investigation to do with atomic structure.

Student Perspectives

In their reports on the oil drop experiment, my students at the second–year university level are expected to comment on the challenges in performing the experiment and the role that their preconceptions of 'e' played as a guiding idea in their analysis of the data and relate that to what Millikan must have experienced. Before they perform the lab, they are told the story of Fletcher and Millikan in a condensed fashion and then view the Fletcher movie.

After students have been introduced to the historical case, have seen the Fletcher movie, and have completed performing the oil drop experiment and doing the calculations, many of them show remarkable insights into Millikan's dilemma. For example, one student wrote:

> If we had used the data for every drop we observed, our result would have not agreed with the accepted value at all. I suppose that Millikan must have depended quite heavily on his preconception of the value of e, assuming his apparatus was similar to ours. If Millikan did not have a testable basis for rejecting drops (either a measurable problem with the equipment, or perhaps a statistical model that allowed him to reject values between the quantized steps), I can not see how the experiment would give one confidence that charge quantization had been observed.

Evident in the students' writing is the same frustration that Kruglak pointed out with obtaining 'e' values in between the expected quantized values. As we know, Millikan also found some of these, only not as frequently as our students do. It is not difficult for students to identify with Millikan's situation and see why he selected data based on his preconceptions. It should be noted that students are not left to simply deduce the fact that their data choice is guided by their presuppositions—they have been directed with appropriate questions that challenge them to be reflective about Millikan's work. With the appropriate background and guidance, students will show insights such as another student shared with me:

> By having a preconceived notion of what *e* should be, we knew what to expect, and disregarded observations that were not expected. In saying so, I believe Millikan depended on his preconceived notion, as much as we did. It is likely, that when Millikan noticed a quantization trend of the charges, he selected only those drops that would illustrate the phenomena and excluded those few that distort it. Millikan does [sic] not change the results, but merely reduced the marginal error of his experiment. By doing so, he was able to illustrate his discoveries for us to understand undoubtedly. Had he not, who knows when we would finally acknowledge charge quantization?

Based on their own experiences and with the knowledge of relevant aspects of the history of science, students readily conclude that the discovery of charge quantization would not have been possible by the traditional scientific method based on the idea of inductive reasoning. Niaz (2005) has written about this very issue, pointing out that speculative or cutting edge research cannot be successful by applying the traditional scientific method, based on the idea of inductive reasoning, but must be guided by presuppositions in order to be successful.

Besides the historical and philosophical issues, the practical issue of successfully making the measurements is a major concern, as experienced by instructors teaching the experiment (Kruglak 1972). One can, likely, not find better insights into the problem than by consulting a thoughtful student account of the challenges. One student expressed the experience, in a refreshingly candid fashion, in the following way:

> In just the beginning of the experiment we have already faced its most excruciating task: the challenge of finding a 'good' drop. Even though a galaxy of drop [sic] is observed, it is difficult to find a drop that would fall between ten to twenty seconds, and would rise in a time greater than five seconds. Moreover, we have learned to hold our rejoice [sic] when we spot a good candidate. The drop would initially fit the criteria, but would ionize after a while, and we find ourselves back from the start. It is yet another challenge to hope that the drop would remain unionized [sic] until we have taken enough data from it. Also, it is another difficulty to be able to isolate the chosen drop among the other drop [sic] in the view. It is easy, and regretful to lose the chosen drop, and to face the initial challenge once again. The experiment requires a person with a generous source of patience, the ability to be immobile when necessary, the strength to be able to keep an eye open for as long as possible, and some gift of luck. It is [a] relief, however, to be able to work with a team.

Such student insights will prove to be invaluable in improving the instructional situation. One issue brought forward is that drops seem to ionize in mid–stream. This possibility needs to be investigated, since if it is the case, and the cause can be discovered, one major source of frustration would be minimized.

Discussion

As I pointed out at the start, the Millikan experiment has taken on mythic proportions, and there is a gap between its legendary greatness and current student experience. However, as I have shown, an appropriate insertion of history of science changes the situation completely. Students readily identify with Millikan and Fletcher and with proper guidance show remarkably profound insights.

So far, I have concentrated on humanizing the scientists behind the great discoveries, illuminating the challenges in interpreting data, and elaborating on the difficulties in obtaining good data. There is yet one more aspect which students fail to comprehend fully—that the oil drop experiment is not about drops falling through air but about the fundamental nature of electricity. One of Millikan's earliest accounts of his measurement of 'e' may be useful in giving students further insight into the matter (Chicago Daily Tribune 1910). Here Millikan gives his frank analysis of the significance of his and Fletcher's work when interviewed on May 24, 1910:

It follows that an electric current which is simply a charge in motion consists of a movement of these atoms of electricity through or over the conducting body. … I am not broaching any new theories but have established beyond doubt several which, until I was able to isolate the electrical charge, never were proven. In isolating the ion I was able to see by a method of my own just what it did when separated. Once isolated, the charge could be measured for the first time and was found to be 5.13x10-10 electrostatic units. Mr. Fletcher and I have been able to give a tangible demonstration of the correctness of the view advanced many years ago that an electric charge is not a homogeneous something spread uniformly over the surface of a charged body, but that it has a definite granular structure. It consists, in fact, of a definite number of specks or atoms of electricity, exactly alike, peppered over the surface of the charged body. (Chicago Daily Tribune 1910, p. 7)

I pointed out earlier that students are not normally encouraged to think about what evidence for an elementary electrical charge is provided by the oil drop experiment and what this charge has to say about the nature of the electron. The oil drop experiment provided evidence for the electron as being real. Along these lines, Millikan, ever the master of overstatement, argued, in his Nobel Lecture, that

if a man had seen a football which someone told him was the electron he would be far less certain that what he had seen corresponded to reality, than is the man who has become familiar with the foregoing [oil drop] experiment. By its aid he can count the number of electrons in a given small electrical charge with exactly as much certainty as he can attain in counting his fingers and his toes. (Millikan 1924, p. 55)

In the lecture, Millikan speaks not only of realism about the electron, but of the historical origins of the electron concept and of the oil drop experiment. Science students rarely have the opportunity to consider such matters. Using a case study such as outlined here to guide instruction significantly changes what students take with them from having done the experiment.

Four essential aspects have been identified that will need to be dealt with in reviewing the student version of the Millikan oil drop experiment. These are (1) humanizing the experiment's originators, focusing on the important and overlooked role of Harvey Fletcher, (2) exposing the difficulty in obtaining results in speculative experiments if the traditional scientific method is followed rather than allowing presuppositions to guide data analysis, (3) dealing with the difficult and frustrating nature of the experiment, apparently somewhat less so in Millikan's time than with our current students, and (4) considering that the oil drop experiment establishes various aspects of the fundamental nature of electricity. These aspects and the insights surrounding them, arising both from the history of science and from current student work, should serve as an effective basis for the pedagogical renewal of the Millikan Oil Drop Experiment.

Acknowledgements: The researching and writing of this paper was made possible, in part, through funding provided by The University of Winnipeg and the NSERC CRYSTAL at The University of Manitoba.

Notes

[1] Where not otherwise stated, the historical account is based on: H. Fletcher, 1982; S. Fletcher, 1992; J.R. Goodstein, 1991; Millikan, 1950.

References

Allan, J.B.: 1996, 'Harvey Fletcher's role in the creation of communication acoustics', *Journal of the Acoustical Society of America* 4(1): 1825–1839.

Binnie, A: 2003, 'The Laby Experiment', in U. Hoepli, (ed.), *Volta and the History of Electricity*, Università degli Studi di Pavia, Milano: 377–386.

Chicago Daily Tribune: 1910, 'Electric Secrets Located at Last', article in the Chicago Daily Tribune, May 25, 1910.

College of St. Catherine: 2006, Millikan Oil Drop Experiment. [Available on–line: http://www.stkate.edu/physics/phys112/curric/millikan.html].

Crease, R. M: 2002, 'Critical Point: The Most Beautiful Experiment', *Physics World* **15**(9): 19–20.

Duckworth, H.E.: 1992, 'The British Association for the Advancement of Science Meets in Winnipeg in 1909', Luncheon Address at the Manitoba History Conference, May 14, 1992.

Fletcher, S.H.: 1992, 'Harvey Fletcher', in *Biographical Memoirs* **61**, National Academy Press, Washington D.C.: 165–189.

Fletcher, H.: 1963, 'A Talk with Harvey Fletcher', [motion picture], Bell Telephone Laboratories. Available on–line, http://auditorymodels.org/jba/BOOKS_Historical/FletcherVideo/, accessed June 22, 2006.

Fletcher, H: 1982, June, 'My work with Millikan on the oil–drop experiment', *Physics Today*, 43–47.

Franklin, A.: 1981, 'Millikan's Published and Unpublished Data on Oil Drops', *Historical Studies in the Physical Sciences* **11**: 185–201.

Goodstein, D.: 2001, 'In defense of Robert Andrews Millikan', American Scientist **89**: 54–60.

Goodstein, J.R.: 1991, 'Millikan's School: A History of The California Institute of Technology', W.W. Norton & Company, New York.

Hanson, R: 1997, 'History of the Physics Department' On–line document available from http://www.physics.uni.edu/history.shtml, retrieved June 7, 2006.

Hodson, D: 1993, 'Re–thinking Old Ways: Towards A More Critical Approach To Practical Work , In School Science', *Studies in Science Education* **22**: 85–142.

Holton, G.: 1978, 'Subelectrons, Presuppostitions, and the Millikan–Ehrenhaft Dispute', *Historical Studies in the Physical Sciences* **9**: 161–224.

Holton, G.: 1988, 'On the Hesitant Rise of Quantum Physics Research in the United States', in S. Goldberg and R. H. Stuewer, (eds.), *The Michelson Era in American Science 1870–1930*, American Institute of Physics, New York: 177–205.

Izmir Institute of Technology: 2006, 'Modern Physics Laboratory Manual', [Available on–line: http://uio.iyte.edu.tr/PDF/classes/physics/eng/phys_305_eng.pdf].

Kruglak, H: 1972, 'Another look at the Pasco–Millikan oil–drop apparatus', *American Journal of Physics* **40**: 768–769.

Lunetta, V. N: 1998, 'The School Science Laboratory: Historical Perspectives and Contexts for Contemporary Teaching', in B. J. Fraser & K. G. Tobin (eds.), *International Handbook of Science Education*: 249–262.

Matthews, M. R: 1994, *Science Teaching: The Role of History and Philosophy of Science*, Routledge, New York.

Millikan, R. A: 1913, 'On the Elementary Electrical Charge and the Avogadro Constant', *The Physical Review* **2**: 109–143.

Millikan, R. A: 1924, 'Nobel Lecture'.

Millikan, R.A., 1950, *The Autobiography of Robert A. Millikan*, Prentice Hall, New York.

Niaz, M., & Rodríguez, M.A: 2005, 'The Oil Drop Experiment: Do Physical Chemistry Texts Refer to its Controversial Nature?', *Science & Education* **14**: 43–57.

Niaz, M: 2000, 'The Oil Drop Experiment: A Rational Reconstruction of the Millikan–Ehrenhaft Controversy and Its Implications for Chemistry Textbooks', *Journal of Research in Science Teaching* **37**(5): 480–508.

Niaz, M., 2005, 'An Appraisal of the Controversial Nature of the Oil Drop Experiment: Is Closure Possible?', *British Journal for the Philosophy of Science* **56**: 681–702.

Petrosino, A. J: 1998, *At–Risk Children's Use of Reflection and Revision in Hands–On Experimental Activities*, Unpublished PhD Dissertation, Vanderbilt University.

Rodríguez, M.A., & Niaz, M: 2004, 'The Oil Drop Experiment: An Illustration of Scientific Research Methodology and its Implications for Physics Textbooks', *Instructional Science* **32**: 357–386.

Teaching Physics to in-service primary school teachers in the context of the History of Science: the case of the fall of bodies

PANAGIOTIS KOKKOTAS; PANAGIOTIS PILIOURAS; KATERINA MALAMITSA; IOANNIS VLACHOS; KATERINA PLAKITSI; MIRONAS MAUROGIANNAKIS; EFTHYMIS STAMOULIS

Pedagogical Department of Primary Education, National and Kapodistrian University of Athens, Greece

Abstract: Our paper concerns the presentation of an in-service primary school teachers' training program which is based on the potential role that history of science has for promoting the learning of physics. This training program has been developed in the context of Comenius 2.1 that is a European Union program. The program we developed in University of Athens is based on socioconstructivist and sociocultural learning principles with the intention to help teachers to appropriate basic knowledge about the topic of the fall of bodies and to make explicit through the exploitation of authentic historical science events on the above topic (Aristotle's, Galileo's and Newton's interpretations on the fall of bodies) the contemporary views about Nature of Science (NOS), Nature of Learning (NOL) and Nature of Teaching (NOT). In the implementation of the program we exploited a variety of teaching and learning strategies (e.g. group work, making posters, concept maps, simulations) that exploit authentic historical science events in the topic of the fall of bodies.

Introduction

Our paper concerns the presentation of an in-service primary school teachers' training program which is based on the potential role that history of science has for promoting the learning of physics. This training program has been developed in the context of Comenius 2.1 that is a European Union program. The title of the project is "The MAP prOject": A science teacher training program based on Science Education and history of science". Science education as well as history of science were the two main pillars towards the development of this training program in which five European Universities cooperated.

In general the aim of "The MAP prOject" was:

- To make a productive synthesis of the contemporary theoretical insights and research outcomes both of history of science and science education in order to produce situated in practice in-service science teachers' training strategies.
- To develop and evaluate innovative training situations for in-service science teachers by using case studies from history of science.

The MAP prOject is an acronym formed from the names of the European cities where the five co-operating institutions are situated, that is: Thessaloniki (The), Madrid (M), Athens (A), Pavia (P), Oldenburg (O).

Thessaloniki *(Pedagogical Department of Primary Education, Aristotle University of Thessaloniki)*

Madrid *(Didactic of Experimental Sciences, University Complutense of Madrid)*

Athens *(Pedagogical Department of Primary Education, National and Kapodistrian University of Athens)*

Pavia *(Department of Physics "Alessandro Volta", Università di Pavia)*

prOject *(**O**ldenburg-Physics Education/History and Philosophy of Science, Institute of Physics, Carl von Ossietzky University)*

In the context of the program, all partners tried to bridge their expertise in different aspects of history of science and/or science education (Kokkotas et al 1998; Kokkotas 2003a; Bevilacqua & Giannetto 1998; Matthews Bevilacqua & Giannetto 2001; Riess 2000; Seroglou & Koumaras 2000; Seroglou 2002; Moreno 2001; Heering 2000) in order to optimize the ways that history of science and science education can be interrelated for the development of in-service science teachers' programs that exploits the work and the research outcomes both of these.

In this paper we present the procedure we followed to develop part of the program in the University of Athens. Three types of activities led us to the attainment of our goal:

a) The elicitation of science teachers' views, concerning NOS, NOL, and NOT and a thorough review of international literature concerning these issues.

b) Design and development of an in-service science teachers' training curriculum (goals, content, strategies, educational material, assessment process) that is based in the exploitation of authentic historical science events on the topic of the fall of bodies (Aristotle's, Galileo's and Newton's interpretations on the fall of bodies).

c) Teaching the in-service training program in the University of Athens.

History of Science and Science Education

In the last years historians and philosophers of science and educators have repeatedly drawn attention to the enormous potential role history of science has for promoting the learning of science (Matthews 1994). Research has shown that incorporation of history of science into science instruction and teachers' in-service training are effective in leading students and teachers to a better understanding of the science concepts and the NOS (Seroglou & Koumaras 2000; Nott & Wellington 1998. Many contemporary curricula emphasize the relations between science education and history of science (e.g. Project 2061). History of science can enable students to get an insight into the principles of science and an appreciation of the values of the scientific endeavor. We believe that bridging history of science with science education in science teachers' training programs that teachers are active and function in a collaborative inquiry mode of meaning making, will create the conditions teachers to appropriate gradually contemporary meanings about NOL and NOT.

Tracing the contemporary assumptions about NOS, NOT and NOL and their implication in science teachers' professional development

For us professional development of science teachers goes beyond the term 'training' with its implications of learning skills, and encompasses training strategies of helping teachers not only learn new skills but also develop new insights in pedagogy and in their own practice. We regard that teaching practice has many dimensions, five of the most important of them are (Bartholomew Osborne & Ratcliffe 2004):

- Teachers knowledge and understanding of the nature of science
- Teacher's conceptions of their own role
- Teachers' use of discourse
- Teachers' views of learning goals
- Teachers' views about the nature of classroom activities

All these dimensions have to do explicitly or implicitly with three basic aspects of science education namely NOS, NOL, and NOT. A contemporary teacher professional development program should be based and stress the contemporary assumptions about these three aspects.

In the context of "The MAP prOject" we made a research in the four countries, using as research instrument a questionnaire, in order to investigate science teachers' views about NOS, NOT, NOL, and about issues that concern the fall of bodies. Our aim was to elicit science teachers' knowledge and views relative to history of science, to science education and their scientific knowledge about the theory of the gravitational force, to analyze science teachers needs and to search thoroughly the international contemporary science education literature, in order to established the basis for the optimum development of the training curriculum which was the culminating goal of "The MAP prOject".

The presentation of the instrument

The instrument/questionnaire had four parts the first of which referred to the demographic information of our teachers. The other parts of the questionnaire had on the whole 21 questions. The second one, was consisted of eight questions, it concerned the tracing of teachers' views about NOS and was based (with small alterations) on the (VOSTS) Views on Science-Technology-Society questionnaire (Aikenhead Ryan & Fleming, 1989; Ryan & Aikenhead 1992). The VOSTS, as Aikenhead et al support, is an inventory of student viewpoints about science, and about how science is related to technology and society. Indeed, the VOSTS was based in authentic Canadian high school students views about the above issues. From this research tool we chose eight questions that are related mainly with issues about NOS (e.g. "What is science?", "What is scientific method?").

The third part concerned the tracing, with an indirect way, science teachers views about NOL and NOT. The seven questions of this part were taken and properly adapted from a Science Questionnaire that was used in 2000 National Survey of Science and Mathematics Education. The 2000 National Survey of Science and Mathematics Education was designed to provide up-to-date information and to identify trends in the areas of teacher background and experience, curriculum and instruction, and the availability and use of instructional resources. The forth part aimed at eliciting the views of science teachers about "fall of bodies" knowledge. The seven questions have been designed in the basis of the notion of alternative ideas of the constructivist literature about the fall of bodies (Driver et al 1994).

The research instrument (see a representative sample of the questionnaire in annex 1) has been made from the University of Athens and it was initially applied in a pilot study in 40 Greek primary science teachers. There was the appropriate amendment of it and in its final version it was sent to all partners of "The MAP prOject" in order to use it for data collection.

The changing views about the NOS and science teachers' professional development

Science education is not a static field but a dynamic one. It changes in direct relation to the developments in the society which it serves. One of the most important issues that led to the change of the direction in science education is the gradual change of our views that concern the NOS. (Kokkotas 2003b).

It is argued that a contemporary accepted view about this issue is that "science is dynamic, changing and tentative. Science is not a static collection of facts. We can not take current scientific knowledge to be complete and final" (Bell Lederman & Abd-El-Khalick 2000). Lemke (2001) supports that:

> "Historians, sociologists, and cultural anthropologists came increasingly to see that science had to be understood as a very human activity whose focus of inter-

est and theoretical dispositions in any historical period were, and are, a part of the dominant cultural and political issues of the day".

Teaching the NOS in schools and training teachers about it have attracted the attention of the science education research community (e.g. Hodson 1988; Matthews 1998; McComas Clough & Almazroa 1998; Abd-El-Khalick & Lederman 2000; Jenkins 1996).

Studies have shown that high school science students and in-service teachers' views of the NOS are not consistent with current accepted definitions of it (Lederman 1992; Ryan & Aikenhead 1992; Driver et al 1996; Leach et al 2000; Lederman Wade & Bell 1998). For example, most teachers and students believe that all scientific investigations adhere to an identical set and sequence of steps known as the scientific method (McComas 1996), and does not recognize the fact that scientists' disciplinary training and commitments, as well as their personal experiences, preferences, and philosophical assumptions do influence their work (Akerson Abd-El-Khalick & Lederman 2000). The results of our research, in the context of "The MAP prOject" about NOS, helped us develop a more informed primary teachers' training program about the fall of bodies. We present the results and our interpretation of two questions concerning teachers' views about NOS.

In the first question of the questionnaire teachers should read the statements and then choose one statement that fits better to their opinion.

Table1 *Question 1: What, in your view, is science?*	N	%
Science is a truth-seeking process. It is not a collection of unquestionable "truths."	43	19,3 %
Science is an objective, logical, and repeatable attempt to understand the principles and forces operating in the natural universe.	81	36,3 %
Science can prove anything, solve any problem, or answer any question.	1	0,4 %
Science is primarily concerned with understanding how the natural world works.	70	31,4 %
Science involves dealing with many uncertainties.	16	7,2 %
Science is a self -correcting discipline. Such corrections may take a long time but as scientific knowledge accumulates the chance of making substantial errors decreases.	2	0,9 %
I don't understand.	5	2,2 %
I don't know enough about this subject to make a choice.	16	7,2 %
None of these choices fits my basic viewpoint.	3	1,3 %
Total	223	100 %

In table 1 we can see that 36,3% of all respondents regarded that *"science is an objective, logical, and repeatable attempt"*, 31,4% of them believed that *"science is primarily concerned with understanding how the natural world works"* and a percentage of 19,3% of them believed that *"science is a truth-seeking process"*. This table shows that most of the respondents seem to adopt views about NOS that are not consistent with current accepted definitions of it. These results are in accordance with other researches (e.g. Bartholomew et al 2004; Lederman et al 1998; Leach et al 2000).These studies claim that a significant proportion of teachers have no recognition of the tentative nature of scientific knowledge, and they hold outmoded positivist or empiricist views of the NOS.

In another question teachers should read the statements and then choose one statement that fits better to their opinion about the nature of scientific method.

Table 2 *Question 3. When scientists investigate nature, it is said that they follow the scientific method. The scientific method is:*	N	%
Controlling experimental variables carefully, leaving no room for interpretation.	10	4,5 %
Considering what scientists actually do, there really is no such thing as the scientific method.	3	1,4 %
Testing and retesting - proving something true or false in a valid way.	19	8,6 %
Postulating a theory then creating an experiment to prove it.	40	18,0 %
Questioning, hypothesizing, collecting data and concluding.	133	59,9 %
A logical and widely accepted approach to problem solving.	5	2,3 %
None of these choices fits my basic viewpoint.	6	2,7 %
I don't know enough about this subject to make a choice.	6	2,7 %
Total	222	100%

In table 2 we can see that the majority of respondents are in favour of the opinion that *"scientific method is questioning, hypothesizing, collecting data and concluding"* (59,9%). The other responses have less respectable percentages apart from the view that *"scientific method is postulating a theory, then creating an experiment to prove it",* which attracted 18,0% of the responses. Both of these views show that teachers have a very specific opinion about scientific method, which may be regarded as something steady. On the contrary more informed and contemporary accepted views about the NOS and the scientific methods have attracted small percents of the respondents. For example *"Testing and retesting - proving something true or false in a valid way."* have been selected by 8,6% of the respondents. Bell, Lederman, & Abd-El-Khalick (2000) support that "there is no one, single Scientific Method. The overly simplified hypothetico-deductive method that is frequently given as the only example of scientific methodology in the initial chapters of textbooks is not the only way science progresses". McComas, Clough, & Almazroa (1998) claim that there is no way to do science, and therefore, there is not universal step-by-step scientific method.

Although the sampling among the participating countries in our project and the groups of the teachers of different professional status, does not allow us to generalize our conclusions, as they concern science teachers views about NOS, few points are apparent and firm.

1. Teachers adopt a variety of views about NOS. The most of them, as other internationally researches have also indicated (e.g. Bartholomew et al 2004; Lederman et al 1998; Leach et al 2000), have no recognition of the tentative nature of scientific knowledge, and hold positivist or empiricist views of NOS.
2. The most of the teachers regard scientific method as something steady, as a universal step-by-step scientific method.
3. The vast majority of science teachers hold the view that "scientific discoveries result from a logical series of investigations". Furthermore, the majority of the respondents seem to ignore that the history of science reveals both an evolutionary and revolutionary character of the scientific knowledge.

The result of our study is in accordance with other studies that have shown that high school science students and in-service teachers' views of the NOS are not consistent with current accepted definitions of it. According to McComas 1996 the most teachers and students believe that all scientific investigations adhere to an identical set and sequence of steps known as the scientific method. Other researchers have argued that teachers do not recognize the fact that scientists' disciplinary training and commitments, as well as their personal experiences, preferences, and philosophical assumptions do influence their work (Akerson Abd-El-Khalick & Lederman 2000).

In conclusion, our research demonstrated teachers' need for better understandings of the contemporary accepted views about NOS. So in the development of the training program we followed the principle of the explicit instruction about NOS engaging teachers to study and compare the views, on the fall of the bodies, of some important philosophers and scientists such as Aristotle, Galileo and Newton and the exploitation of authentic historical science events to present the social and cultural nature of science.

The changing views about the NOL and the NOT and science teachers' professional development

The last years there was a shift from perspectives that adopt individual constructivist assumptions (Tobin 1993; Von Glasersfeld 1995) to socioconstructivist, sociocultural and discursive ones (Solomon 1993; Driver 1995; Lemke 2001, Mortimer & Scott 2003; Wells 1999). The major thrust of the critique coming from situated and sociocultural approaches, is that reasoning is a phenomenon that shows considerable variability depending on the cultural context in which it occurs, the meaning of the task at hand and the tools available (Cole 1996; Lave 1996; Rogoff 2003).

The socioconstructivist and sociocultural shift in views about NOL and NOT creates major conflicts with the fundamental teaching beliefs of many science teachers. Despite the overwhelming push towards teaching methods involving collaborative research and inquiry, there is little evidence that these practices are happening in science classrooms. In many studies, on the role of the teacher in relation to the new socioconstructivist and sociocultural approaches, has been established the conviction that most of the teachers aren't familiarized with constructive and sociocultural learning principles and function mainly as dispensers of knowledge and less as facilitators, are closed and authoritative and not open and dialogic (e.g. Nystrand 1997; Newton et al 1999).
Many scholars have promoted a social constructivist view of learning in teacher development (e.g. Tobin 1993; Bell & Gilbert 1996). Others have proposed a sociocultural one (e.g. Matusof & Hayes 2002; Wells 2002). Sociocultural driven teacher development programs are based on the assumption that learning is a community process of transformation of participation in sociocultural activities (e.g. Rogoff Matusov & White 1996). A socio-cultural view of learning suggests a situated work embedded model which is more likely to engage science teachers in activities and interactions that are directly related to their practices (Fletcher 2002). Situated action has the potential for teachers to re-examine their practice within the context of their every-day work enabling them to evaluate outcomes as they relate to their teaching experiences and their students' learning needs.
In the followings we shall present some of the results of our research that concern science teachers views about NOL and NOT.

In question 8 that concerns teachers' views about NOL (see appendix 1) the most of the respondents (67,7%) agreed with the socioconstructivistic assumption that *"Learning is an active process in which learners construct new ideas or concepts based upon their current/past knowledge"*. A percentage of 14,1% of the respondents adopted the cognitive constructivist statement that *"Learning is a process of building cognitive structures"*. These results indicate that respondents are at least theoretically informed about constructivist approaches and not hesitate to adopt such assumptions about NOL. However research has shown, that many times the constructivist theoretical assumptions that the teachers adopt, are in opposition with their more traditional pedagogical practices (e.g. Tobin 1993). In reverse, we concluded from the results that science teachers weren't familiarized and didn't adopt statements such as "Learning is a search for meaning" (0,9%), "Learning is an act of membership in a community of practice" (1,4%), "Learning is a co-construction (or reconstruction) of social meanings emerging from socially negotiated and discursive activity" (5,9%), and "Learning is a process of social apprenticeship" (1,4%), which all of them are statements that belong to a sociocultural perspective. This implies that in our training programs we should create the conditions for the epistemological and ontological issues about NOL to emerge. A basic strategy to achieve this goal is to design a training program that it is based on sociocunstructivist and sociocultural principles. We support that it is difficult if not impossible for science teachers to teach with the appropriate ways if they don't experience in the praxis this teaching and learning strategies in meaningful training procedures.

In question 8 (see appendix 1) the vast majority of respondents ("fairly well prepared" - 45,4 % plus "very well prepared" - 28%) support that they are well prepared to *"Take students' prior understanding into account when planning curriculum and instruction"*. Research in science teachers' practices does not confirm these statements of science teachers' responses (Kokkotas 2003a). But the majority of respondents ("not adequately prepared" - 26 % plus "somewhat prepared" - 38,6%) supported that they were not prepared to *"Make connections between science and history of science"*. Additionally, in the question 13 the vast majority of the respondents declared that either sometimes (39,6%) or seldom (34,0%) used events of the history of science when they teach science. This means that history of science does not seem to be a popular issue among the participants in the research. Also, in question 10, the vast majority of respondents ("somehow prepared" - 29% plus "fairly well prepared" - 45,4 % plus "very well prepared" - 28%) supported that they were well prepared to *"Have students work in cooperative learning groups"*. These views of teachers are in opposition to a series of situated in praxis studies that indicate that most of the teachers don't have the appropriate cooperative competencies (e.g. Lazarowitz & Hertz Lazarowitz 1998). Also the majority of respondents ("not adequately prepared" - 22,4 % plus "somewhat prepared" - 38,3%) supported that they were not prepared to *"Guide a class of students using investigative strategies"*. Finally the majority of respondents ("not adequately prepared" - 47,7 % plus "somewhat prepared" - 23,1%) declared that they were not prepared to *"Use computers for simulations and demonstration of scientific principles"*.

Question 14 intended to elicit science teachers background knowledge about the history of the evolution of theories of the fall of bodies (Aristotle's, Galileo's, Newton's and Einstein's views), and other important issues such as how prepared they feel to use strategies that relate science education with the history of science. The majority of the science teachers ("not adequately prepared"- 44,7% plus "somewhat prepared" - 26,5%) supported that they were not well informed about aspects that concern "Aristotle's view of the fall of bodies". This has some consequences in the way they teach science, as it is difficult if one doesn't know the "paradigms" of the fall of bodies to implement strategies that relate science education with history of science (Matthews 1994). Science teachers' supported that they where well informed about aspects that concern "Newton's view of the fall of bodies". We believe that this is something normal since for years they had been trained in physics in the Newtonian "paradigm".

Conclusively, as it concerns science teachers' views about NOL and NOT, the results of our study indicate that:

- Science teachers weren't well prepared to use pedagogical strategies such as: make connections between science and history of science, guide a class of students using investigative strategies, and use computers for simulations.
- The answers of the science teachers indicated a conceptual centred orientation of teachers' views about science learning instead of the contemporary meaning making one.
- Science teachers declared that they give emphasis in the objective their students to learn about the history and the nature of science, although this assertion is in opposition with many research results internationally.
- Science teachers weren't well prepared to use strategies such as: debates - argumentation activities, role playing, simulations, concept maps and utilization of events of the history of science.

In conclusion, our research demonstrated teachers' need for better understandings of the contemporary accepted views about NOL and NOT. In the development of the training program we followed the principle of the explicit instruction about NOL and NOT engaging science teachers in a collaborative inquiry process in which they interact with peers around topics concerning the fall of bodies, in meaningful activities with the guidance of the instructor. We support that if we want teachers' to transform their views and their ways that teach science based in contemporary views about NOS, NOT, and NOL, they should be involved in a professional development program that it will have the characteristics similar to these contemporary views.

Also in the design of the training program we take into account some broadly accepted thesis and directions that should characterize every contemporary training program. The review of literature in science education research has shown that the designing of teachers' in-service training programs has to be in accordance with the following fundamental principles (Kokkotas, 2003a):

a. The aim of teachers' in-service training is by using teachers' views and practices to achieve appropriation of knowledge and competencies in some important teaching and learning aspects of science.
b. In-service science teachers following a professional development program are learners who actively construct their own theories about teaching and learning. This is achieved through their personal teaching experience and determined by their attitudes and beliefs.
c. The quality, breadth, and flexibility of teachers' practices they use in the classroom are tightly connected with their professional development.
d. In-service science teachers who attend professional development programs improve their teaching and learning processes by interacting with their colleagues and expanding experience on the processes of teaching and learning.
e. The teaching and learning processes receive continuously support from teachers who attend professional development programs, and offer opportunities allowing them to interact with colleagues and exchange experience on these processes.

Our proposal: Science teacher professional development as a process of collaborative/dialogic inquiry

Each of the models for teacher education is aligned implicitly or explicitly with a particular theory or theories of learning and has different implications for the nature of a teacher professional development program, because different models predict different roles for teacher and trainers, different training materials, different curriculum organization, and different time-frames.

The traditional teacher-centered model in which knowledge is "transmitted" from the trainer to the trainee has rapidly being replaced by alternative models of teacher development (constructivist and sociocultural ones) in which the emphasis is on guiding and supporting teachers as they learn to construct their understanding of the culture and communities of which they are a part (Duffy & Cunningham 1996). In the process of shifting our attention to the constructive activity of the teacher, it is recognized the need to anchor learning in real-world or authentic contexts that make teacher professional development meaningful and purposeful. The current emphasis is to embed knowledge and competencies appropriation within a framework of teacher professional development, collaborative programs, and interactive research within a community of learners.

We believe that professional development should be seen as a social process of enculturation in a work practice. So we propose a model of professional development that will be based on participation and not in an acquisition metaphor (Bruner 1996). For the designing of science teachers' training program we adopted socioconstructivist and sociocultural learning principles and we were based on the potential role that history of science has for promoting the learning of science. Our aim was the training program to be characterized from the following:

- To make explicit through the exploitation of authentic historical science events the contemporary views about NOS (explicit instruction about NOS)
- To include a variety of teaching and learning strategies (e.g. group work, debates - argumentation, concept maps, role playing, making posters, creating interviews, simulations) that exploit authentic historical science events in the topic of the fall of bodies.
- To facilitate learning/training and development through collaborative inquiry activities among trainer and in-service primary science teachers (learning is an inherently social-dialogical activity). An oriented in collaborative inquiry training program should involve teachers in conversations about common experiences and in development of collaborative projects (e.g. collaborative construction of a poster).
- To involve science teachers in collaborative activities in order to develop their own educational instructional material (e.g. worksheets). We believe that when teachers are given the opportunity to research their own practice, collaboratively and with support, they establish what works for them and their students. This could become a creative transformative process that is participant-driven.
- To contextualize training, learning, and joint productive activity in the experiences and competencies of in-service science teachers (knowledge is embedded in practice) and to anchor training/professional development in real-world or authentic problems of science teaching, that make science teacher development meaningful and purposeful (learning is embedded in the activities and practices in which it occurs).

Our overall aim was teachers to re-examine their role in the classroom and try to transform their teaching methods, procedures and approaches. We believe that every science teacher training procedure should pursue the gradually shifting of the views of teachers from traditional to contemporary aspects about NOS, NOT, and NOL.

An overview of the training program

The program we developed in the University of Athens was based on socioconstructivist and sociocultural learning principles with the intention to help teachers to appropriate basic knowledge about the topic of the fall of bodies and to make explicit through the exploitation of authentic historical science events on the above topic (Aristotle's, Galileo's, Newton's and Einstein's interpretations on the fall of bodies) the contemporary views about NOS, NOL, and NOT.

In the implementation of the program we used a variety of teaching and learning strategies (e.g. debates - argumentation, group work, making posters, concept maps, simulations) that exploit authentic historical science events in the topic of the fall of bodies.

In the context of the training program a series of eight sessions (lessons) was produced and implemented. Its session (lesson) was accompanied by an appropriately designed work sheets. The eight sessions are the following:

- The changing views about the NOL and NOT
- The changing views about the NOS
- Aristotle's views about the fall of bodies
- Galileo's aspects of the fall of bodies
- The fall of bodies and the NOS in the cosmological models proposed by Aristotle and Galileo
- Newton's views about the fall of bodies
- Design a worksheet for teaching the topic of the motion of planets around the Sun
- Space trips

In the training meetings the focus was on teachers' changing participation in activities that actively involved them in discussions and actions that concern issues about NOS, NOT, and NOL through the exploitation of authentic historical science events on the topic of the fall of bodies. In this view, learning was a collaboratively and socially constructed entity, rather than an individual possession. Teacher learning was a collaborative inquiry process as teachers interacted with peers around topic, in activities with the guidance of the instructor who has had particular experience in the area.

A training/educational scenario concerning the study of Aristotle's views about the fall of bodies

In order to make explicit the procedure of the training program, and with the experience of the implementation of this program for two academic semesters, we proceed in the description of a specific training meeting. More specifically, we will make an elaborated description of a possible training/educational scenario that could concern the study of Aristotle's views about the fall of bodies and the chances that are created for science teachers with the appropriate guidance of the instructor to negotiate issues about NOS, NOL, and NOT. In appendix 2 is presented the corresponding worksheet that in-service science teachers can have in their disposition during the training procedure. The pedagogical and didactical procedure we propose in this training lesson concerns the implementation of a worksheet that could include the utilization of the following strategies:

- Reading texts from the history of science relative with the life and the work of Aristotle
- Science teachers collaborative work with guidance of the instructor
- Using simulations to present or represent Aristotle's views about the fall of bodies
- Creative ways of communication and presentation of the collaborative work of science teachers teams in the plenary session [poster design, slide creation, journalistic reportage, and an interview with the philosopher (Aristotle)].

The duration of the session could be about 120 minutes teaching. The activities that are included in the worksheet are designed for collaborative work so it is preferable if not necessary science teachers to work in groups.

In order to facilitate the learning process the instructor has to be appropriately prepared for the lesson. Instructor has to choose suitable reading texts from Aristotle's Physics. For example, in the case of Greek science teachers, we use a Greek book (Paleopoulou-Stathopoulou & Koukopoulou-Arnellou 1999) that is referred in the cosmological thinking of 2500 years' period from ancient years until now.

In the first and the second activities of the worksheet instructor could start a discussion about the Aristotelian interpretation of the fall of bodies after science teachers or students having studied the corresponding texts in small groups. It is needed appropriate texts from history of science's books that referred to Aristotle's view about the causes of the fall of bodies and his view about the positions of the Earth and the Sun in the universe to be selected. A synopsis of Aristotle's view about the "cosmological model" and the "theory of bodies' motion" that are included in the book of Paleopoulou-Stathopoulou & Koukopoulou-Arnellou would be useful for the instructor in order to coordinate the discussion of the teams in the plenary session that will follow the group discussions. [For example: *Natural Motion:* is the motion towards the "natural place" of a body. The cause of natural motion of the bodies is their substance. *Impetuous Motion:* is the motion towards any direction except of that of the "natural place" of a body. The cause of impetuous motion is an external force. The motion lasts as long as the body is in contact with the cause. The forces are exercised only when there is a contact point. The speed of the body is proportional to the force.] The presentation in a transparency of the synopsis of Aristotle's view about the fall of bodies is proposed to be presented to the science teachers at the end of the session because initially they should express their views and opinions freely and spontaneously. During the first and second activities trainees should be enabled to record their views, to discuss, and to argue about the target topic. During the discussion on the group work and the plenary session the instructor should offer the appropriate guidance to them.

In the context of third activity, trainees, based on the historical texts and the discussion that took place previously with instructor's guidance, write down the specific Aristotelian "laws" about the fall of bodies (for example: "The bodies move towards their "natural place" depending on their substance").

In the forth activity trainees are invited to compare the interpretation of the fall of bodies that Aristotle suggests and the one they have been taught in school. A farther step is to make a discussion about the rationale of the Aristotle's interpretation. The process of interpretation will offer to instructor a fruitful field to investigate students' views about the NOS.

When the groups have completed the four activities instructor could invite them to present their work to the above activities in the whole class by choosing one of the following ways: poster design, slide creation, report presentation, and interview with the philosopher (Aristotle).

References

Abd-El-Khalick, F. & Lederman, N. G.: 2000, 'The influence of history of science courses on students' views of nature of science', *Journal of Research in Science Teaching* 37, 1057-1095.

Aikenhead, S.G., Ryan, G.A., Fleming, W.R.: 1989, '*Views On Science-Technology Society, Form CDN.mc.5*', Department of Curriculum Studies College of Education.

Akerson, V. L., Abd-El-Khalick, F. & Lederman, N. G.: 2000, 'The influence of a reflective activity based approach on elementary teachers' conceptions of the nature of science', *Journal of Research in Science Teaching* 37(4), 295-317.

Bartholomew, Osborne & Ratcliffe: 2004, 'Teaching Students ''Ideas-About-Science'': Five Dimensions of Effective Practice', *Science Education*, 88(5), 655-682.

Bell, B. & Gilbert, J.: 1996, *Teacher development: A model from Science Education*, Falmer Press, London.

Bell, R.L., Lederman, N.G. & Abd-El-Khalick, F.: 2000, 'Developing and acting upon one's conceptions of the nature of science: A follow-up study', *Journal of Research in Science Teaching* 37, 563-581.

Bevilaqua, F. & Giannetto, E.: 1998, 'The History of Physics and European Physics Education'. In B. J. Fraser & K. G. Tobin (Eds.), *International Handbook of Science Education,* Kluwer Academic Publishers, pp. 981-999.

Bruner, J.: 1996, *The culture of education*, Harvard University Press.

Cole, M.: 1996, *Culture in mind*, Harvard University Press, Cambridge, MA.

Driver, R.: 1995, 'Constructivist approaches to science teaching'. In L. Steffe & J. Gale (Eds.), *Constructivism in Education*, Hillsdale, N. J.: LEA, pp.385-400.

Driver, R., Leach, J., Millar, R. & Scott, P.: 1996, *Young people's images of science,* Open University Press, Buckingham England.

Duffy, M.T. & Cunningham J. D.: 1996, 'Constructivism: Implications for the design and delivery of instruction'. In L D.H. Jonassen (Ed.), *Handbook of Research on Educational Communications and Technology*, Scholastic, New York.

Fletcher, M. A.: 2002, 'Reciprocity in professional partnerships: Outcomes for everyone'. In J. Searle & D. Roebuck. (Eds.), *Envisioning practice - implementing change: 10th Annual International Conference on Post-Compulsory Education and Training,* Australian Academic Press, Brisbane.

Heering, P.: 2000, 'Getting Shocks: Teaching Secondary School Physics Through History', *Science & Education* 9, 363-373.

Hodson, D.: 1988, 'Towards a philosophically more valid science curriculum', *Science Education* 72(1), 19-40.

Jenkins, E.: 1996, 'The "nature of science" as a curriculum component', *Journal of Curriculum Studies,* 28(2), 137-150.

Kokkotas et al.: 1998, 'Developing the concept of the material nature of air through laboratory experiences at the Primary school level', *Practical Work in Science Education, International Conference,* Copenhagen, 20-23 May, 1998, pp. 122-135.

Kokkotas, P.: 2003a, *Didactics of Science Education (Part II) - Cotemporary Trends for Teaching Science: Constructivism in Science Teaching and Learning,* Athens (in Greek language).

Kokkotas, P.: 2003b, 'The nature of science as a determinant factor of shaping of science curriculums'. In K. Skordoulis & Kr. Chalkia (eds), Proceedings *of 2nd Pan-Hellenic Conference entitled: The contribution of History and Philosophy of Science in the Science Education,* Leader Books, Athens, pp. 87-98.

Lave, J.: 1996, 'Teaching, as learning in practice', *Mind, Culture, and Activity* 3, 149-164.

Lazarowitz & Hertz Lazarowitz: 1998, 'Cooperative learning in the science curriculum'. In B. Fraser & K. Tobin (Eds), *International Handbook of Science Education,* Kluwer Academic Publishers, pp. 449-469.

Leach, J., Millar, R., Ryder, J. & Séré, M-G.: 2000, 'Epistemological understanding in science learning: the consistency of representations across contexts', *Learning and Instruction,* 10 (6), 497-527.

Lederman, N. G., Wade, P. D. & Bell, R. L.: 1998, 'Assessing understanding of the nature of science: A historical perspective'. In McComas, W. (Ed.), *The nature of science in science education: Rationales and strategies,* Kluwer Academic Publishers, pp. 331-350.

Lederman, N. G.: 1992, 'Students' and teachers' conceptions of the nature of science: A review of the research', *Journal of Research in Science Teaching* 29 (4), 331-359.

Lemke, J. L.: 2001, 'Articulating communities: sociocultural perspectives on science education', *Journal of Research in Science Teaching* 38, 296-316.

Matthews, M. R.: 1994, *Science Teaching. The Role of History and Philoshophy of Science,* Routledge, New York - London.

Matthews, M.: 1998, 'The nature of science and science teaching'. In B. J. Fraser & K. G. Tobin (Eds.), *International Handbook of Science Education,* Kluwer Academic Publishers, pp. 981-999.

Matthews, M.R., Bevilacqua, F. & Giannetto, E. (eds.): 2001, *Science Education and Culture: The Role of History and Philosophy of Science*, Kluwer Academic Publishers.

Matusof E. & Hayes R.: 2002, 'Building a community of educators versus effecting conceptual change in individual students: multicultural education for pre-service teachers'. In G. Wells & G. Claxton, (Eds.), *Learning for life in the 21st century: Sociocultural perspectives on the future of education,* Blackwells, London.

McComas, W. F., Clough, M. P. & Almazroa, H.: 1998, 'The Role and Character of the Nature of Science in Science Education'. In W.F. McComas (Ed.), *The Nature Of Science In Science Education. Rationales and Strategies*, Kluwer Academic Publishers, pp. 3-39.

McComas, W.: 1996, 'Ten myths of science: Reexamining what we think we know about the nature of science', *School Science and Mathematics*, 96, 10-16.

Moreno, A.: 2001, ' "Weighing" the Earth: a Newtonian Test and the Origin of an Anachronism', *Science & Education* 10(6), 515-543.

Mortimer, E. & Scott, Ph.: 2003, *Meaning Making in secondary science classrooms,* Open University Press, Maidenhead-Philadelphia.

Newton, P., Driver, R. & Osborne, J.: 1999, 'The place of argumentation in the pedagogy of school science', *International Journal of Science Education* 21, 553-576.

Nott, M., & Wellington, J.: 1998, 'Eliciting, interpreting and developing teachers' understandings of the nature of science'. *Science & Education* 7(6), 579-594.

Nystrand, M., Gamoran, A., Kachur, R. & Pendergast, C.: 1997, *Opening dialogue. Understanding the dynamics of language and learning in the English classroom.* Teachers College Press, New York.

Paleopoulou-Stathopoulou, R. & Koukopoulou-Arnellou, G.: 1999, *Cosmologica: The acquaintance of the cosmological thinking of 2500 years' period.* Athens: Typothito (in Greek language).

Rieß, F.: 2000, 'History of Physics in Science Teacher Training in Oldenburg', *Science & Education* 9, 399-402.

Rogoff, B.: 2003, *The cultural nature of human development*, University Press: Oxford.

Rogoff, B., Matusov, E. & White, C.: 1996, 'Models of teaching and learning: Participation in a community of learners'. In D. Olson & N. Torrance (Eds.), *Handbook of education and human development: New models of learning, teaching, and schooling*, Blackwell, London.

Ryan, A.G. & Aikenhead, G.S.: 1992, 'Students' preconceptions about the epistemology of science', *Science Education* 76, 559-580.

Seroglou, F. & Koumaras, K.: 2001, 'The Contribution of the History of Physics in Physics Education: A Review', *Science & Education,* 10(1-2), 153-172.

Seroglou, F.: 2002, 'Science for the Education of the Citizen: Development and Evaluation of a Teacher-Training Programme. A Transdisciplinary Approach of the Teaching of Pendulum'. *Proceedings of the International Pendulum Project,* Sydney, October 16-19, 2002, pp. 231-237.

Solomon J.: 1993, 'The Social construction of children's scientific knowledge'. In P.J. Black & A.M. Lucas (Eds.), *Children's informal ideas in science,* Routledge, New York.

Tobin, K. (ed.): 1993, *The Practice of Constructivism in Science and Mathematics Education,* AAAS Press, Washington DC.

Von Glasersfeld: 1995, 'A constructivist approach to teaching'. In Steffe, L. & Gale, J. (Eds.), *Constructivism in Education*, Hillsdale, N. J.: LEA, pp.3-16.

Wells, G.: 1999, *Dialogic inquiry: Towards a sociocultural practice and theory of education,* Cambridge University Press, New York.

Wells G.: 2002, 'Inquiry as an orientation for learning, teaching and teacher education'. In G. Wells & G. Claxton (Eds.), *Learning for Life in the 21st Century: Sociocultural Perspectives on the Future of Education, Blackwell* Publishing, *London* pp. 197-210.

Appendix 1

A sample of the questionnaire

Part 2: The nature of science
1. What, in your view, is science?
Please read from 1 to 9, and then choose (darken) one statement that fits better to your opinion.

①	Science is a truth-seeking process. It is not a collection of unquestionable "truths."
②	Science is an objective, logical, and repeatable attempt to understand the principles and forces operating in the natural universe.
③	Science can prove anything, solve any problem, or answer any question.
④	Science is primarily concerned with understanding how the natural world works.
⑤	Science involves dealing with many uncertainties.
⑥	Science is a self -correcting discipline. Such corrections may take a long time but as scientific knowledge accumulates the chance of making substantial errors decreases.
⑦	I don't understand.
⑧	I don't know enough about this subject to make a choice.
⑨	None of these choices fits my basic viewpoint.

3. When scientists investigate nature, it is said that they follow the scientific method. The scientific method is:
Please read from 1 to 9, and then choose (darken) one statement that fits better to your opinion.

①	Controlling experimental variables carefully, leaving no room for interpretation.
②	Considering what scientists actually do, there really is no such thing as the scientific method.
③	Testing and retesting — proving something true or false in a valid way.
④	Postulating a theory then creating an experiment to prove it.
⑤	Questioning, hypothesizing, collecting data and concluding.
⑥	A logical and widely accepted approach to problem solving.
⑦	I don't understand.
⑧	I don't know enough about this subject to make a choice.
⑨	None of these choices fits my basic viewpoint.

Part 3: Teaching and Learning in science classrooms

8. Please darken the statement that you agree the most.	
①	Learning is a search for meaning.
②	Learning is the acquisition of new behavior.
③	Learning is a process of building cognitive structures.
④	Learning is an act of membership in a "community of practice".
⑤	Learning is a process of peoples' changing participation in sociocultural activities of their communities.
⑥	Learning is a co-construction (or reconstruction) of social meanings emerging from socially negotiated and discursive activity.
⑦	Learning is a process of social apprenticeship.
⑧	Learning is an active process in which learners construct new ideas or concepts based upon their current/past knowledge.

10. Please indicate how well prepared you currently feel to do each of the following in your science instruction.

(Darken one oval on each line.)	**Not Adequately Prepared**	**Somewhat Prepared**	**Fairly Well Prepared**	**Very Well Prepared**
a. Take students' prior understanding into account when planning curriculum and instruction.	①	②	③	④
b. Make connections between science and History of Science.	①	②	③	④
c. Have students work in cooperative learning groups.	①	②	③	④
d. Use the textbook as a resource rather than as the primary instructional tool.	①	②	③	④
e. Teach groups that are heterogeneous in ability.	①	②	③	④
f. Guide a class of students using investigative strategies.	①	②	③	④
g. Recognize and respond to students cultural diversity.	①	②	③	④
h. Use computers for simulations and demonstration of scientific principles.	①	②	③	④

13. How frequently do you use the following teaching strategies in your class?

(Darken one oval on each line.)	Very Frequently	Frequently	Sometimes	Seldom	Never
a. Laboratory activities	①	②	③	④	⑤
b. Debates – Argumentation activities	①	②	③	④	⑤
c. Problem solving	①	②	③	④	⑤
d. Drama activities	①	②	③	④	⑤
e. Field trips in Museums	①	②	③	④	⑤
f. Group work	①	②	③	④	⑤
h. Role Playing	①	②	③	④	⑤
i. Telling Stories	①	②	③	④	⑤
j. Open ended investigations	①	②	③	④	⑤
k. Guided investigations	①	②	③	④	⑤
l. Simulations	①	②	③	④	⑤
m. Concept maps	①	②	③	④	⑤
n. Library search	①	②	③	④	⑤
o. Utilization of events of the History of Science	①	②	③	④	⑤

14. Think about your plans for a physics course.
You plan to teach the subject of "falling bodies".
A. Please indicate to what extent you have a stable background knowledge on the following.

(Darken one oval on each line.)	None	Minimal Emphasis	Moderate Emphasis	Heavy Emphasis
The history of the evolution of theories of falling bodies.	①	②	③	④
About the nature of science.	①	②	③	④
Physics of falling bodies.	①	②	③	④
Strategies that relate the science education with the History of Science.	①	②	③	④

B. Please indicate how well informed you currently feel about aspects that concern history of the evolution of falling bodies.

	Not Adequately Prepared	Somewhat Prepared	Fairly Well Prepared	Very Well Prepared
Aristotle's view of the fall of bodies.	①	②	③	④
Galileo's view of the fall of bodies.	①	②	③	④
Newton's view of the fall of bodies.	①	②	③	④
Einstein's view of the fall of bodies.	①	②	③	④

Appendix 2

Worksheet 3
Aristotle's views about the fall of bodies
In order to answer the following questions of the worksheet you need to study first Aristotle's contribution in Physics constitution. For your information attached to the worksheet is a relevant text.

1. What is Aristotle's view about the causes of the fall of bodies? What role does the material of the falling body play?

……………………………………………………………………………………………………
……………………………………………………………………………………………………
…………

2. According to Aristotle, what are the positions of the Earth and the Sun in the Universe?

……………………………………………………………………………………………………
……………………………………………………………………………………………………
…………

3. The fall of bodies was studied by Aristotle, who formulated specific "laws". According to the attached text, which are they?

……………………………………………………………………………………………………
……………………………………………………………………………………………………
…………

4. What is the difference between the interpretation of the fall of bodies that Aristotle suggests and the one you have been taught in school? Discuss in your group and argue about the rationale of Aristotle's interpretation.

……………………………………………………………………………………………………
……………………………………………………………………………………………………

After you have finished the above work each group is expected to present its answers to the above questions in the whole class in one of the following ways:

- Poster drawing/construction that will be hung up and presented
- Slide creation for overhead projector
- Report presentation
- Presentation of an interview with the philosopher (Aristotle).

Weighing the Earth, weighing the Worlds
From Cavendish to modern undergraduate demonstrations

PIERRE LAUGINIE

Groupe d'Histoire et de Diffusion des Sciences d'Orsay (GHDSO).
Université Paris-Sud, centre d'Orsay, bat 407. F-91405 Orsay cedex, France.

Abstract: An historical overview is given from the first XVIIIth century experiments "to determine the density of the Earth" to the recent determinations of the newtonian "big G" constant. The dominant role played in this story by the gravitational torsion balance — including recent measurements — is emphasized through a detailed study of Cavendish's gravitational balance and its important technical, methodological and editing innovations. In the framework of our "historical experiments" approach of Science and History of Science Teaching, the use of a modern pedagogical version of the Cavendish balance in a university classroom is presented and commented at great length. The relationship of the "weighing the Earth" challenge with Astronomy is finally shortly evoked.

Introduction

To weigh the Earth: why and how? A problem posed by geologists during the XVIIIth century: getting a first information about our Earth's deep interior from its mean density. Simultaneously, the Earth will become a mass standard for the weighing of the whole solar system. The concern will turn towards a "universal" issue related to the gravitational constant "big *G*" only at the end of the XIXth century.

Following the first XVIIIth century attempts with the "mountain" method, Cavendish's gravitational torsion balance implemented new standards of rigor, accuracy and methodology so that his 1798 report could have been considered as the first *modern* scientific paper (Everitt, 1977).

Miniaturized pedagogical variants of the original torsion balance — as said "a scale to weigh the Worlds" — are nowadays in use in universities, a very typical example for our experimental approach of the History of Science: accepting simplifying opportunities brought by modern techniques and materials, while retaining the essentials of the original concepts. A detailed account of our working way will be given. Such teaching allows revealing unusual students skills and developing qualities more in agreement with actual scientific working.

Meanwhile, improved versions of the gravitational torsion balance have been in use throughout the XIXth and XXth centuries as research tools (improved Earth mass and *G* determinations, gravimetry); and again, at the very beginning of the XXIst century, recently yielding a one order of magnitude leap in G-accuracy. A remarkable precision tool over two centuries! The relationship of the Earth mass or G determinations with Astronomy will be finally briefly evoked.

Motivations

Throughout the XVIIIth century, the mean density of the Earth has been a very important concern, especially for geologists. Knowing nothing about the deep interior of our common home was a very irritating question! An "irritating" and also very difficult problem: indeed, the following centuries will show that it is easier to learn about the Sun's interior than the Earth's one. Thus, since the mean density of the Earth's surface – the Earth's crust – was known, from rocks and other matters, to lie between 2 and 2.5 times the density of ordinary water, any approach of the mean density of the interior would yield a first information about the Earth's internal constitution. This was a highly significant challenge! It is clear that, determining the Earth's mean density requires, at least implicitly, the Earth's mass, but this was not the concern at the time: which interest would be the exact number of tons of matter in the Earth? Even more so, our modern gravitational "big G" constant – which does not even appear in Newton's writings – was completely out of concern; though G is a measure of the intensity of the gravitational interaction, it will be brought to the forefront only at the end of the XIXth century. Recent papers by B.E. Clotfelter and by A. Moreno have specially put the emphasis on that point (Clotfelter, 1987; Moreno, 2001).

However, a question is sometimes asked, especially by our students: wouldn't it exist some "astronomical" way to derive the Earth mass? The answer is known from Kepler's third law and Newton's gravitational law: since all bodies are "falling" on the Sun with the same acceleration, the period of a planet is depending only on the Sun's mass and its distance to it and not on the planet mass (as far as this last one is negligible relatively to the Sun one). Let us write, once and for all, those laws in their usual *modern* form:

- Newton's gravitational law: the mutual attraction F between two point masses M and m placed at a distance d from each other is:

$$F = G\frac{Mm}{d^2} \qquad (1)$$

  where G is the "universal" gravitational constant. Newton has shown that the same law works for homogeneous spheres, with their centres d apart.

• Kepler's third law: the orbital period T of a planet about the Sun and its mean distance a to the Sun (the half-diameter of the ellipse) are related to the Sun's mass M_S according to:

$$\frac{T^2}{a^3} = \frac{4\pi^2}{GM_S} \qquad (2)$$

Nevertheless, at the XVIIth and XVIIIth centuries, *such formulas were never written*: G never appeared, the attraction was only said to vary like the masses and the reciprocal of the squared distance; the squared period of the orbit was only said to vary like the cube of the distance to the Sun; all calculations and deductions were drawn from comparisons and proportions between related quantities. From such comparisons, Newton was able to deduce the *relative values* of the Earth and Sun masses on one side, and of the Sun and planets with satellites on the other side (Kepler's law was applied, for example, to the Jupiter-satellite system). He could thus get the relative values of the masses of the Sun, Jupiter, Saturn and Earth, from which their mean densities which were found respectively as: 100; 94.5; 67 and 400 (Newton, Principia, Part II, Prop. VIII). We stress Newton's interest, not only in the planets masses but specially in their densities as revealing something of their internal constitution. We note also that, since all those quantities are *relative* values, determining the mass (or density) of our planet would yield the other absolute masses and densities, a "weighing" of the solar system. Thus, getting a mass standard for the solar system, and consequently for the Universe, required a terrestrial "weighing" of our planet.

NEWTON'S PREDICTION

In the same chapter of the Principia, Newton made a surprising prediction about the Earth's mean density:

> "Thus, since the Earth is usually at its surface approximately twice heavier as water, and lower in the mines it is found 3 or 4 times, and even 5 times heavier; it is likely that the total amount of matter in the Earth be five or six times larger than if it were entirely made of water, the more so that it has previously been shown that the Earth is almost five times more dense than Jupiter." (Newton 1714, Part III, Prop. X, p. 373)[1,2].

Though this prediction finally proved to be outstandingly correct, it was however a prediction, not a measurement. Newton himself, in the Principia, proposed two methods: the deviation of a plumb-line in the vicinity of a mountain, and the direct measurement of the attraction between two heavy spheres, but he severely doubted of the feasibility of any of such experiments.

In both type of experiments, a given test-mass m would be submitted to the attraction of a large mass M. Suppose the attractive force F between m and M be measured: then, comparing it to the proper weigh p of the test-mass (its attraction from the Earth), and correcting from the reciprocal of the squared distances, will lead to the Earth's mass M_T (its density will follow). Precisely:

$$\frac{p}{F} = \frac{M_T}{M} \cdot \frac{d^2}{{R_T}^2} \qquad (3)$$

where R_T is the Earth's radius and d the m-M "distance". From which M_T follows.

In the "mountain" experiment, the test-mass m is the bob of a plumb-line; the large mass M is provided by a nearby mountain; the attraction of the mountain is measured from the (very small) deviation of the plumb-line; the mass M is estimated from the supposed constitution of the mountain; a correct account of the distances requires a detailed survey of the region and a numerical integration over the whole mountain.

In the "Spheres" experiment, the test-mass and the large mass are lead (or any heavy metal) spheres; their mutual attraction has to be directly measured in the "laboratory" (see below for the assembly); according to Newton's laws, the distance d is just the distance between their centres and is thus very well determined. The whole experiment is performed in a small room. Though this method may appear preferable, people were so much convinced that the mutual interaction force was too much tiny to be measured, that the "mountain" method was first tried out.

THE XVIIIth CENTURY FIRST MEASUREMENTS. BOUGUER, MASKELYNE AND HUTTON:

The first significant attempts in the mid-XVIIIth century were those of Bouguer on Chimborazo, a volcanic mountain in Peru, in 1749. Bouguer, a French astronomer, was a member of the famous French Academy expedition in the Andes to determine the length of the degree of meridian under the Equator; as a side-experiment, he tried to measure the deviation of the plumb-line due to the attraction of the Chimborazo mountain. Though a deviation was observed, the results were not conclusive (insufficiently accurate instruments, unknown internal constitution of the mountain, extremely bad weather and working conditions). However, Bouguer had shown the way.

The experiment was resumed in the years 1774-1775 by Sir Nevil Maskelyne, Astronomer Royal in Greenwich, and Charles Hutton who will be in charge of the detailed survey of the region. They used a mountain of adequate shape and of homogeneous constitution, the Schiehallien hill in Scotland, "extremely solid and dense, and seemingly composed of an entire rock" (contrary to Chimborazo). A preliminary re-

port was published by Maskelyne in the *Philosophical Transactions* in 1775, based on a first rough survey. He concluded to a mean density of the Earth between 4 and 5 times that of water and added: (a result) "totally contrary to the hypothesis of some naturalists, who supposed the Earth to be only a great hollow shell of matter". However, two more years were necessary to C. Hutton in order to achieve the complete detailed 3-dimensional survey of the region over several miles. It is worth mentioning how innovative this last work has been: new techniques of measurement and numerical integration were implemented and, for the first time, the results appeared as a contour map. The complete results were published by Hutton in a 100-pages article in the *Philosophical Transactions* in 1778 (Hutton, 1778) and he concluded to an Earth's mean density of 4.5 "assuming the density of the hill equal to that of common stone". A shortened version of this article, incorporating new comments, appeared in the "Abridged" version of the the *Philosophical Transactions* in 1809 (Hutton, 1809, pp. 408-423). Taking account of a new evaluation of the constitution of the hill by John Playfair (cited in: Hutton, 1809), the Earth's mean density is then estimated to "nearly 5 times the weight of water". Since this article – omitting the tiresome listings of all measurements and calculations – describes completely the principles, the whole set of procedures and computation techniques, and the final results, it is this paper which should be recommended within the framework of an historical approach of the "Earth's density" problem at the university level.
Finally, *the situation at the beginning 1780's was as follows*: an accurate implementation of the "mountain" method yielding a first knowledge about the Earth's deep interior: since the surface density is only around 2–2.5, the Earth's core necessarily incorporates heavy elements. However, the final result is entirely depending on an estimate of the mountain's mean density, which, in any eventuality, will remain an estimate; the method requires time and much work for an extremely detailed and accurate survey over a several miles region. Though the "mountain method" will be performed again through the XIXth century (near Marseille in 1814 and near Edinburgh in 1856), it appears at the time as having reached its limits and something else should be undertaken. We shall only mention a third method using the variation of the acceleration of gravity *g* either at the top of a mountain or at the bottom of a mine (see Clotfelter, 1987; Moreno, 2001). This method is of less historical importance and the remainder of this article will be devoted to the "Spheres" method.

The "Spheres method" and the Michell-Cavendish experiment

Thus, around the 1780's, the Reverend John Michell, a "multidimensional" scientist, astronomer, physicist, geologist, will, the first, build an instrument in view of a direct laboratory measurement of the attraction between two lead spheres. In order to measure such a minute attraction, he chose the torsion balance technique, a quite new technique at the time, which, according to Cavendish, he invented independently of Coulomb[3]. The choice was excellent: it will allow measuring such minute forces which appeared quite non-measurable beforehand. However, Michell could not perform the density of Earth experiment with the balance he had built because of numerous occupations, disease and finally death. The physicist John H. Wollaston inherited the instrument, and finally passed it to Cavendish who almost completely reconstructed it (for example the diameter of the large spheres was increased from 8" to 12", increasing their mass by a factor 3.37); he will apply to it high standards of methodology, rigor and accuracy unknown beforehand. The balance is thus known as "Cavendish's gravitational balance", even if the role of John Michell should always be recalled in the genesis of the experiment.

CAVENDISH'S GRAVITATIONAL TORSION BALANCE:

The Cavendish torsion balance and the measurement process are described on Figure 1.

The small spheres or "balls" are attached at both ends of a beam, itself suspended to a torsion wire. When the large spheres or "weights" are approached in a fixed position close to the small ones, their mutual attraction is deduced from the deviation of the beam towards a new equilibrium position.

The whole assembly is placed is a special room closed by strong walls. It is operated from outside in order to avoid vibrations and thermal effects from the operator (air convection, the most important). The deviation of the beam is checked on a small ivory rule with a vernier allowing a 1' angular resolution. The picturesque of the thing is that, since the scale is observed from outside through a small telescope crossing the wall, it is lit up with a projector, the light source of which is a candle (note that no use is made of the Poggendorf's method which will be invented only in 1826. See fig. 3).

The torsion constant of the wire is deduced from the swinging period of the beam. Because of a very long period (7 and 15 min, depending on the wire used) and very small damping, Cavendish cannot wait hours to reach the equilibrium which would be more and more subject to external perturbations: the equilibrium point is deduced from the first three extremes. Let us note the global dimensions: 6 feet (1.83 m) for the beam, 730 g for the balls, 158 kg for the weights. With this instrument, the ratio between the (measured) attraction undergone by a "ball" from the nearby "weight" and its proper weight (eq. 3) was as one part in 50 millions. Thus, a tremendous assembly, so much by its size than by the overcome difficulties!

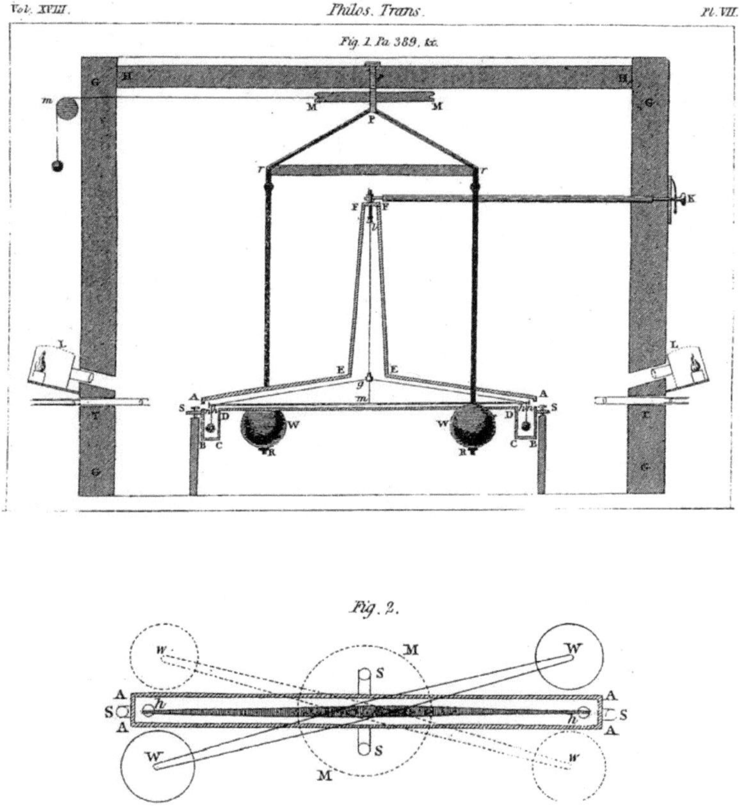

Figure 1. Cavendish's gravitational balance (original drawing).
In a mahogany case, two leaden “balls” or “test masses”, 2” or 5.08 cm in diameter and 730 g each, are hung at each extremity of a wooden arm, 6 feet or 1.83 m long. The arm is suspended in an horizontal position by a slender wire about 1 m long, its oscillation period is roughly either 15 mn or 7 mn, depending on the wire which was used. The two large leaden “weights” or attractive masses, 1 foot or 30.48 cm in diameter and 158 kg each, can be approached close to the balls, “so that their attraction would conspire in drawing the arm aside”. From the deviation of the beam when equilibrium is reached, the gravitational attractive force between the balls and the weights is deduced. Those deviations were approx. 1° and 1/4° respectively for both wires. The lower picture is a top view of the assembly. The balls are labelled *h,* and the weights *W* in the former position and *w* (dashed lines) in the symmetrical one (Cavendish, 1798, Tab. XXIII-XXIV, p. 526; 1809, pl. VII).

CAVENDISH'S 1789 ARTICLE:

Cavendish's 1798 article has been considered by Francis Everitt as the first *modern* scientific paper (Everitt, 1977): a complete table of results, including 29 independent sets of measurements (see fig. 2); a detailed account of the measurements with dates, conditions and special difficulties. This is new: for example, in his 1785–Mémoire aiming to establish the inverse square law for Electrostatics with a torsion balance, Charles Coulomb published only one set of 3 measurements (with the distance divided by a factor 2, and again a factor 2). Clearly, what was produced was an example of a "good" result among (likely) many others (Coulomb, 1785)[4]. With Cavendish, *all* is published. We can again remark that the final result (last column in fig. 2) is the mean density of the Earth, not its mass or anything else.

Cavendish goes directly into the heart of the matter through a *modelization*, i.e. a simplification of his assembly. And he refers to the continuation of the paper for the whole corrections brought to the model, always on quite physical grounds, and gives a very detailed account of the external perturbations. Some may be cited: the obstinate search for eventual magnetic influences; the effect of the non-uniformity of the gravitational field of the weights resulting in a slight frequency shift of the oscillatory regime — which will give rise to one of the modern *G*-determinations (Luther & Towler, 1982) — and the wonderful way in which it is solved; the nice demonstration of the attraction of the balls by the near wall of the case considered as part of a massive plane; the choice of a symmetrical operation (as shown on Figure 1, lower part), thus eliminating some errors due to a slight uncontrolled asymmetry in the zero-deviation position; the highlighting of the thermal effects due to different thermal inertia between the large weights and the rest of the assembly. We shall come back below to some aspects of this article in the pedagogical section of this paper.

WAS CAVENDISH'S EXPERIMENT A "TEST" OF NEWTON'S GRAVITATIONAL LAW ?

We strongly emphasized the fact that neither Maskelyne nor Cavendish did intend to test the inverse square law of the gravitational interaction, well established by Newton more than a century ago and proved with the high precision of astronomical observations. No more was Cavendish looking for the value of our gravitational constant *G* which nowhere appears in his report, nor elsewhere, at the time. However, it is sometimes considered that the *experiment* itself would be a test for the newtonian law, independently of astronomical observations, and this deserves discussion. Two answers can be brought:

1. did Cavendish bring a direct laboratory test of the "universality" of the gravitational interaction, i.e. of the existence of a mutual attraction, not only between celestial bodies, but between any bodies, including the common ones on the Earth? The answer is unquestionably yes. Let us notice that the Maskelyne and Hutton experiment had already brought a first answer, though limited to very large geological bodies.
2. did Cavendish's experiment bring a test of Newton's inverse square law? The answer is unquestionably *no*! Indeed, in order to test the inverse square law, one would have to *vary the distance* between the interacting spheres. On the contrary, not only Cavendish used a fixed distance (8.85" or 22.5 cm from center to center) in his reported measurements, but, moreover he *makes use* implicitly of the inverse square law in order to determine the Earth's mass or density (see eq. 3 above). And this use cannot be avoided! Thus, the inverse square law is *accepted*, not demonstrated.

Cavendish himself was quite aware of this "acceptance". Let us mention here a noteworthy final remark in his paper, which evidences again how scrupulous he was. With a remarkable intuition, Cavendish puts the question of an eventual modification of the inverse square law at small distances (8.85" or 22.5 cm in his case):

> "Another objection, perhaps, may be made to these experiments, namely, that it is uncertain whether, in these small distances, the force of gravity follows the same law as in greater distances. There is no reason, however, to think that any irregularity of this kind takes place, until the bodies come within the action of what is called the attraction of cohesion, and which seems to extend only to very minute distances" (Cavendish, 1798, p. 522; 1809, p. 406; shortened version, 1959, p. 91).

Then, Cavendish reports to have tested the eventual influence of "the attraction of cohesion" by putting the "weights" quite close to the case, with a negative result[5]. This question of the validity of the inverse square law is nowadays again quite topical, but Newton's law has now been successfully tested down to distances less than 1 mm — again with a torsion balance! (Hoyle, Schmidt et al, 2001) — and the relevant distances in some modern theories are at least two orders of magnitude lower than in Cavendish's experiment.

We should add that Maskelyne himself considered his own experiment as a test of the inverse square law. How could it be? The argument was that, if the law had been a reciprocal one instead of inverse square, the magnitude of the effect would have been so small as to be completely non-measurable. The same is true for Cavendish: with an

inverse relationship, the ratio of the interaction force to the ball's weight (eq. 3), instead of one part in 50 millions, would have been of the order of one part in 10^{15}! However, this cannot be considered as a "demonstration" of the inverse square law.
Finally, when the question of a "test" of Newton's gravitational law is asked, how do you understand it? In my mind, I would immediately think to the inverse square law, because the sole existence of a universal attraction is a physical fact, not a "law", i.e. not a quantitative relationship between physical quantities. And, as far as Cavendish is concerned, my answer would be *no*! However, I expect to have given a full overview of this somewhat controversial question.

CAVENDISH'S QUALITY MEASUREMENTS:
Cavendish reports 29 results for the mean density of the Earth (fig. 2, last column): a first series of 6 measurements obtained with a "too smooth wire" corresponding to a swinging period of roughly 15 mn[6]; a second series of 23 measurements were obtained with a stiffer wire (period now roughly 7 mn). Both series are said to yield a mean value for the Earth density of 5.48, while the modern value is close to 5.517. Excellent! Let us cite him: "By a mean of the experiments made with the wire first used, the density of the earth comes out 5.48 times greater than that of water; and by a mean of those made with the latter wire, it comes out the same" (Cavendish, 1798, p. 521; 1809, p. 405).
However, those results will suffer some critics and scepticism from successors. Around the 1800's, the confidence in Maskelyne and Hutton results (4.5 to 5) was so high that, for example, A. Libes will require further measurements in order to discriminate the results (Libes 1813). Furthermore, F. Bailey, who will undergo hundreds of measurements some 40 years later with the same technique, expresses doubts about the probative force of the Cavendish result towards his contemporaries: rather a *specimen* of a method than an actual scientific result (Baily 1842; Heilbron 1994). The argument lies in part on the global dispersion of the data (the difference between extreme values, 1/14, relative value, for the 23 measurements of the second series according to Cavendish) which was usually considered, at the time, as characterizing the uncertainty. It lies also on the fact that the global mean of the 29 reported results is evidently 5.45, not 5.48 as Cavendish claimed!
Did Cavendish make an error? On that last point, our understanding of Cavendish's report is different from Baily's.

520 *Mr.* CAVENDISH'S *Experiments to determine*

CONCLUSION.

The following Table contains the Result of the Experiments.

Exper.	Mot. weight	Mot. arm	Do. corr.	Time vib.	Do. corr.	Density.
1	m. to +	14,32	13,42	′ ″	-	5,5
	+ to m.	14,1	13,17	14,55	-	5,61
2	m. to +	15,87	14,69	-	-	4,88
	+ to m.	15,45	14,14	14,42	-	5,07
3	+ to m.	15,22	13,56	14,39	-	5,26
	m. to +	14,5	13,28	14,54	-	5,55
4	m. to +	3,1	2,95		6,54	5,36
	+ to −	6,18	-	7,1	-	5,29
	− to +	5,92	-	7,3	-	5,58
5	+ to −	5,9	-	7,5	-	5,65
	− to +	5,98	-	7,5	-	5,57
6	m. to −	3,03	2,9	-	-	5,53
	− to +	5,9	5,71		-	5,62
7	m. to −	3,15	3,03	7,4 by mean.	6,57	5,29
	− to +	6,1	5,9			5,44
8	m. to −	3,13	3,00		-	5,34
	− to +	5,72	5,54	-	-	5,79
9	+ to −	6,32	-	6,58	-	5,1
10	+ to −	6,15	-	6,59	-	5,27
11	+ to −	6,07	-	7,1	-	5,39
12	− to +	6,09	-	7,3	-	5,42
13	− to +	6,12	-	7,6	-	5,47
	+ to −	5,97	-	7,7	-	5,63
14	− to +	6,27	-	7,6	-	5,34
	+ to −	6,13	-	7,6	-	5,46
15	− to +	6,34	-	7,7	-	5,3
16	− to +	6,1	-	7,16	-	5,75
17	− to +	5,78	-	7,2	-	5,68
	+ to −	5,64	-	7,3	-	5,85

Figure 2. Cavendish's results (*facsimile* original paper).

This is Cavendish's original listing. Clearly, the sole purpose is the density of the Earth (last column). The first six results, obtained with a very slender wire which "was not stiff enough", with a "time of vibration" of ca. 15 min, have a mean of 5.31; the 23 following ones obtained with a second wire with a "time of vibration" of ca. 7 min, have a mean of 5.48. However, Cavendish writes: "By a mean of the experiments made with the first wire used, the density of the earth comes out to 5.48 times greater than that of water; and by a mean of those made with the latter wire, it comes out the same" (Cavendish, 1798, p. 520; 1809, p. 405).

Let us examine the results: the mean of the six initial measurements with the smoother wire appears to be 5.31 (consequently lowering the global mean to 5.45, less good than 5.48, as observed by many commentators who noticed "Cavendish's error"). Cavendish asserts explicitly 5.48 in the text. Somebody is mistaken: Cavendish himself or a copyist who would have incorrectly reproduced the table? We likely shall never know. Then, for the 23 following measurements with the stiffer wire, the mean is 5.48 — "again", Cavendish says — and this time, he is right. Clearly, since the torsion wire is the fundamental part of the apparatus, there are two distinct instruments. Consequently, the average *cannot be calculated on the whole set* as Baily and others did, but only *separately on each series of results*: we must follow Cavendish himself who, very judiciously, considered separately each series of data (even if, by chance, he seemed to get the same mean result for both). And, according to Cavendish himself (see note 6), the first instrument can be considered as an imperfect "prototype" while the second one is the final assembly. Thus, if one value had to be accepted from Cavendish's work, it is 5.48, just as he said.

Furthermore, may we suggest an hypothesis about the origin of the error? The reported six data of the first series are:

5.50; 5.61; 4.88; 5.07; 5.26; 5.55 (cf. fig. 2)

The mean of this series is 5.31. Surprisingly, Cavendish asserts very strongly that it is 5.48. Then, we can remark that the 4.88 value deviates much more than all other data of both series, which remain in the range 5.07-5.85. Suppose that the Cavendish experimental value would have been actually 5.88 instead of 4.88: then the mean of the first series comes out to 5.48, exactly as Cavendish wrote! At our knowledge, this fact has never been previously noticed. Thus, the Cavendish's text would recover its full coherence. Obviously, we are going beyond the as-published text, and since we have not Cavendish's manuscript at our disposal, we leave this as hypothesis. However, it appears to us quite likely, and, within this framework, Cavendish would be fully right!

A METROLOGICAL EXPERIMENT:

Finally, we should emphasize that the "weighing the Earth" experiments are Metrology experiments. This is conceptually quite different, for example, from Coulomb trying to evidence an – evidently preconceived – inverse square law for Electrostatics (Coulomb, 1785; Heering, 1994). In this last case, measurements are invoked in order to infer or to prove a physical law. On the contrary, in Metrology, they have to yield, with the highest possible accuracy, the value of a physical parameter or constant. Joule's determination of a mechanical equivalent of heat (Joule, 1850) is, in this respect, an example similar to Cavendish's density of the Earth: no preconceived ideas could help them finding results only 1% off the modern value. This has been well underlined by J. Heilbron (1994, p. 160)[7].

"Weighing the Earth" and Science teaching: the gravitational balance as a pedagogical tool

HISTORICAL EXPERIMENTS AND SCIENCE TEACHING:

Following years of classic "academic" teaching, we have been involved in the 1990's in a more efficient and experimental approach of basic Physics, especially of the principles of classical Mechanics at the 1st and 2nd year levels at the university in Orsay. I used to tell our students: "Theory is at your fingers tips", and finally this was taken as an illustrative title for this pedagogical project: in French, "La Théorie au bout des Doigts". A good idea of what we were trying to do can be found in a paper which appeared later, by M.P. Silverman: "Science as it is taught should more closely resemble science as it is done by professional scientists" (Silverman, 1995).

History of Science was particularly involved in several of our topics: the Foucault pendulum and Earth's motion, the velocity of light and historical models of light, the electrostatic torsion balance, the "brachistochrone" and, of course, Cavendish's gravitational balance. This resulted in an original and efficient cooperation between physics lecturers and historians of science in Orsay: introducing History of science elements in science teaching, but also to introduce experimental work on "great experiments in Physics" as part of an History of Physics approach. This was also further proposed in "continuing education trainings" to high schools teachers. This original way of "concrete experimental History" received an enthusiastic welcome from the participants.

Replicating historical measurements is a full research activity, aiming to bring to the History of Sciences elements which cannot be found in written documents; it is namely a speciality of the Oldenburg group in Germany (see: Heering, Riess & Sichau 2000; Blondel 1994). However, replication is not an easy subject within usual classroom environments, either in high schools or universities. Then, what can be done? Accepting an intermediate level, reproducing historical measurements in the classroom – with reduced constraints but in full respect of the basic concepts of the original experiment – is a good way to link science to its history and to help the students acquiring useful skills. Our main, and strict, criterion is: provided that the assembly be completely "transparent" as regards the original concept and working, modern materials or some miniaturization can be accepted. The original paper is always provided; students have to read it, at least its most significant parts, and include in their laboratory report a comparison between the modern pedagogical version and the original one. Thus, our approach lies just at an intermediate level between the "full research approach" in progress in Oldenburg, and the traditional academic teaching. The Cavendish experiment will be an example of this approach.

I would finally lay stress on some points (among many others) which I consider important in a scientific cursus:

- those experiments are difficult, they sometimes require time, a lot of attention, patience, manual skills and rigour. This is part of students training, but rather unusual in academic cursus; just like in real labwork, specially if we cast our mind back to the historical environment at the time, the experiment may fail — this happens! there are "bad days" — and, just as in the case of the replications of Coulomb and Joule experiments (Heering, 1994; Blondel, 1994) much is also to be learned from failures. Again, unusual!
- very important is to allow sufficient time for work[8]. One day is the right amount of time for most of our experiments. Furthermore, the students are then placed in more realistic lab conditions, the more so that a full success is not necessarily required at the end of the day. In the case of two-months sessions (one day per week), they can choose to begin again next time, just like in actual scientific life.
- a last main point is the mix of theoretical and practical topics throughout the work. Generally, a first information is given by the lecturers at the beginning of the day in order to allow the students to start working rapidly. They have to temporarily accept some facts and undemonstrated theoretical laws. Further, during their work, lecturers give several short talks introducing theoretical background and historical aspects. No detailed written working plan is given, the sole documents are the manufacturer's instructions for use, and historical papers. The oral relationship between lecturers and students is given the first place.

We believe those practical skills to enrich a study of "great experiments" in Physics through a better understanding of the actual conditions of their achievement, finally yielding an improved understanding of what is science in its historical and social context.

BEFORE THE TORSION BALANCE. MASKELYNE AND HUTTON:

Though lecturers in mathematics rather than physicists are concerned, we would like to mention the study of the 1778–article by C. Hutton as an excellent introduction to a course on integration. We already said how innovative it was, both in measurements, computation and integration techniques: how the mountain was divided in a large number (2000) of cylindrical pillars; how the contribution of each one was accounted, with justified approximations, reducing each pillar to its axis; how the whole set of pillars of common height were gathered in a unique subset along a level line (Hutton invented the contour map). Much is to be gathered for an introductory course!

WORKING A PEDAGOGICAL VERSION OF CAVENDISH'S BALANCE:

Still nowadays, the "Weighing of the Earth Experiment" reported by Henry Cavendish in 1798 (Cavendish, 1798; 1809) remains fascinating to our eyes: weighing the Earth, really? Just in this lab? Absolutely fascinating!

- *The balance*: our gravitational torsion balance, shown on fig. 3, was bought from the Leybold Company (Germany). Following Boy's demonstration at the end of the XIXth century, the dimensions have been strongly reduced as compared to Cavendish's original: 10 cm for the beam, 15 g and 1.5 kg for the balls and weights respectively. The motion of the beam is detected through the Poggendorf method using a laser ray and yielding a spot on a distant wall. However, despite those differences, the essentials of the original experiment have been kept: apart from miniaturization, the instrument is physically and conceptually transparent and similar to the original: no "black box", the wire, the beam and the balls are easily seen, the weights are hand-operated from outside, all is easily understandable. More important, the swinging period — the crucial parameter — is similar to Cavendish's one: roughly 10 minutes, while it was 15 and 7 minutes for Cavendish, with his first and second wires respectively. The result is that the deviation of the beam at equilibrium under the attraction of the weights is about 0.5°, while Cavendish's ones were respectively approx. 1° and 0.25°. Thus, the students *work essentially in similar physical conditions* to Cavendish's ones, — including the same type of difficulties he encountered — an essential feature of our concept of historical experiments.
- *What the students have to do:* normally, the right amount of time for the gravitational balance experiment with undergraduate students is one day. This allows time enough for a full preliminary study – experimental and theoretical – of the torsion pendulum through the free oscillations of the beam (without the weights). This also allows time for requiring that the students *themselves* proceed to the adjustment of the initial zero-position of the beam, a preliminary operation sometimes requiring one hour or even two. When only a half-day is available, the balance is pre-adjusted by the lecturers and the behaviour of the torsion pendulum briefly recalled. However we shall describe below a full-day experiment (see fig. 3 and 4):

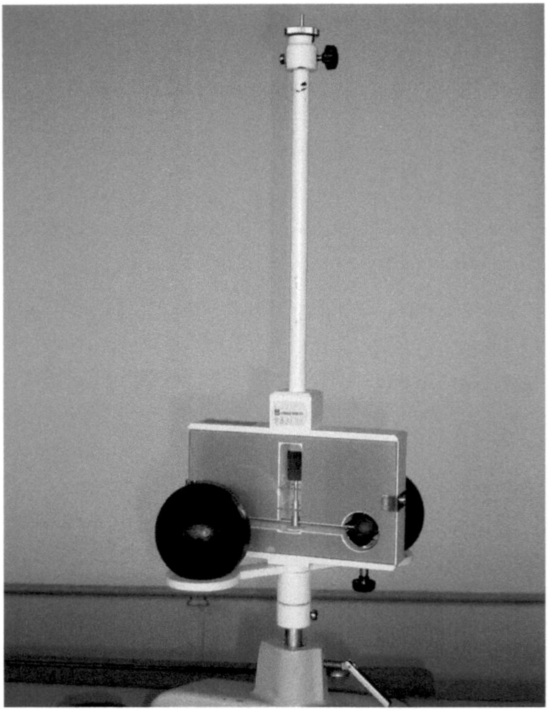

Figure 3. A pedagogical version of Cavendish's gravitational balance.

The size is considerably reduced: the beam (in a small rectangular glass case) is 10 cm long, the small "balls" weigh 15 g and the "weights"~1.5 kg, instead of 730 g and 158 kg respectively in Cavendish's original (see fig. 1). The torsion wire is a beryllium-copper ribbon, approx. 250 mm in length, cross-section .010 x .15 mm. The Poggendorf method is used for detection of the motion of the beam: a light ray, from a fixed source, is reflected on a mirror attached to the beam and yields a spot on the room wall 4 to 10 meters away. Despite those differences, the oscillation period, which is the main physical parameter — here 10 min — remains similar to the Cavendish original (15 or 7 min, fig. 1), while the deviation of the beam at equilibrium (0.5°) is also of the same order (see fig. 1). The ratio of the force experienced by a ball from the nearby weight, to its proper weight, is ~1part in 200 millions (instead of 1 in 50 millions in the original). This model was bought in the 1990's from the Leybold Company (Germany). We also tested another model, later proposed by Pasco (USA) and also in use in universities: the sole important differences are in the small balls: 38.3 g, and the beryllium-copper ribbon cross-section: .017 x .15 mm, resulting in a period of 8 min; the beam length (10 cm), the "weights" (1.5 kg) and the length of the beryllium-copper ribbon (approx. 26 cm) are similar to the previous Leybold model.

– *the first item* is the zero-adjustment with the knob at the top of the metal tube surrounding the torsion wire. A long and difficult process when the system is completely misadjusted! Remember that the whole system is extremely sensible, the swinging period very long (10 min) and the damping so small that hours are required for the beam definitively stopping at the equilibrium position. This position has to be estimated from one or two complete oscillations. The difficulty does not reside in getting the spot zero-position exactly in the middle of the wall (which is not required) but in the extreme sensibility of the beam to minute rotations of the knob. Let us note that the suspended equipment has to be left freely hanging during at least 12 hours (one night) before any measurement, in order to remove internal stresses in the torsion wire.
– *the next item* is the study of the free oscillations of the beam during at least three or four complete periods. The data are obtained checking the spot position on the wall (4 to 10 m distant) every 30 seconds, and further reported in a computer to get oscillations curves similar to those shown on fig. 4. Then, the equation of the damped torsion pendulum can be introduced, and the students (in good cases) can be invited to fit the curve to an exponentially damped sine curve. Indeed, the main cause of damping is the air friction on the balls; the period is so long, and the deviation so small (maximum some degrees) that the maximum velocity of the balls motion (for the maximum amplitude allowed by the case, i.e. approx. 0.1 radian) is some tenths of microns per second. For such low velocities, and taking account of the diameter of the balls, the air friction damping remains well within the viscous regime (friction force proportional to the velocity) resulting in an exponential damping. Good fits have been obtained in "good cases", i.e. when the beam motion is not affected by external disturbances. This justification has its interest by itself.
– the second part of the day is devoted to *the gravitational measurements*. The weights are now placed in either of both active positions, just against the case (say position "*W–W*" of fig. 1) and the oscillations of the spot again carefully reported during several swinging periods yielding, for example, the circle points curve in fig. 4. Then, the weights are placed in the second symmetrical active position (position "*w–w*" of fig. 1) and the oscillations again reported (solid points in fig. 4). The equilibrium position in each situation is deduced from a generalization of Cavendish's "mean" technique : if x_1, x_2 *and* x_3 are the first three extremes (first oscillation), an estimate of the equilibrium is: $\frac{1}{2}\left(\frac{x_1+x_3}{2}+x_2\right)$. Since our damping is larger than Cavendish's one (due to smaller balls), it is better to repeat this mean on the following oscillations and finally take the "mean of the means". The equilibrium positions are shown by two

horizontal solid lines on fig. 4. The experimental part of the work is achieved.
– subsequently, the students come to the *interpretation of their data.* Since we are in the XXIst century, not the XVIIIth one, the experiment is generally presented as a determination of the gravitational constant G (which requires only measuring the attraction between the balls and the weights, see eq. 1); the density of the Earth, requiring the further knowledge of the acceleration of gravity g is then deduced; furthermore, using Kepler's third law (eq. 2), they can deduce the density of the Sun and of Jupiter! However, when the experiment is performed in a more historical framework, the opposite way can be proposed, giving the first place to the density of Earth determination similarly to the XVIIIth century habit (G is then a secondary consequence). Now, *how to proceed?*

As shown on fig. 4, the distance δ between both equilibrium positions is well measurable. Its ratio δ/L to the wall distance L, yields four times the deviation θ of the beam from the neutral or zero position to one of its equilibrium position with the weights being active (twice due to the use of the Poggendorf technique, and twice again because of double deviation resulting from both symmetrical operations). Then, the momentum of the forces F between the weights and the balls at the equilibrium has to be equalled to the antagonist couple of the torsion wire:

$$2Fb = C\theta \qquad (4)$$

where F (depending on G) is given from eq. 1, b is half the beam length and C is the torsion constant of the wire. Knowledge of C (see below) yields finally the gravitational constant G.

– *determining the torsion constant* C of the wire: C is deduced from the period T of the free oscillations of the beam, according to the classical result: $C = 2\pi\sqrt{I/C}$ where I is the moment of inertia of the beam. This last one is approximately determined taking only account of the lead balls at its extremities. It is worth mentioning here a simplifying method due to Cavendish himself, which is described below (see: *lessons to be learned...*): it allows measuring the period only in the presence of active weights, i.e. directly on the curves of fig. 4 (though the oscillations are not strictly free), and using in eq. 4 only the initial distance between the balls and weights centres (4.65 cm). Thus, only two records of data are necessary, with the weights in both active positions, avoiding a separate study of the free oscillations when not enough time is available.

– *corrections*: simplifying assumptions which are made along the calculation are justified within the global accuracy of the experiment. However, there is an important correction which cannot be neglected: not only does a given ball experience an attraction from the nearby weight (4.65 cm apart), but also from the opposite more distant weight (~10 cm, see fig. 1, lower part). This last attraction contributes negatively to the force momentum. This contribution is easily calculated since all distances are well known. Finally, it results in a positive important correction in the G final result of + 7.5 %. Cavendish himself was aware of this problem, but the correction was, in his case, negligible because of the large distance between a ball and the opposite weight (6 feet). The problem was, on the contrary, very sharp in Boys experiment in 1895, in which the dimensions had been drastically reduced to values similar to ours; he solved it in a fully technical way: each couple "ball – weight" was placed at quite different altitudes at each end of the beam, thus strongly increasing the distance between a given ball and the distant weight (Boys, 1895; 1898). But it was a highly sophisticated assembly!

– *accuracy and difficulties*: the standard accuracy of our gravitational balance is within 5 %. However, the practical accuracy to be expected in a normal teaching building subject to vibrations, noise, lots of passers-by, and important thermal variations is rather 10 %, sometimes up to 15 or 20 %. Thermal variations are specially harmful since they induce convection air flow inside the case, resulting in a motion of the beam. All those external disturbances can induce slow drifts of the zero initial position; the curves shown on fig. 4 can then become asymmetrical, and the estimation of an equilibrium position very uncertain. Should we take an instant rough estimate or wait? In every such cases, just like in actual scientific work, a decision has to be taken on observational and discussion grounds[9].

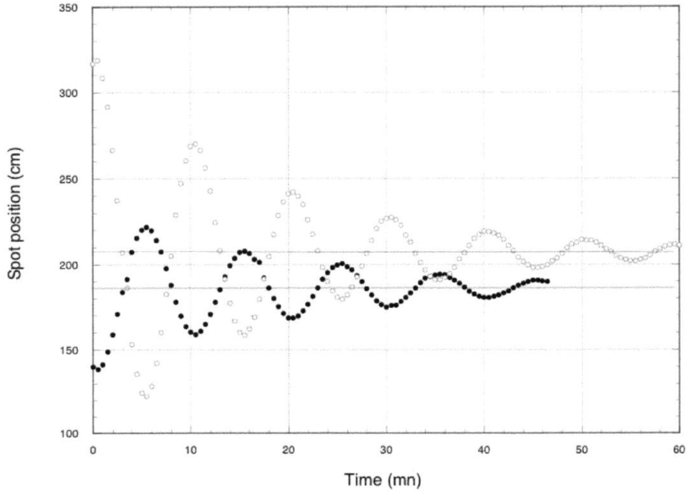

Figure 4. An example of students results.

The oscillations of the beam are detected through the motion of a laser spot (Poggendorf technique, see fig. 3) on a wall at a distance of 7.25 m. Each curve is related to one of the two symmetrical active positions of the "weights" (see fig. 1). In this "pretty good case", the oscillations were quite regular, and the damping fitted well to an exponential (viscous friction of the balls on air). The two horizontal solid lines represent the equilibrium position of the beam in each case, as deduced from an analysis of the successive maxima and minima on the graph. Both equilibrium spot positions differ by δ=21.6 cm, a quite measurable quantity. The gravitational constant G was deduced to be $6.33 \cdot 10^{-11}$ (int. units), a value within 5 % of the modern accepted value ($6.67 \cdot 10^{-11}$). Current accuracy in a university classroom environment is approx. 10 %, specially depending on quiet conditions and even temperature.

LESSON TO BE LEARNED FROM CAVENDISH'S ARTICLE:

Particularly good lessons are to be learned from Cavendish's article, the main qualities of which have been related above (cf. "Cavendish 1798 article"). The students are provided with the original paper and, depending on the context, invited to read either the 1959–shortened version (calculations omitted), or at least the first pages describing the goal, the historical context, detailed description of the assembly and the way to use it; and the last pages with the final results and author's comments. Without being exhaustive at all, let us mention some topics:

- *an introduction to dimensional analysis:* it is striking how all useless constants are eliminated (contrary to our modern textbooks!): no formulas, all calculations are reduced to ratios and proportions between relevant quantities. The density of the Earth is deduced without using nor even mentioning the G gravitational constant, a "modern" parameter unknown at the time. For example, the acceleration of gravity (necessary to evaluate the weight of the balls) is accounted only through a relevant *length*: "the length of the pendulum vibrating seconds in this climate", which will also be used as a standard for the oscillation of the arm[10]. We can admire the choice of a volume unit well adapted to this problem of spheres, the *spherical foot*; and you can easily guess (it is not written) that it is the volume of a sphere whose diameter is just one foot. For example, "each of the weights... is equal in weight to 10.64 spherical feet of water". With this unit, the volume of any sphere is simply the cube of its diameter! This approach is similar to our modern use of special units well adapted to a given class of problems, for example the "atomic units" in quantum Mechanics. Systematically, only the fundamental quantities *length, mass, time* do appear explicitly in the calculations, a very good introduction to dimensional analysis[11].
- *an original approach to the slightly non-harmonic oscillator:* when the weights are settled in an active position close to the case side, each ball performs its motion in the strongly non-uniform gravitational field of the nearby weight. As a consequence, the motion of the beam is no longer strictly harmonic and the usual relationship between the swinging period and the torsion constant is modified. A first order calculation of this "frequency shift" — which cannot be completely neglected — is easy. However, Cavendish invented a wonderful solution: once the beam has moved from its neutral position to its new equilibrium one (due to attraction from the weights), the distance between the balls and the weights has been slightly shortened and one should take account of this shortening. However, Cavendish has shown that if both this shortening and the frequency shift are neglected, then both errors cancel exactly in first order. Then, only the original distance is used and the swinging period is measured with the weights in their active position, close to the case side. No measurement is necessary with the weights in neutral position (i.e. in the symmetry plane of the balance): two measurements are made, with the weights at first in the W–W position (cf. fig. 1) and secondly in the w–w one. Furthermore, this symmetrical way of operation allows some other errors due to an imperfect initial alignment to cancel in first

order. This is a wonderful way of solving a non-harmonic problem, exempting the future users of any calculation, and a very good example for the students.

- *an exercise of statistical data analysis:* Baily's scepticism about Cavendish's results was mentioned above (cf. "Cavendish's quality measurements"), lying in part on the global dispersion of the data (the difference between extreme values, 1/14, relative value) usually considered, at the time, as characterizing the uncertainty. Avoiding any anachronism, we can invite our students, as a *modern* statistical analysis exercise, to check the accuracy of Cavendish's data. We know that the more numerous the data are, the larger the global dispersion. Estimating the precision from the global dispersion, as at Bailey's time, would systematically prove wrong the one who made the most numerous measurements! On the contrary, we can use the classical statistical analysis in order to yield *a posteriori* a modern appreciation on the quality of Cavendish's data. We remember how Cavendish himself considered two distinct series of measurements, corresponding to two different torsion wires. Which series to choose ? The second one, of course, which is more numerous and was obtained, according to Cavendish himself, with a more reliable instrument. Then, the uncertainty on the mean of this 23-samples series of density of Earth values, taking account of the number of samples and defined within 2 standard deviations about the mean (5.48), amounts to ~1.50 %, and the modern value (1.517) is well within the uncertainty. An excellent result!

The gravitational balance through History

FURTHER IMPROVEMENTS IN G MEASUREMENTS:

- *The XIXth century:* following Cavendish, further numerous measurements in the XIXth century — F. Baily, A. Cornu and J. Baille (Baily, 1842; Cornu, 1873) and specially C.V. Boys in 1895 — will definitively give right to Cavendish and its torsion balance technique in the historical controversy with Maskelyne. It is Charles Vernon Boys who, a century after Cavendish, brought the main improvement to the gravitational torsion balance: he demonstrated the great advantage of a strong *miniaturization* (1/18 in size) of the Cavendish balance: the forces will be strongly lowered, but the wire – which has to support the whole arm – can be chosen much slender. Furthermore, the control of air thermal convection is much easier in a smaller case (Boys, 1895; 1897). This progress was allowed from the new ability to build very thin quartz wires from molten silica by the "bow and arrow" technique[12].

However, mention should be given of a third technique implemented at the end of the 1880's by Poynting and by Von Jolly: the use of an ordinary scale. Let a scale be in equilibrium; then a large spherical mass is approached on one side, over or below the plate, thus disturbing the equilibrium. From which *G*, or the Earth's mass, can be deduced (Poynting, 1894). The 5.75 tons sphere of Von Jolly, made of lead bricks, which is shown in the Deutsches Museum in Munich is worth seeing!

Only at the end of the XIXth century the main concern turned from the density of the Earth to "big *G*", the universal gravitational constant or Newton's constant (though it did not appear explicitly in Newton's writings). Boys is an enthusiastic supporter of this new point of view: in his 1897 paper, he strongly emphasizes the importance and universality of the *G* constant as regards the contingency of the mass of our Earth, an "ordinary" planet among others. He claims his experiments do not aim at determining the Earth's mass or density. A completely reversed situation! Boy's *G* determination ($6.657.10^{-11}$, int. units) will remain the reference during several decades. In the domain of applied science, let us mention Eötvös, in Hungary, who will launch worldwide a new technique, Gravimetry, with a modified differential torsion balance able to measure local *g* variations.

- *The XXth and XXIST centuries:* there will not be any decisive improvement in *G* determination during the main part of the XXth century. The concern in big *G* resumed at the beginning of the 1980's and provoked a new development of the field. Very recent measurements have allowed a one order of magnitude jump in *G* accuracy at the very beginning of the XXIst century. A very remarkable fact, is that, following an obvious lack of agreement between some experimental results, the uncertainty in the CODATA value for *G* has been *increased* during the 1980's, a unique fact in the history of Metrology! Let us cite some recent studies:

  — in 1982, the "frequency shift technique" is performed for the first time with a torsion balance, taking advantage of the modification of the swinging period of the beam by the non-uniform gravitational field of the weights, a fact well described by Cavendish himself, from which *G* follows (Luther & Towler, 1982).

  — a new and very original use of the torsion balance by Terry Quinn at the *Bureau International des Poids et Mesures* (BIPM) in Sèvres, near Paris. Renewing an ancient observation, the torsion wire is loaded at 2/3 of its breaking load (use of a very heavy beam, contrary to all others); then, the main torsion energy is of gravitational, not elastic, origin (due to the fact that the beam slightly rises up when it is twisted); the mechanical properties of the wire, which may vary with time, have a negligible role in this situation (Quinn, 1997; 2001).

— a very sophisticated torsion balance operated with a zero deviation technique (no torsion) by the Eöt-Wash group in the university of Washington: the beam is slowly rotated with a feed-back system so that, despite the weights, the wire does not suffer any torsion at any time. *G* is deduced from the variations of the driving torque (Gundlach & Merkowitz, 2000).

— finally, a renewal of interest for the "ordinary" balance, by a group from Zurich: a mass of 13.6 tons (one cubic meter!) of mercury is flowed from an upper spherical tank to a lower one, on both sides of one of the scale plates (Schlamminger et al, 2002).

— since Cavendish himself evoked the eventuality of a modification of Newton's inverse square law at short distances, it is worth citing the experimental demonstration of the validity of the law down to 0.2 mm, by the Eöt-Wash group, using the torsion balance (Hoyle, Schmidt et al., 2001). Let us recall that such an eventuality is related to some modern theories of Gravitation (specially the search for "extra" dimensions and the "fifth force").

Finally, the last three determinations yield for the meaningful digits of *G*: $6.674.10^{-11}$ for the Eöt-Wash and Zurich groups, and $6.675.10^{-11}$ (int. units) for the BIPM. Further improvements are still in progress.

"WEIGHING THE EARTH" AND ASTRONOMY

"Weighing the Earth" is related to Astronomy in a slightly more complex way than one could *a priori* imagine.

1. *Weighing the Earth, weighing the Worlds*: we have shown above how, through the use of Newton and Kepler's laws, the knowledge of the Earth's mass and density could immediately yield the mass and density of the Sun and of the planets with satellites: the Earth's mass as a mass standard for the solar system. For the other planets, their masses can, in principle, be deduced from the study of the mutual perturbations between them and between them and the Earth, provided the knowledge of their distance. That is why, finally, the Cavendish balance was named "the balance for weighing the Worlds", quite in accordance with Newton's Principia.

2. *"Weighing the Earth" and astronomical distances*: the mutual interactions between the planets, as revealed from the irregularities of their orbital motion, are

depending on both their masses and their distances. Thus, one might infer that the new knowledge of an absolute mass scale at the end of the XVIIIth century could yield a new method to deduce the distances.

Actually, things have been a bit more complex. Two main names should be quoted for the "perturbation method": Laplace for the mathematical grounds and Le Verrier for applying it to the planets studies. But did Le Verrier make an explicit use of the Earth's mass? No; though the observed irregularities are implicitly depending on the absolute masses, it is from reasonings on those sole irregularities that he could, for example, predict the existence of the planet Neptune in 1846 (Le Verrier, 1846).

Actually, a main concern was an improved determination of the Sun's distance, or, equivalently, the "solar parallax"[13]. Le Verrier could apply the perturbation method to the solar parallax from the consideration of the precise Earth and Moon orbits, and of the *relative* masses of the *Earth and Moon* and of the *Earth and Sun* (Le Verrier, 1872). Thus the experimental Earth mass did not enter his calculation. Then, we could imagine that, at least, this last knowledge might have brought a useful cross-check with Le Verrier's calculations. However, the answer, again, is *no*. The deep reason is that the uncertainty in Cavendish's (and others, at least until Boys–1895) determination of the Earth mass is just of the order of the Moon mass (Moon mass is 1/80 of the Earth's one)[14]. Thus, the "weighing of the solar system" had not, through the XIXth century, such an influence on the astronomical distances as it could have been expected. Only during the first half of the XXth century, the absolute masses will directly enter the calculation of an asteroïd distance in view of improving the accuracy of the solar parallax (Toulmonde, 2004).

Conclusion

This wonderful story of carrying out the weighing of the Earth has highlighted the strong efforts of the scientists of the ending XVIIIth century to get absolute values of the densities of the Earth and other planets as a first information about their internal constitution. Both the Cavendish gravitational torsion balance and Cavendish's 1798 article are illuminating examples for our students within the framework of our experimental approach of History of Science. Working a modern pedagogical version of Cavendish's balance is a very efficient way to introduce our students to realistic scientific working conditions. Much has to be learned from either the working of the experiment itself, or from the attentive reading of Cavendish's article.

In Boys's 1897–article, the interest of the Earth mass – a "particular" planet – was strongly denied as compared to the "universality" of the G constant. We shall not fully follow Boys on that point: not only the Earth mass and density were obtained, but also (through Newton and Kepler laws) those of the Sun – a "particular star", but how important for us! – and finally a complete weighing of the Solar system. It is not so often that we can offer a modern student of our universities the opportunity of saying, like Cavendish, when back at home: "to-day, I weighed the Earth!" while simultaneously offering him such an opening onto the Universe. This is highly motivating!

Notes

[1] *"Unde cum communis suprema quasi duplo gravior sit quam acqua et paulo inferius in fodinis quasi triplo vel quadruplo aut etiam quintuplo gravior reperiatur; verisimile est quod copia materiae totius in Terra quasi quintuplo vel sextuplo major sit quam si tota ea acqua constaret; prae sertim cum Terram quasi quintuplo densoriem esse quam Jovem jam ante ostensum sit."* (Newton 1714, Part III, Prop. X, p. 373).

[2] Newton wrote: "Earth is almost five times more dense than Jupiter", though he found himself (see Prop. VIII) a ratio of 400/94.5 = 4.23, closer to 4 than to 5. In his 1756-French translation, Mme du Châtelet corrected "five times" in "four times".

[3] "Mr. Coulomb has, in a variety of cases, used a contrivance of this kind for trying small attractions; but Mr. Michell informed me of his intention of making this experiment, and of the method he intended to use, before the publication of any of Mr. Coulomb experiments" (Cavendish 1798, p. 470; 1809, p. 389). The torsion balance would thus have been invented independently in France and in England during the 1780's years.

[4] A discussion of Coulomb's results is out of the scope of this paper, however let us mention that they have been recently extensively discussed by several authors in (Blondel/Dörries 1994). Included is the report of a replication of Coulomb's balance by P. Heering.

[5] This modification in the distance between the weights and the balls cannot be considered as an attempt in order to test the inverse square law, as it is sometimes heard. The normal distance between the surface of the weight and the side of the case was 0.2" (Cavendish 1798, p. 472; 1809, p. 390) while the centre to centre distance between the weight and the ball was 8.85" (ibid. 1798, p. 510; 1809, p. 399). In the present attempt (experiments 9-10-11 and 15), "the balls were made to rest as close to the sides of the case as they could", thus reducing the centre to centre distance down to 8.65" or a bit more; such a reduction of only 2.25 % in the spheres distance is in any eventuality insufficient to perform any test of the variation of the attraction versus distance. On the contrary, Cavendish avoided any "distance" effect by varying it only in a negligible amount; the goal in this last test was only to evidence the eventual existence of an "attraction of cohesion" between the weight and the side of the case. The result was, of course, negative.

[6] "Its stiffness was such, as to make the arm perform a vibration in about 15 minutes. I immediately found, indeed, that it was not stiff enough, as the attraction of the weights drew the

balls so much aside, as to make them touch the sides of the case; I however, chose to make some experiments with it, before I changed it".

[7] We should mention an error in Heilbron's article (p. 156) while he is comparing the Coulomb and Cavendish torsion balance measurements: the rough mean of the, as-published, whole set of Cavendish's results for the density of the Earth is not 5.31 as Heilbron writes, but 5.45. The 5.31 value is the mean of the six first measurement made with the smoother wire. Furthermore, Heilbron is led to an overestimation of the dispersion of the results, 1 part in 10 according to him, instead of one part in 14 according to Cavendish's article. Actually, from Cavendish's text itself, the 1/14 dispersion value is clearly related to the *second series* of 23 measurements, not to the whole, and it is correct. This is quite in agreement with Cavendish's habit to consider *separately* both series of measurements made with two different torsion wires (as we would do nowadays), and to consider the second series as the main one.

[8] I am afraid to observe how, more and more in our universities (at least the French ones), the lecturers have no longer time to teach, and students no time to learn. No time to do one's work. Which world? Where are we going?

[9] When sufficient time is available, it is recommended to perform a third experiment with the weights in the neutral position, i.e. in the symmetry plane of the instrument (this is different from free oscillations without the weights because the zero can be disturbed when putting the weights on their support). A test of no zero-drift is then that this third equilibrium position lie in the middle of the two preceding ones.

[10] Let us note that, since this pendulum length can vary throughout the world surface, Cavendish adds in a footnote, with a high sense of rigor, the following comment: "In strictness, we ought, instead of the mean diameter of the Earth, to take the diameter of that sphere whose attraction is equal to the force of gravity in this climate; but the difference is not worth regarding" (Cavendish 1798, p. 510; 1809, p. 399).

[11] Avoiding any explicit use of derived secondary quantities else than length, mass and time, is a long story: in the Middle-Age, the velocity is defined either as the time required for a given length, or as the length covered in a given time. But nobody would divide a length by a time! Descartes and Galileo themselves suffered some difficulties with the concept of velocity. The dimensional analysis concepts are far from being evident, and this holds also for our modern students.

[12] A spectacular technique! The bow was bent back, the tail of the arrow was put in a molten quartz bath and the bow was released: the arrow flew in the air, driving a very homogeneous quartz wire behind it, some microns in diameter and the length of which could reach 30 meters! The experiment has been successfully repeated in the 1990's at the Science Museum in London.

[13] The solar parallax is the mean angle subtended by the Earth's radius as viewed from the centre of the Sun. From it and the Earth's radius, the Sun's mean distance follows immediately.

[14] I am indebted to Michel Toulmonde, from Paris Observatory, for an illuminating discussion of this point (cf.: Toulmonde 2004).

References

Baily, F.: 1842, *Phil. Mag.* **21**, 111.

Boys, C.V.: 1895, 'On the newtonian constant of Gravitation', *Phil. Trans. of the Royal Society* **186**, 1–72.

Boys, C.V.: 1897, 'La constante de la gravitation' (a reduced version of the original paper, in French) in: *Revue générale des sciences pures et appliquées* **8**, 46-54.

Blondel, C. & Dörries, M. (eds.): 1994, *Restaging Coulomb — Usages, controverses et réplications autour de la balance de torsion*, Leo S. Olschki, Firenze.

Bradley, J.: 1728, *Phil. Trans. of the Royal Society* **35**, 637.

Cavendish, H.: 1798, 'Experiments to Determine the Density of the Earth', *Phil. Trans. of the Royal Society of London* **88** (17), 469–526. Reprinted (full text): 'To Determine the Density of the Earth' in: 1809, *The Philosophical Transactions of the Royal Society of London, from their commencement, in 1665, to the year 1800; Abridged, with notes and biographic illustrations*, C. Hutton, G. Shaw & R. Pearson (eds), **XVIII** (Anno 1798), 388–408. Reduced version with comments (calculations omitted) in: 1959, *Great experiments in Physics*, M.H. Shamos (ed.), Holt, Rinehart and Winston, N.Y., chap. 6, pp. 75-92; reprinted: 1987, Dover. French translation by CHOMPRÉ, M.N.M.: 1815, 'Expériences pour déterminer la densité de la Terre', *Journal de l'École Polytechnique, Paris*, tome **X**, 17th cahier, 263–320.

Clotfelter, B.E.: 1987, 'The Cavendish experiment as Cavendish knew it', *Am. J. Phys.* **55**, 210–213.

Cornu, A.: 1873, 'Détermination nouvelle de l'attraction et de la densité de la Terre', *Comptes-rendus hebd. Acad. Sci., Paris*, **76**, 954–958.

Coulomb, C.: 1785, 'Premier Mémoire sur l'Electricité et le Magnétisme', in: 1788, *Mémoires de l'Académie Royale des Sciences*, année 1785, 569–577.

Everitt, C.W.F.: 1977, 'Gravitation, Relativity and Precise Experimentation', *Proc. of the First Marcel Grossmann Meeting on General Relativity*, North-Holland, Amsterdam, 545–615.

Gundlach, J.H. & Merkowitz, S.: 2000, 'Measurement of Newton's Constant Using a Torsion Balance with Angular Acceleration Feedback' *Phys. Rev. Lett.* **85**, 2869–2872.

Heering, P.: 1994, in: *Restaging Coulomb — Usages, controverses et réplications autour de la balance de torsion*, Blondel, C. & Dörries, M. (eds.), Leo S. Olschki, Firenze, pp. 47–66.

Heering, P., Riess, F., & Sichau, C. (eds): 2000, *Im Labor der Physikgeschichte*, bis, Oldenburg.

Heilbron, J.L.: 1994, in: Restaging *Coulomb — Usages, controverses et réplications autour de la balance de torsion, Blondel*, C. & Dörries M. (eds.), Leo S. Olschki, Firenze, pp. 151–161.

Hoyle, C.D., Schmidt, U., Heckel, B.R., Adelberger, E.G., Gundlach, J.H., Kapner, D.J. & Swanson, H.E.,: 2001, 'Submillimeter Test of the Gravitational Inverse-Square Law: A Search for Large Extra Dimensions' *Phys. Rev. Lett.* **86**, 1418–1421.

Hutton, C.: 1778, 'An account of the Calculations made from the Survey and Measures made at Schihallien, in order to ascertain the Mean Density of the Earth', *Phil. Trans. of the Royal Society of London* LXVIII, 689–788.

Hutton, C.: 1809, 'An account of the Calculations made from the Survey and Measures made at Schihallien, in order to ascertain the Mean Density of the Earth' (abstract), in: *The Philosophical Transactions of the Royal Society of London, from their commencement, in 1665, to the year 1800; Abridged, with notes and biographic illustrations*, C. Hutton, G. Shaw & R. Pearson (eds), XIV, 408–423.

Joule, J.: 1850, Philosophical Transactions of the Royal Society of London, **140**, 61. Version with modern comments in: 1959, *Great experiments in Physics*, M.H. Shamos (ed.), Holt, Rinehart and Winston, N.Y., chap. 12, pp. 166–183; reprinted: 1987, Dover.

Le Verrier, J.: 1846, 'Recherches sur les mouvements d'Uranus', Comptes-rendus hebd. Acad. Sci., Paris, 21, 1050-1055 ; 'Sur la planète qui produit les anomalies observées dans le mouvement d'Uranus. Détermination de sa masse, de son orbite et de sa position actuelle', *ibid.* **22**, 907–918.

Le Verrier, J.: 1872, Comptes-rendus hebd. Acad. Sci., Paris, 75, 165–172.

Libes, A.: 1813, in: Traité complet et élémentaire de Physique, Paris, vol. 3, p. 182.

Luther, G.G. & Towler, W.R.: 1982, *Phys. Rev. Lett.* **48**, 121.

Maskelyne N.: 1775, *Phil. Trans. of the Royal Society of London* **LXV**, p. 494 and p. 500.

Moreno, A.: 2001, 'Weighing the Earth; a Newtonian Test and the Origin of an Anachronism', *Science and Education* **10**, 515–543.

Newton, I.: 1714, *Principia*, second edition.

Poynting, J.H.: 1894, T*he mean Density of the Earth*, C. Griffin & Co, London.

Quinn, T.J.: 1997, 'Novel torsion balance for the measurement of the newtonian gravitational constant', *Metrologia* **34**, 245–249.

Quinn, T.J.: 2001, 'A new Determination of G using two Methods', *Phys. Rev. Letters* **87**, 111101-1–4.

Schlamminger, S., Holzschuh, E. & Kundig, W.: 2002, *Phys. Rev. Lett.* **89**, 161102.

Silverman, M.P.: 1995 'Self-directed learning: A heretical experiment in teaching Physics', *Am. J. of Physics* **63**, 495.

Toulmonde, M.: 2004, 'La parallaxe du Soleil', *L'Astronomie* **118**, 285–287.

Alternative perspectives and conceptual change: integrating pre-scientific knowledge into teaching-learning sequences in school science

SHU-CHIU LIU

University of Oldenburg (Physics Education & History and Philosophy of Science), Germany

Abstract: The present work is concerned about the integration of pre-scientific knowledge into teaching-learning sequences in school science with a specified theoretical basis from a conceptual change perspective. The central point of this theoretical basis is, to enhance students' understanding of the intended scientific knowledge alternative ideas and perspectives should be provided and serve as "bridging" representations while students are faced with a new set of ideas which often conflicts their personal view. One of the most powerful representations of this kind is, as argued based on empirical evidence, particular pre-scientific knowledge which has similar conceptual grounds to students'. Empirical evidence discussed in the paper originates from studies on students' concepts regarding two scientific topics: "the model of the universe" and "the nature of heat". A theoretical framework and instructional approaches of integrating pre-scientific knowledge into teaching-learning sequences aiming to foster students' conceptual change are thus derived from these investigative results.

Introduction

Science teaching and learning can be and is often considered from a conceptual change perspective. Conceptual change is generally defined as learning that changes the learner's existing conceptions (or preconceptions, alternative conceptions) to, in the context of science education, the intended scientific view. Learning for conceptual change goes beyond mere accumulations of new facts and ideas or learning of a new skill. In the process of conceptual change, the learner acts to revise her existing conceptions and even construct a new conceptual framework that she uses to solve problems and explain phenomena in the world.

There are two fundamental ideas regarding teaching and learning for conceptual change to be summarized: First, students construct their own knowledge about how the world works before they enter the science classroom. And this kind of knowledge is often different from the scientific knowledge we present to them (for an extensive bibliography, see Duit, 2004). Second, students' view is difficult to change because it is normally compatible with their perceptual experience and adequate in everyday problem-solving. Therefore, if instruction does not engage their earlier concepts and help students recognize the shortcomings of these concepts, they will just try to memorize new facts and ideas without really understanding what they are about. Moreover, conceptual change instruction requires a thorough understanding of the topic or phenomenon in question as the information delivered in school science is often full of fragmented and rigid knowledge that neither presents the nature of science nor interests students. If the learning outcome as suggested by teachers and educators is not clear and well-defined, how can we expect students to undergo conceptual change and reconstruct a fruitful and meaningful conception?

Thus, to consider science teaching and learning from a conceptual perspective should involve 1) uncovering students' preconceptions about a particular topic or phenomenon, 2) clarifying the subject matter of that topic or phenomenon, and 3) using various techniques to help students change their conceptual framework.

It is the primary goal of science teaching to initiate and accelerate the process of conceptual change. It is then of significance to clarify two sets of knowledge, students' and the intended, in order to develop materials and activities to assist students in conceptual change. This idea construes the central theme of the research model upon which the two empirical studies to be discussed in the paper are based. Some results of the two studies suggest the use of pre-scientific knowledge as alternative perspectives to bridge the two views present in the science classroom. It is argued that pre-scientific knowledge, if presented with an emphasis on the perspectives from which and the context in which it is generated and evolved, can be a powerful tool to have students become aware of their conceptions and their underlying meanings and further differentiate them from the intended knowledge in the science lesson. This would coincide with the contemporary conceptual change theory which highlights "conceptual appreciation" and "conceptual differentiation" as learning outcome. It claims that emphasis of science instruction must be laid on helping students appreciate diverse perspectives and methods regarding a given problem, including their own conceptions, and moreover, differentiate one from another. For it shares some similar perceptive grounds with students' knowledge and provides conceptual roots of today's scientific view, pre-scientific knowledge should have its potential to bring about such learning outcome.

Model of Educational Reconstruction

As discussed in the previous section, teaching and learning for conceptual change requires a thorough understanding of students' view and the intended scientific view in order to develop materials and activities which can assist students in revising the former and even changing it to the latter. In line with this perspective, the theoretical framework proposed by several German researchers (Kattmann *et al.*, 1996), the model of "Educational Reconstruction" (Figure 1), can serve as an underpinning of research-based instructional design.

The essential point of this theoretical framework is to ground instructional design with a deep knowledge of both students' view and the subject matter. Thereby alternative perspectives can be derived in attempt to bridge the two views. Based on the view of an integrated constructivism, three components are essential in this theoretical framework: students' representations, epistemic aspect of subject matter and Integration of the two based on educational theories.

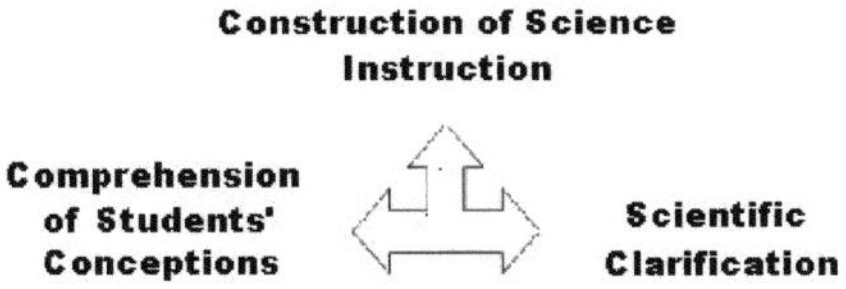

Figure 1. The model of "Educational Reconstruction" (Kattmann et al., 1996).

In this framework, both the students' representations and the subject matter are considered to be problematic and thus need to be clarified. It is argued that, first, science is not something that exists out in nature, but is rather a tool in the mind of the knower; the scientific knowledge should not be treated as "truth," but rather presentations of the consensus held by the scientific community. The scientific contents taught at school are like a reconstruction of these presentations made by the educators and teachers. At this point the science educational issues bear a fundamental question: What is actually the thing called science which students ought to learn at school? Second, it is well acknowledged that students already hold certain ideas (their personal knowledge) before they enter the science classroom, and these ideas make sense to them and in turn can bring about change. As research shows that these preconceptions in fact play a crucial role in the learning process, they should be taken into serious consideration when developing the curriculum. Thus, the model of educational reconstruction treats the student's preconceptions and the body of scientific knowledge as resources of the same value in reconstructing the initial instructional module.

It proposes a double reference in developing and elaborating teaching-learning sequences. This double reference- on the one hand to students' personal knowledge about a specific topic and the conceptual framework that organizes this knowledge and to scientific knowledge on the other – allows researchers to gain insight into the core issues that differentiate students' conceptualization of the phenomena and scientific accounts presented in the classroom. It also explicitly or implicitly suggests that students' view is often a self-sufficient set of knowledge that should be activated, not devalued, in the teaching and learning process, and can serve as an indicator of what and how science should be taught. Students' conceptions and their frameworks are seen as an important parameter having the same weight with the subject matter in the process of educational reconstruction and both are essential for the achievement of the goals of science teaching and learning. The model is specially characterized by its dynamic interactions between the components: "knowledge gained in one of the components influences the activities and the interpretation of the results of the other components in a dynamic process" (Meheut & Psillos, 2004, p.526).

Study 1: Students' alternative models of the universe

A cross-cultural study designed in accord with the model of "Educational Reconstruction" is carried out in Taiwan and in Germany to discover students' concepts about the earth and the sky. There are totally 64 students of Grade 3-6 (ages 8-12) involved in the study. These students are asked to describe and explain the basic features of the earth and sky objects and events in an interview setting including the use of role-play. Prior to the interview, none of the students had received formal instruction on this topic, yet had been generally exposed to related information through mass media, familiar adults and peers.

Previous studies (e.g.,Vosniadou & Brewer, 1992, 1994) have revealed convincing evidence on students' metal models of the earth, which are often a hybrid version of the flat earth, as one "feels" on earth, and the spherical earth, as it is in reality. The discussed study is thus intended to go beyond this space scale and to find out how students conceive of the obvious cosmic bodies, including the earth, sun, moon and stars, and their relations, and whether a structural feature is also evident.

As elaborated in Liu's report (2005), throughout the interview, the model-like structure of the student's knowledge is apparent. The interviewed students seem to give their accounts substantially based on a model of the universe, when they describe basic features of the earth, sun, moon and stars, such as the shape, the relative position and motions, and also explain the cause of some familiar sky event, e.g., the phases of the moon. This model is, to various extents, different from the scientific model of the universe we adults know as true, and varies among students. A limited number of such models are distinguished and categorized. Most evident is the tendency that students' models exhibit either an earth-centered or a sun-centered view. Figure 2 and 3 show two models which are most dominant among students in the study. They are respectively representative of the earth-centered and sun-centered view. Students who hold the model as illustrated in Figure 2 typically emphasize the position and the role of the earth. They believe that the sun and the moon, and sometimes even the stars, circle around the spinning or static earth, and the earth has some special features, e.g., gravity, that do not exist with other cosmic bodies. When a sun-centered model is in mind, students often considered the sun being static and much bigger than any other objects (like the one in Figure 3).

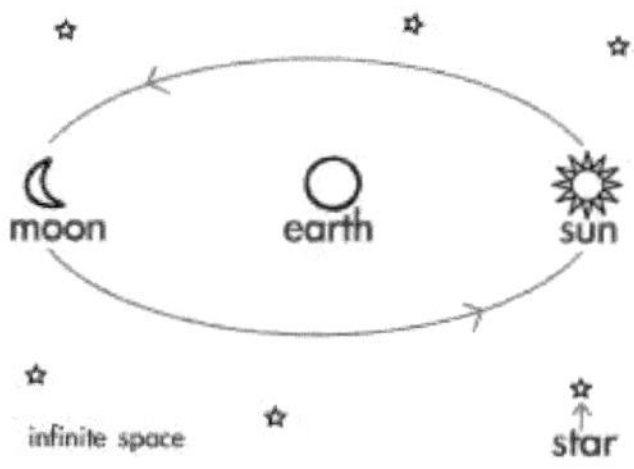

Figure 2. A earth-centered model.

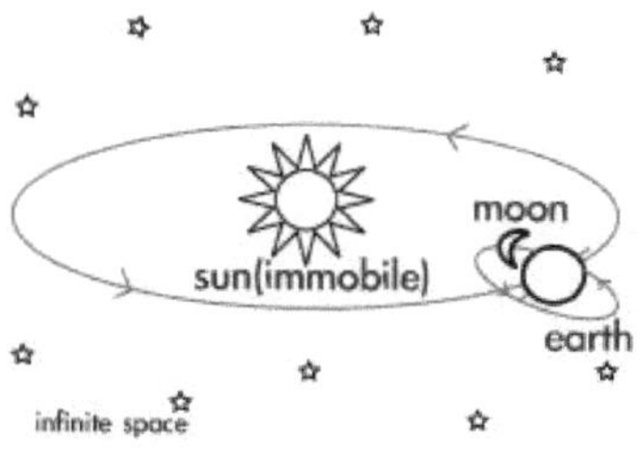

Figure 3. A sun-centered model.

It is also evident that the earth-centered models are often coupled with a hybrid model of the earth, in which the perceptive experience of the flat landscape and the received information of the spherical earth are reconciled. An example is given by a third-grader who describes a spherical earth where people live inside on the two layers of surface rid of the risk to fall off to space, as illustrated in Figure 4. Some students do not fully understand why the earth is round while the everyday observation tells us that the earth is flat. Therefore, they reconstruct a model as such to integrate these two views.

Another characteristic of the earth-centered models is that an imaginary horizon is predominant. This imaginary horizon assigns an absolute direction of up and down, and therefore, for example, the sun is in the day located only "above" the earth, as it must go from one side "up", through the outside circle of the earth, and then move "down" on the other side. This is understandably derived from the prevailing everyday experience, such as the horizon on the sea, flat landscape, and sun rising and setting.

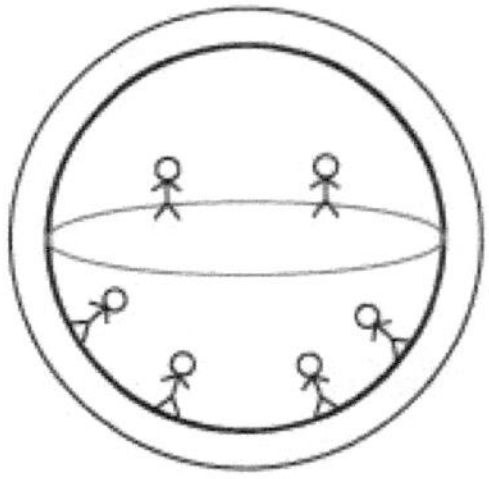

Figure 4. A hybrid earth model presented by a girl in Grade 3.

Almost all students in the study believed in an infinite universe, which seems to contradict their earth-centered or sun-centered view - something infinite in size would not have a center. Yet, the further interrogation shows that students' models actually speak to an "observational universe" instead of the infinite cosmos. For the students, the earth and the obvious sky objects, such as the sun, moon and stars, seem to be located in a confined space, whereby their model of universe is conceptualized. Beyond this space is the unknown and invisible world, sometimes described as "nothingness".

Interview data also shows that the student who holds no physical reasons for familiar phenomena tends to construct more primitive models than others. The information provided by students in the earth-centered group is substantially descriptive. When asked to offer explanatory accounts (reasons for a given phenomenon, such as the phases of the moon), considerably more students in the earth-centered group say "I don't know." than those in the sun-centered group. Furthermore, comparing the explanatory ideas presented, its accuracy seems to correspond to the models. For example, a number of students with earth-centered models explain the phases of the moon as a result of earth's shadow, cloud's obstruction, or sun's shadow, while the majority of the students with sun-centered models points out that it has something to do with the relative movements and positions of the earth, sun and moon.

Instructional considerations and the historical approach

The previously discussed results indicate the predominant role everyday experience plays in students' conceptualization of the world. When they are exposed to the new and authoritative information that conflicts their original beliefs and further explanations on this conflict are not given, students tend to reconstruct their own models which become a mixture of different views. In order to explain the conflict, it is necessary to provide students opportunities to see how different opinions and ideas can arise as a give topic is viewed from different perspectives. As this point, pre-scientific knowledge can be beneficial.

Following the results on students' models of the universe, it can be argued that, first, instruction should provide accounts of perspectives from which the earth or the sun is regarded as a central point. The earth-centered view is compatible with the perceptual experience that where we are located is huge and static, and the sky objects are small and moving. To see the sun as a central object is a step towards the kind of perspective beyond what is obvious. The historical models proposed by Ptolemy and Copernicus can be therefore illustrated in such a way that students may learn something about the relationships between models and perspectives from which they are viewed (e.g., to look at the earth from a tiny point on it understandably leads to a feeling of the earth being in some way flat) and also the importance to distinguish what is obvious and what is real (e.g., obviously the sun comes up from the horizon, but in reality, it's the earth which moves down). The intriguing philosophical and social context where these two models emerged and evolved can furthermore serve as a vehicle for an image of science beyond sets of ideas and facts.

The second point is that we can use pre-scientific knowledge to offer students genuine approaches to supporting models. For example, Aristotle provided interesting evidence on spherical earth based on observation to shops which appear or disappear on the horizon. As showed in Figure 5, if one observes a ship coming towards the seashore, it is always its mast that first comes to the view, and gradually more become visible until it is in full view. This is presumably uncomplicated to explain or to replicate to high school or even primary school students.

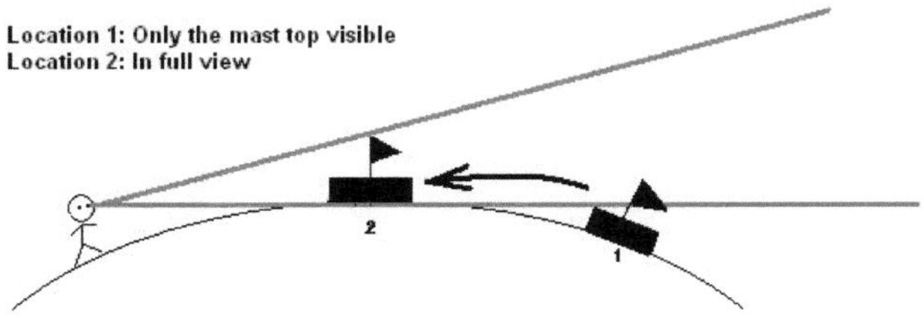

Figure 5. One of Aristotle's reasons for spherical earth: the ship approaching the shore is first seen only with the mast and slowly comes to the full view.

Another good example is how Eratosthenes measured the earth's circumference by assuming the earth to be spherical and using relatively simple calculations based on his measures of length of sun shadow at two locations. His experiment can provide the idea of spherical earth with another plausible reason in that his results can be only explained given the earth is spherical. This experiment can be appropriate for replication for high school students with thoughtful instructions and help them see the linkage between the model and evidence and moreover mathematical applications.

As well acknowledged in history of science, early scientists constructed their models based on very careful and attentive observations to the sky, we can therefore argue that students should be given opportunities to have real contact with the sky. If they observe the sky carefully, some naïve ideas may be demolished relatively easily. Students would, for example, find it difficult to keep arguing that phases of the moon is caused by the earth's shadow, cloud's covering, or sun's shadow, and that the sun and moon are located on two opposite sides of the earth.

Study 2: Students' understanding of the concepts about heat

Another study is concerned about students' understanding of the concepts about heat. 252 Taiwanese secondary students (Grade 10-12; 15-18 years old) participated in the study which uses a mixed method to investigate the conceptual landscape of students' knowledge about heat. It is intended to identify in what way and to what extent students grasp the meanings of the basic scientific concepts in relation to heat after a few years of science instruction at school, including topics in question.

These students responded to a written test, which contains several closed and an open-ended questions. The questionnaire is a revision of a previous study with a different group of students from the same age range and the response items to the closed questions are derived from these students' responses. The questions cover definitions of terms (e.g., "What is heat?") and explanations and predictions of phenomena (e.g., "Why does metal generally feel colder than wood?" and "What happens to the particles of the metal stick whose one end is being heated?") without or with elementary mathematical application. To correctly answer those questions requires a basic understanding of heat, heat transfer, thermal equilibrium, the function of thermometer, etc. Factual knowledge about thermal concepts does not suffice because explanations of a thermal phenomenon often require an integration of several concepts. Students need to know, for example, the differentiation of feeling hot/cold and being really cold/hot, and the ideas that heat is transferred from hot to cold, and that capacity of transferring heat varies depending on the material if a scientifically correct response to the question of "Why metal generally feels colder than wood?" is to be given.

The results indicate that the participant students are familiar with a number of basic thermal terms: heat, temperature, heat transfer, thermal equilibrium, specific heat, conductivity, etc. However, their accounts show various levels of disintegration and confusion. Many terms are used in an incorrect and confusing manner. It seems that many students fail to grasp the underlying meanings of the concepts and as a consequence simply memorize what is taught to them. To be more precise, students talk about heat in relation to several different things, e.g., temperature, hotness, and internal energy (Figure 6). A considerable number of students agree to this statement "Heat is a kind of energy that can make temperature rise. An object with more heat must be hotter." It indicates that heat is related to higher temperature, something hot, and something inside a body. The definition of heat seems to be a mixture of heat source, hotness and internal energy.

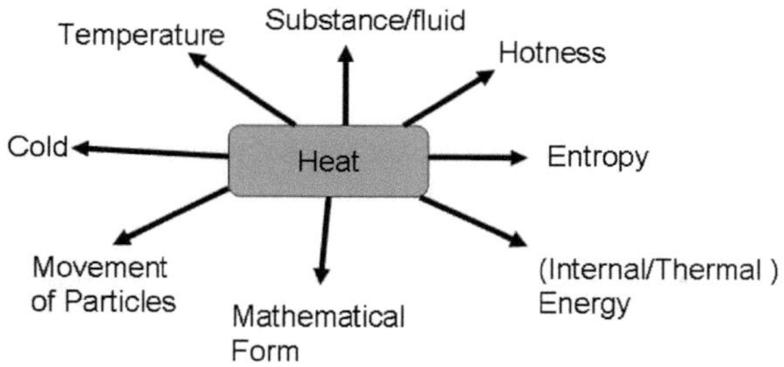

Figure 6. Students relate heat to various things.

When students are asked to choose, if any, correct information about the graph in Figure 7, a number of them take the statement: "A has 5000 cal of heat, while B has 4500." They seem to apply an altered version of the learnt mathematical formula $\Delta H = M \times s \times \Delta T$ (where M is the mass of material, ΔH is the amount of heat energy transferred to or out of the material, and ΔT is the change of the material's temperature) to this problem by interpreting ΔH as heat possessed by the material, and ΔT as the temperature of the material. This again indicates that students have a concept of heat which is a function of temperature and is something inside a body. Heat seems to be often confused with internal energy and can be calculated using some simple formula. Also, as radiant heat is a part of the science lesson and seems to be compatible with the idea of heat as internal energy, students also tend to give accounts that connect heat to internal energy and radiant heat. Comments such as that "Heat energy is released by all substances at over absolute zero." and "(Heat is) the energy that exists inside a body, even for a body at 0°C." (Figure 8) are given by 12-graders in defining the concept of heat.

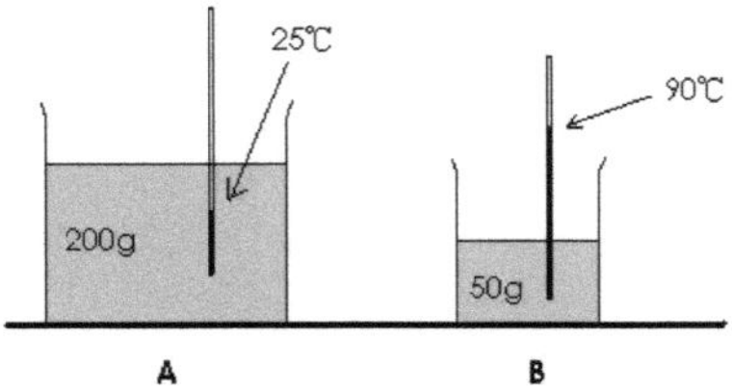

Figure 7. Students are asked to choose correct statement(s) based on this graph.

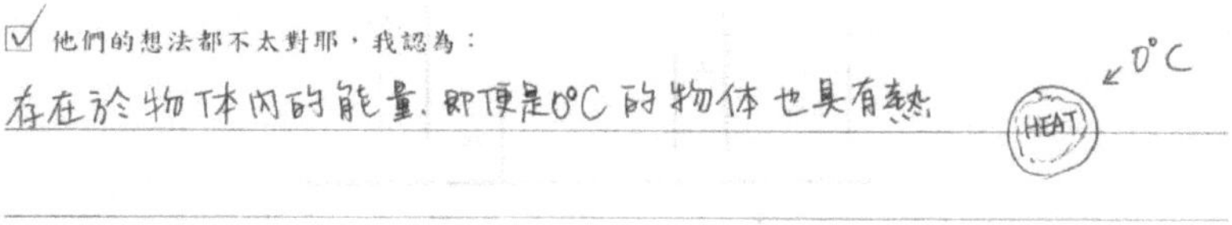

Figure 8. A 12th-grade student pointed out that heat is a quality possessed by the object, even at 0°C.

Many students also maintain a material view of heat. A question which requires students to describe actions of the particles in the metal stick as shown in Figure 9 revealed that nearly a third of the participant students believed that "particles move away carrying heat which flows in from the source." It clearly shows that students hold the idea of particles behaving like a carrier and transporting heat in a manner like a flow (Figure 10). A student drew a picture as illustrated in Figure 11, where heat is depicted as a current inside the stick heated on one end.

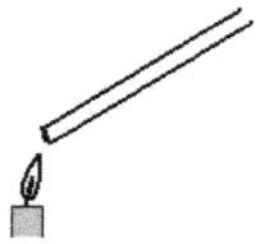

Figure 9. What happens to the particles of the metal stick?

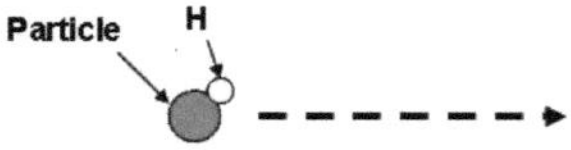

Figure 10. A popular idea: The particle carries "heat" (H) and moves like a flow.

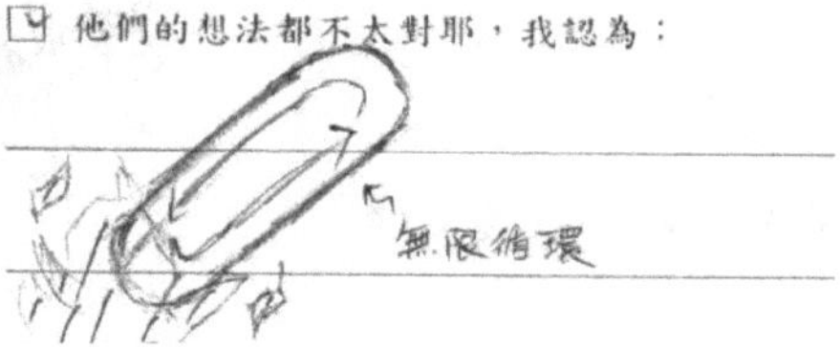

Figure 11. Heat transfer depicted as a current by a 12th grader.

Instructional considerations and the historical approach

It seems to us that students explain heat in terms of and in relation to different concepts. These terms are, however, used in an ill-defined and confusing manner. Temperature, hotness/cold, internal energy, radiant heat, heat flow, etc. seem to be frequently mentioned, explicitly or implicitly, in students' chosen or written statements without a sufficient distinction between one and another. This result indicates that students do not grasp the underlying principles of these concepts and consequently lack a conceptual map to organize these concepts. One of the problems may arise from the coexistence of different meanings of heat.

For example, heat in relation to hotness and energy is both manifest in everyday context, but students do not understand why and how these two very different explanations are correct for one single thing. Students do not understand in what context or from which perspective a specific explanation is given. More precisely, they do not understand that the underlying meanings of these explanations have something to do with microscopic or macroscopic level, external or internal quality, and a quality or an effect of the thing. Thus, instruction should lay emphasis on specifying the perspectives from which and the context in which heat is explained in a specific way. At this point, it should be valuable to present historical reconstruction of the science of heat in order to convey elements and contexts of foundation and development of theories that can support students in the process of conceptual change.

We can mirror students' conceptions onto the pre-scientific ones and present the historical arguments with sequences and context. Students' ideas to relate heat to hotness, to substance/fluid, to motion, or to temperature are not novel in view of the history of science (e.g., heat as non-material element, Phlogiston, caloric, and kinetic theory). The pre-scientific knowledge does exhibit some similar perceptive grounds to students'. Moreover, pre-scientific knowledge is intrinsically contextual. The process of theory change often indicates some features which are particularly educationally meaningful. For example, the arguments and experiments of friction leads to the decay of caloric theory, and when heat as motion comes to discussion, the question about what is actually moving leads to the microscopic level of explanations of heat. Pre-scientific knowledge, if presented in a contextual and sequential manner, can likely help students realize why they think the way they think, and see the underlying principles of the different meanings.

Closing remarks

The issues dealing with the inclusion of the history of science in school science have been in discussion for decades. Many of the science educators and teachers would agree with the value of the historical aspect of science in teaching and learning, albeit its actual practice seems to remain scarce. The problem in general lies in the implementation level. First of all, historical materials as located in historiography are difficult for science education practitioners to access as they often contain complicated descriptions using technical terms and mathematical forms which may be no longer in use or have different meanings today. Second, there is always the problem as to how these materials can be utilized and appropriately selected in the classroom. It is often cautioned that lacking thorough considerations, lessons with elements of history of science may lead to simplification of the nature of science and distortion of the image of scientists and their work (Brush, 1974; Metz, 1999). Thus, considerations about using historical material in science instruction must be taken carefully and thoughtfully in reference to students' ideas and the subject matter in question.

At this point, let me turn back to the theoretical framework that underpins the two discussed studies. It stresses the idea that teaching-learning activities should have their grounds in a deep knowledge about students' view and the knowledge to be taught. While these two views often show a clear division, alternative perspectives should be brought in to bridge them and promote the wanted conceptual change. It can be argue that, based on these two studies, pre-scientific knowledge can serve as a bridging representation between students' view and scientific knowledge to be taught, as it in some sense help to clarify the both knowledge. For early beliefs and ideas show a perceptive ground in which they are generated and this ground is also present in students' view. For early beliefs and ideas, along its attached context, provide some clues on the ways in which the subject matter as presented in the classroom comes to being.

Moreover, their relation is bilateral (Figure 12). Students and scientific view also help to clarify the instructional use of historical materials. Without the knowledge about students' view and the subject matter we would not be able to decide on what historical material should be important or useful and how these materials should be organized in order to bring about students' conceptual change. We should bear in mind while developing teaching-learning sequences that students need the kind of representations in which they see the connections of concepts and perspectives or context, and which they can see some underlying principles that discern and relate the knowledge presented to them.

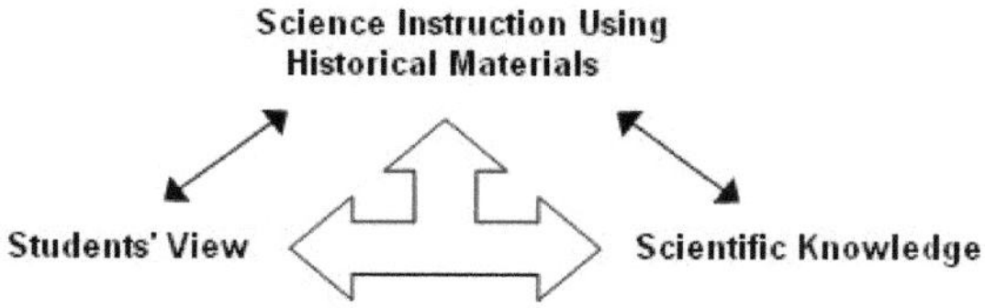

Figure 12. The use of historical material in science instruction rooted in the clarification of, and for the purpose of clarifying, students' and scientific view.

References

Brush, S. G. (1974). Should the history of science be rated X? *Science, 183*, 1164-1172.

Duit, R. (2004). *Bibliography - STCSE: Students' and teachers' conceptions and science education*: IPN - Leibniz Institute for Science Education at the University of Kiel, Germany.

Kattmann, U., Duit, R., Gropengiesser, H., & Komorek, M. (1996). *Educational reconstruction- bringing together issues of scientific clarification and students' conceptions.* Paper presented at the Annual Meeting of the National Association of Research in Science Teaching (NARST), St. Louis.

Liu, S. C. (2005). Models of "the heavens and the earth": An investigation of German and Taiwanese Students' alternative conceptions of the universe. *International Journal of Science and Mathematics Education, 3*(2), 295-325.

Meheut, M., & Psillos, D. (2004). Teaching-learning sequences: Aims and tools for science education research. *International Journal of Science Education, 26*(5), 515-535.

Metz, D. J. (1999). Integrating the nature and history of science and science instruction. In L. Sokal & R. Bruno-Jofre (Eds.), *Proceedings of the 1999 Graduate Students Symposium* (pp. 85-100). Winnipeg: The University of Manitoba.

Vosniadou, S., & Brewer, W. F. (1992). Mental models of the earth: A study of conceptual change in childhood. *Cognitive Psychology, 24*, 535.

Vosniadou, S., & Brewer, W. F. (1994). Mental models of the day/night cycle. *Cognitive Science, 18*, 123.

Learning about Light in Grade 4: What Happened to the Illuminating Stories from The History of Science and Technology?

BARBARA A. MCMILLAN

Faculty of Education, University of Manitoba, Winnipeg, MB, Canada

Abstract: The research reported here focuses on impediments to using historical narratives and stories in an instructional unit developed to address government mandated learning outcomes for science and the common conceptions held by children. The research took place in a Grade 4 classroom and involved a five-week series of thirteen connected lessons, a review, and a unit test. The lessons, in draft form, included (1) stories of light investigations of Archimedes and Isaac Newton designed to guide students in their investigatory work and to provide a context for substantiating new concepts, (2) narratives of technological innovations of David Brewster, Galileo, Johann Gutenberg, Robert Hooke, and Antonie van Leeuwenhoek designed to introduce students to individuals in science and technology and to the objects they built and/or the purposes for which the objects were utilized, and (3) collaborative learning activities, first-hand investigations, and teacher-student conversations designed to support learning and assist students in attaining knowledge, skill and attitude outcomes. The results obtained show that the10% of a teaching day for Kindergarten through Grade 6 science, recommended by Manitoba Education, is inadequate for implementing essential phases (engagement, exploration, explanation, elaboration, and evaluation, with closure) of a meaningful science lesson in which elements of the history of science or technology are incorporated. In addition to the constraint of time, factors that inhibited the implementation of the prepared historical narratives and stories included weather conditions, teacher knowledge, and the science knowledge and skill levels of the students. Other factors that inhibited the educative potential of the implemented but abbreviated narratives and stories were the social environment of the classroom and the psychosocial resilience of children in the face of adversity. The implications of the findings are discussed.

Introduction

Readers interested in the use of history in science teaching and learning are aware of the reasons for including historical material that have been expressed by Winchester (1989), Solomon and colleagues (1992, p. 410), Matthews (1994, p. 50), Justi and Gilbert (1999, pp. 287-288) and, more recently, Fowler (2002) and Solbes and Traver (2003, pp. 707-709). Although the current literature, following publication of the National Science Education Standards (National Research Council 1996), could lead one to believe that the most appropriate use of science history is in the development of an appreciation of the nature of science (see Allchin 2002, 2003 and 2004; Bentley

2000; Bellocchi 2004; Clough & Olson 2004; de Berg 2006a and 2006b; Harding & Hare 2000; Howe & Rudge 2005; Irwin 2000; Lin & Chen 2002; Lonsbury & Ellis 2002; McComas 2005; McKinney & Michalovic 2004; Niaz 2001; Olson, Clough, Bruxvoort & Vanderlinden 2005; Rudge & Howe 2004; Tao 2003; Venville & Milne 1999; Yip 2006; among others), there are those like Monk and Osborne (1997) who would like the inclusion of the history of science to 'directly contribute to students *learning of science concepts* and satisfy that principal objective' (p. 405). Academic writing in this latter category falls neatly into two groups: authors who present ideas and/or resources for historical experiments, vignettes, role plays, confrontations, dialogues, stories, thematic narratives, case studies, or entire science units; and authors who present the results of classroom investigations where resources with historical components have been used. While not a complete inventory[1], examples of the former include Allchin 2001; Begoray & Stinner 2005; Binnie 2001; Carson 2004; Cotignola, Bordogna, Punte and Cappannini 2002; Gauld 2004; Galili 2001; Galili & Tseitlin 2003; Hagen, Allchin & Singer 1996; Hakim 2002; Höttecke 2000; Justi & Gilbert 2000; Kettle 2001; Klassen 2004 and 2006; Kipnis 1996 and 2001; Machamer & Hepburn 2004; Martins & Silva 2001; McGinnis & Oliver 1998; Matthew, Gauld & Stinner 2004; McMillan 2004 and 2005; Metz 2004 and 2006; Metz & Smerchanski 2003; Mihas & Andreadis 2005; Nussbaum 1998; Raftopoulos, Kalyforratou & Constantinou 2005; Stinner 1994; Stinner & Teichmann 2003; Teichmann 1999; van Tiggelen 2006; and Wandersee and Roach 1998. Examples of classroom-based studies include Becker 2000; Cartwright 2005; Cavicchi 2006; Chan 2000; Degenais 2003; Ford 2003; Galili & Hazen 2000; Heering 2000; Kafai & Gilliland-Swetland 2001; Kalman & Aulls 2003; Kettle 2005; Kubli 1999; Lin &Chin 2002; Osewold 2006; Seker & Welsh 2006; Seroglou, Koumaras, & Tselfes 1998; Sneider & Ohadi 1998; Stewart & Rudolph 2001; and Van Driel, De Vos & Verloop 1998.

Fifty-six of the seventy-four papers cited above contain resources for, or studies conducted in, secondary (senior years) and tertiary (university or college) classrooms, including studies that involved pre-service teachers. Seven of the remaining eighteen papers were written to inform science teachers at all levels, four presented historical materials for middle years science teachers, one discussed the results of using the history of science in a middle years science unit, and one described the data collected from interviews of two groups of middle years and senior years students between the ages 12 and 19 and 15 and 20, and a group of adults, 26 to 28 years in age.

The remaining five papers are of particular interest here. They have, in varying degrees, attempted to address the science education of learners in the first years of formal schooling (Kindergarten through Grade 4). Mihas and Andreadis (2005) focus on optical experiments of Al Haythan (Alhazen) and argue for structured sequences of activities that scaffold learning and conform to the cognitive profiles of students. They refer to young learners with the following statement in their abstract: 'These applications give an opportunity for a first presentation of some notions of photometry even to the primary level of education' (p. 675), and mention using a particular experimental set-up of a camera obscura with primary school students (p. 689). Feedback on the consequences for learners is anecdotal and limited to a small number of students in a 'teacher training' program. McMillan (2005) describes four of the scientific investigations of Arctic whaler William Scoresby to explain how passages from the logbooks and journals of a 19th century scientist can help Early Years teachers and children think about the origin of questions and the methods and instruments used by scientists in their attempts to answer questions and construct explanations. Kafai & Gilliland-Swetland (2001) invited children in a Grade 4-5 classroom to work with an online selection of primary sources from The Donald Ryder Dickey Collection. Concurrently, the children worked on a project to create up-to-date primary sources of the wetlands in southern California where Dickey conducted fieldwork in the early 1900s. The authors suggest that several outcomes of the comparative documentary practices and analyses 'proved that historical primary source materials can become part of elementary science classrooms' (p. 364). The visual materials were revealed as 'more intellectually accessible' than Dickey's handwritten field notes and the provided transcriptions (p. 366). Seroglou and colleagues (1998) describe a three-part study that included small groups of students between the ages of ten and fourteen (Grades 4 through 8) in parts 1 and 3. Their results suggest that the early history of electromagnetism, when the thinking of scientists was perceptually dominated, can provide researchers and teachers with a good sense of 'fundamental [albeit alternative] ideas' students and student teachers will have formed at the phenomenological level regarding electrostatics, magnetism, and electric current. Based on this information, the authors designed a series of simple tasks, 'inspired' by historical experiments of Gerome Gardano, William Gilbert, and Michael Faraday, to determine the effect on students' thinking (i.e. conceptual change). Such experiments, performed between the 16th and 19th centuries, had led to greater understanding of electromagnetic phenomena within the scientific community. The evaluation of the tasks performed by ten students, ranging in age from ten to fifteen, suggested that the instructional design 'encouraged conceptual change' (p. 279). The youngest participant, for whom the tasks seemed 'fairly complicated', was reported as having said, "these phenomena do not have a common explanation because the one is completely different from the other…we cannot explain them in the same way…" (p. 279). Finally, Sneider and Ohadi (1998) draw upon the constructivist-historical teaching strategy proposed by Duschl, Hamilton, and Grandy to test its effectiveness in changing the conceptions of gravity and Earth's shape held by 539 students in eighteen upper elementary (Grade 4) and middle school (Grades 5 thought 8) classrooms. Using a GEMS publication, *Earth, Moon, and Stars*, written by Sneider for the Lawrence Hall of Science, they found that the six lessons 'enabled large numbers of students to unravel their misconceptions and construct a more adequate model of the world' (p. 281). The authors report pre-test and post-test data for Grades 4-5, Grade 6, and Grades 7-8 and the finding that the youngest group of students in the study 'responded more positively to the experimental treatment' than the group of oldest students who began the study with fewer misconceptions (p. 281). They suggest that post-test results showing 4th and 5th Grade students to be as knowledgeable as 7th and 8th Grade students could be a consequence of either especially talented teachers in Grades 4 and 5, especially flexible students willing to examine and reflect upon their current thinking, or an inverse relationship between ease of unraveling misconceptions and the period of time these ideas have been part of a belief system.

Each of these five papers ends with recommendations for further research. Few, if any scholars, however, have taken up the studies proposed. One does not have to look far to find reasons for this. Hayati Seker and Laura Welsh, a middle-years teacher, point to the limited period of time for teaching and learning school science, preferred methods of instruction that may not be promoted in the available historical resources, students' lack of enthusiasm for science [and/or the incorporation of historical material], and teachers' reluctance to use resources with concepts in the history, philosophy, and sociology of science that they are unfamiliar with or simply don't understand (2005, p. 56). Kafai and Gilliland-Swetland note the problems associated with primary source materials such as manuscripts, still and moving images, oral histories, and scientific documentation. Many are not available to teachers and students, and those that are obtainable 'may not be presented in ways that are intellectually assessable' (2001, p. 366). The same argument has been made for historical resources in general. Galili and Hazan quote a physics educator who claimed there were 'no adequate historical materials' for school audiences (2001 p. 360). Adequate materials according to this individual would 'equally match the demands of the discipline' and 'be clear enough' for students. Other 'experts' interviewed by Galili and Hazan maintained that there would be no progress toward incorporating the history and philosophy of science in physics education until such learning materials are created (p. 359). Rutherford (2001) concurs. To the impediments of material resources and teacher knowledge he adds 'somewhat incoherent science curricula' that is 'overstuffed' with content requirements established by school system authorities who 'are generally unaware of the value of the history of science in science education...and do not set appropriate learning goals' (p. 571).

In this context, it would be surprising to observe anything more than the textbook favoured dates-and-names and anecdote approaches to teaching school science with the history of science. Any richer integration of history, philosophy and sociology in science lessons would certainly involve an academic researcher or a practicum-based assignment from a science educator in a teacher education program. This is particularly true at the Early Years level. Decades of research have show that Kindergarten through Grade 4 teachers have: 1) taken the minimum requirements for science in their secondary and tertiary programs of study; 2) lack confidence in teaching science; 3) avoid the messiness and noisiness of 'hands-on' activities when management is an issue; and 4) claim to have neither the time nor the resources to organize meaningful first-hand explorations (Palmer 2001; Mulholland & Wallace 2003). When science is made accessible to young learners, either the investigative components are missing and science is a teacher-led discussion associated with language arts activities (Rice and Roychoudhury 2003), or the investigative components are present and Seatter's (2003) 'intellectual teaching acts' are missing. That is, there is no attempt to guide children toward cognitive appropriate levels of scientific understanding. Children 'learn by doing' (Rowell & Ebbers 2004, p. 217), and child-constructed ideas, including misconceptions and 'just plain error' (Matthews 1994, p. 144), are considered good ideas. Yet, as Hodson (1999) makes clear, 'Learning science is not simply a matter of "making sense of the world" in whatever terms and for whatever reasons satisfy the learner. Rather, it involves introduction into the world of concepts, ideas, understandings and theories that scientists have developed and accumulated' (pp. 241-242). Such an introduction demands an intelligent other, whether this individual is a knowledgeable teacher or a representative of the scientific community.

The study presented below describes obstacles encountered when a science unit incorporating history of science and technology was implemented in a Grade 4 class. The description begins with an overview of the study and the research questions. This is followed in Section 3 by a description of the light unit, a condensed version of both a science story and a narrative of a technological innovation to provide an impression of the character of the historical materials the children would have heard and responded to, and a brief discussion of the unit review and test. In Section 4 the research context and research participants are described. This is followed in Section 5 by a presentation and discussion of constraints in implementing the historical components and actions that undercut the educative potential of teaching that exposes children to aspects of science history. In Section 6 the implications of the findings are considered and suggestions for further research are provided.

The Study

The impetus for the project began at the 7th IHPST Conference in 2003 with a conversation in which Nancy Brickhouse bemoaned the lack of classroom-based research in the presentations. All papers in the session we had just attended focused on aspects of the history of science that the presenters believed would benefit science teaching and learning. Not one had been field-tested. I had presented earlier on the scientific work of a nineteenth century whaler and was preparing to use the notes and journal entries of his meteorological studies to develop a section of the Manitoba Grade 5 weather unit. With this comment, I realized that such a unit would be of little research value if it wasn't first implemented and the consequences for teaching and learning determined. As a result, Scoresby's weather observations and measurements were incorporated into a presentation on reading original texts in school science, and the light unit was initiated.

The Grade 4 'Light" unit was chosen for three reasons. First, as an Early Years teacher educator I wanted to design materials for young learners. The authors of the 1993 Benchmark document in the United States had stated that 'the history of science and technology is mostly too advanced for students in the earliest grades' (AAAS 1993, p. 15). The more recent writing of Asoko (2002), Dawes (2004), Duckworth (2005), Metz (1995, 1997), and others suggested that this might not be the case. Rather than pointing to developmental constraints in understanding science concepts, these authors described carefully chosen teaching strategies and activities that scaffold the learning of science. Perhaps historical narratives could also be used to scaffold science learning in Early Years. Second, the light unit is believed by teacher candidates in the Early Years program at the University of Manitoba to be the most demanding of all nineteen science units in the Manitoba curriculum for Early Years. In listing the units in order of perceived difficulty, the top five are normally Grade 4 'Light', Grade 4 'Sound', Grade 3 'Materials and Structures', Grade 3 'Forces that Attract or Repel', and Grade 2 'Position and Motion'. It seemed prudent to develop a unit with a likelihood of being used by Manitoba teachers. The third reason for choosing 'Light' was the opportunity for incorporating elements from the history of science that could be more than Matthews' 'icing on the cake' or 'minimal view' (1994, p.11). While the knowledge outcomes addressed in the unit made it unreasonable to create a history-based unit where disciplinary knowledge would be taught in its 'historical evolution' (c.f. Galili and Hazan's optics course, 2001), historical elements could be used to guide some of the unit's exploratory work and to substantiate new ideas introduced by the teacher. The history of technology would take on a lesser role. Students would be introduced to inventor-scientists responsible for optical objects, including the Gutenberg periscope, Galileo telescope, or Brewster kaleidoscope, that they would be expected to construct.

Research Questions

The study was designed to answer the following research questions:

1. What observable factors affect the implementation of the historical elements incorporated in the science lessons of a Grade 4 light unit?
2. What is the effect on students' understanding of science concepts and investigatory methods of a series of science lessons that incorporate information from the history of science and technology?

The Light Unit

Light is one of four science units at the Grade 4 level (Manitoba Education and Training 1999). 'Habitats and Communities' develops ecological knowledge in the life science strand. 'Sound', along with 'Light', develops knowledge about forms of energy in the physical science strand. 'Rocks, Minerals, and Erosion' develops knowledge about Earth materials in the earth and space science strand. Manitoba teachers are to help ten year old children achieve sixteen specific knowledge outcomes in the 'Light' unit. This knowledge is to be developed through first-hand investigation and text-based research in which questions are answered, problems are solved, and decisions are made (p. 2.9) using the twenty-nine skills and attitudes associated with scientific literacy and design technology in Grade 4 (pp. 3.46-3.47).

The specific learning outcomes (SLOs), listed in Appendix A, were collapsed into five essential questions around which the unit's curricular priorities were framed (Wiggins & McTighe 1998). The questions, with the SLOs addressed, were as follows:

1. What is light? (02, 03, and 09);
2. What are the properties of light? (04 and 07);
3. What are sources of light? (05, 06, and 16);

4. What happens to light when it shines on materials? (08, 10, 11, 12, and 15); and
5. What developments have changed how we see? (13 and 14).

Two hierarchical concept maps of the science concepts built into the learning outcomes were used to organize an instructional sequence that allowed for coherence, connectedness, and the development of progressively more complex ideas and skills. As such, the unit did not begin with energy (SLOs 02 and 03 and Essential Question 1). Energy is a most difficult concept to imagine in the context of light, particularly for Grade 4 students who, as Asoko (2002) argues, 'are unlikely to have given real consideration' to how light, light phenomena, and properties of light might be explained (p. 157). Coming to understand energy requires that children have some scientific understanding of light as a physical reality, and as a physical reality that is unlike the air and the solid and liquid materials they would have studied in previous years of school science. The unit, as outlined below, started with darkness: the absence of light and the elicitation of children's conceptions of what light might be as electric lights in a darkened classroom were switched on. The principal idea presented in each of the unit's lessons is stated and followed by the essential question (EQ) that the lesson and teacher would help children to begin to explore, think about, and eventually answer.

Lesson 1	Absence of light (darkness): eliciting children's understanding of light and darkness; myths and legends of how light came into the world; children's questions about light
Lesson 2	Sources of light (natural and human made) EQ 3
Lesson 3	Sources of light continued
Lesson 4	A beam of light and its path through the air; light interacting with reflective materials; story of Archimedes' death ray EQ2 and EQ4
Lesson 5	Light interacting with opaque, translucent, and transparent materials EQ 4
Lesson 6	Investigating shadows EQ 4
Lesson 7	Refraction of light from air to transparent liquids, and from air to transparent solids EQ 2 and EQ 4
Lesson 8	Rainbows, prisms, and colour; story of Isaac Newton's optical experiments and explanation of the spectral colours EQ 2 and EQ 4
Lesson 9	Eye safety (protecting the eyes from natural and human-made sources of light); designing a safety poster EQ 5
Lesson 10	Technological innovations (display of objects); narratives and stories of Galileo's telescope, Isaac Newton's telescope, Robert Hooke's *Micrographia*, EQ 5

Lesson 11 Light as a form of energy; sources of light that produce light or light and heat EQ 1

Lesson 12 Using the design process to create periscopes and/or kaleidoscopes; narratives of inventors Johann Gutenberg and David Brewster EQ 5

Lesson 13 Unit review (rotational graffiti); Bill Nye the Science Guy's 1995 *Light and Color* video

Lesson 14 Unit test

The lessons were designed using the '5-E Model of Instruction'. The 5-E model is an expanded version of the learning cycle developed by Robert Karplus, director of the Science Curriculum Improvement Study (SCIS), and Herbert Thier in the early 1960s (Karplus 1964; Karplus & Thier 1967). To the three phases of instruction in SCIS lessons (exploration, concept invention, and concept application), Trowbridge, Bybee & Powell (2000, pp. 243-249) added an introductory engagement phase and a concluding evaluation phase. The instructional plans for the light unit included engagement, exploration, explanation, elaboration, and evaluation phases. However, in contrast to the model, the teacher was not to wait for the explanation phase to formally introduce and explain new concepts and useful ideas or to model scientific ways of thinking. Following the teaching strategies advocated by Asoko (2002), Hodson (1999), and Seatter (2003), a conceptual structure or scientific model was introduced prior to the exploration phase to stimulate thinking. It was believed that this 'second phase scaffolding' (Hodson 1999, p. 245) would allow students working together in small groups to use the new ideas to explore and explain the light phenomenon at the core of a lesson and, thus, begin to move beyond the limits of their own intuitive ideas and personal or consensual conceptualizations.

The history elements in the light unit were often used to scaffold children's thinking about natural phenomena and about investigative work in science. While telling the story of Newton's theory of light and colours, for example, the teacher was expected to recreate the three experiments described. The story and demonstrations would subsequently guide children in their own investigations with ray boxes and prisms. While some may find the stories to be a distortion and oversimplification of the historical events, as Martins & Silva (2001) describe Towne's use of Newton's colour theory, one must remember that the lessons have been developed for ten your old students who have numerous experiences with light and its effects, but generally little, if any, scientific understanding of their observations. To speak, for example, of minimum deviation position, incident and emergent rays, divergence, relative refrangibilities, methodological arguments, an/or a theory of refraction would not be intelligible, plausible, or fruitful for young learners who are being helped to know and describe the properties of light and the consequences of light interacting with materials. This is not to suggest that the stories and narratives included in the light unit cannot be improved. A shortened version of 'Newton's Prism and His Explanation of Colour' follows. The sections included utilize the 'interrupted storyline' described in Metz, Klassen, McMillan, Clough & Olson (2006). They were designed to link the introductory components of the engagement phase of Lesson 8 to the activities and discussions planned for the exploration and explanation phases.

A Story of Newton's Prism and His Explanation of Colour

Prehistoric paintings that have been discovered on the rock walls and ceilings of caves, and the images on prehistoric pottery tell us that since the earliest times, humans have been interested in colour (Show illustration of cave painting). *Initially people thought that the light from the Sun was "pure light" and that colour was the result of adding things to this pure light. Some people believed this was what a glass crystal or prism did when sun light passed through it. The material the prism was made from gave the pure light its colour. Other ideas that people had about colour came from artists and artist-scientists like Leonardo da Vinci. These artists based their ideas on the colours they made when mixing pigments to paint and draw with. They recognized that red, yellow and blue were the primary colours and that the primary colours along with black and white could be mixed to make any colour of paint they could imagine. The same idea was applied to different coloured beams of light, but it was an idea that wasn't tested until Newton began to make reflecting telescopes. Newton used mirrors to eliminate a fuzzy ring of colour that surrounded the image formed by the lenses in Galileo's refracting telescope. As much as Newton wanted to build a better telescope to study the night sky, he also wanted to understand the cause of the ring of colour when lenses were used in telescopes. So, like you, he began studying what happens to the light when it shines on a triangular glass prism.*

Newton darkened a room in his house with opaque, wood shutters. (Recreate this experiment in a darkened classroom using a ray box and a triangular prism.) *In one of the shutter slats, he then made a small circular hole that the Sun's light was able to pass through. It made a beam of white light, and Newton carefully placed the triangular faces of his prism in the beam. On the opposite wall he noticed a long patch of coloured light. This patch was violet at one end and red at the other and showed, blue, green, yellow, and orange in between. Newton named this pattern of colours the spectrum. He couldn't explain with certainly where the colours came from, and he couldn't explain why a beam of light would spread out in an elongated patch when the hole in the window shutter was circular. So, he changed the size of the hole in the shutter and tested. He changed the size of the glass prism and tested. He even placed the prism outside the window. None of these changes made a different spectrum on his wall. If he always held the prism in the same orientation to the beam of light, the spectrum was always the same shape and the same number and arrangement of colours.*

As a result, he decided that the spectrum must be a result of some unevenness or blemish in the glass from which the prism was made. He tested this idea by passing the beam of light through the prism, and passing the spectrum created through a second prism turned upside down. If it was a problem with the glass, and the glass was causing the beam to spread out, them passing it thought the second prism would make it spread out even more. Guess what he saw on his wall? Do you think it was an even bigger spectrum? (Recreate this experiment in a darkened classroom using a ray box and two triangular prisms.) *Instead of a larger patch of colours, the spectrum that was passed through the second prism made a spot of white. It was just like the beam of sunlight that shone on the wall through the circular hole in the shutter when the single prism was not placed in its path. Yet the light had passed through two different prisms.*

With this information, Newton convinced himself that the light from the Sun is composed of colours. The prism does not add the colours as the light passes through it. The colours, violet, blue, green, yellow, orange, and red, are in the Sun's light all of the time. The prism works by bending (refracting) each of the colours at a different angle and this spreads the beam into a spectrum.

As a further test of this idea, Newton cut a very small hole in an opaque screen on which the spectrum was projected. (Recreate this experiment in a darkened classroom using a ray box, two triangular prisms, and an opaque sheet of boxboard into which a hole has been cut.) *Only one colour of light was able to pass through the hole. When he placed a second prism in the beam red light, neither white light nor the spectrum appeared, only the colour of the light that passed through the small hole in the screen. The second prism had shown that red light is composed only of red light and that prisms don't add colour.* (The storyline continues with Newton's concept of primary and compound colours and how the interaction of white light and an object makes a colour.).

Narratives

The narratives associated with inventors and their inventions were designed to introduce students to several of the individuals responsible for objects that require light for their operation, extend the ability of humans to see, or ensure protection of eyes. These written texts were as short as four sentences, as in the case of Johann Gutenberg's 1430 periscope designed to see over the heads of pilgrims during a religious ceremony in Aachen, or as long as two paragraphs, as in the case of Sir David Brewster's teleidoscope. An excerpt from the latter will provide a sense of the information children would listen to and discuss with their teacher.

The kaleidoscope is a 'magical' instrument that produces changing patterns of unexpected beauty. It was invented by Sir David Brewster, a Scottish scientist, in 1813. The year 1813 is almost one hundred fifty years after Newton constructed his theory of light and colours. Brewster's invention (show children the teleidoscope and pass it around for each child to look through) *is known today as the teleidoscope. Like the three-mirrored kaleidoscope you made at the Museum, the teleidoscope makes a pattern out of the object or objects it is pointed toward. Almost all kaleidoscopes begin with the same parts Brewster used to make his teleidoscope. These are the three mirrors, or reflective surfaces, and a body that holds the three mirrors in place. A sphere scope* (show the sphere scope and pass it around for the children to look through) *is a teleidoscope with a glass marble or crystal sphere at one end cylindrical or triangular-shaped body. A cell scope* (show the cell scope and pass it around for the children to look through) *uses a closed container at the end of the body rather than a marble or sphere. The container can be filled with small tumbling objects or objects in a liquid that move when the cell scope is rotated. A wheel scope has a wheel attached to the end of the body. The wheel is usually bigger than the body, it can be rotated without rotating the body, and is usually made out of stained glass, crystal, found objects that are mounted on a disk, or hand-drawn and photographic images that are glued to a disk* (show an example of each kind of wheel scope and as each kind is described pass it around for the children to look through). The narrative continues with images of antique and contemporary kaleidoscopes (paying particular attention to the type, the materials of construction particularly the materials in the containers and wheels, and the presence or absence of a viewing lens).This is followed by a discussion of the reflected images that would be created and seen using examples of a three-mirror system (not necessarily an equilateral triangle) and a two-mirror system with the mirrors positioned at acute and right angles.

Unit Review and Unit Test

As stated in the lesson sequence, the unit review included a rotational cooperative graffiti and Bill Nye video. Students were to be divided into four groups and each group was to begin the rotational graffiti with a sheet of poster paper on which a question had been written. Their first task would be to write down as many responses as they could in five minutes. At the end of five minutes, the sheets would be rotated clockwise to the next group, and team members would draw as many responses as they could to the new question on their sheet. This would continue for two additional rounds until each group would have responded pictorially or in writing to each question. Four questions (What is light and where does it come from? How can we make a beam of light bend or change direction? What developments change what we see? What can happen when light shines on different materials?) were designed to help students recall what they would have learned with respect to the unit's five essential questions. The video, *Light and Color*, was to be watched during the unit review because it addressed in unique contexts a majority of the concepts that children would have been investigating.

A unit test was designed as a job application. Students were to show the owners of 'The Spectrum Company' how well they understood light and its properties by providing answers to six questions. In Question 1, they were asked to explain the appearance of the light in a photographic image. In Question 2 students were to add three items to a list of light sources that produce heat and feel hot to the touch and to explain why an item listed did not belong. In Question 3 they were to select the best of three materials to be used in manufacturing the Spectrum Company's new envelopes, to justify the material selected, and explain why the other two materials were not suitable. In Question 4, students were to explain to the manager of the Spectrum Company why his eyeglasses that had fallen into a half-full glass of water were not really broken even thought they appeared to be. In Question 5 the owner of the company asked for an explanation, with a diagram, of the colours created by sunlight shining on a prism. In Question 6 the students responded to a child's question about white light and why it makes objects appear to be red or yellow or any of the colours we see. In other words, 'Why isn't everything white?'

The Research Context and Research Participants

The study was conducted in the Grade 4 classroom of a Kindergarten through Senior 1 (Grade 9) school in suburban Winnipeg. During the 2005-2006 school year, there were 240 enrolled students, a staff of 22 teachers, and 8 support personnel. Fifteen girls and eight boys were assigned to Grade 4. Five of the children were from families that arrived in Canada from South America, Europe, the Arabian Peninsula, or Africa. Two were members of Canadian First Nations populations. The students reflected the diversity represented in the majority of Winnipeg schools and came from homes representing low-to upper-middle socioeconomic status. All schools in the division ran on a six-day teaching cycle. That is, if Day 1 was Monday in Week 1, it would be Tuesday in Week 2 and Wednesday in Week 3, and so on. The school day for the Grade 4 students began at 8:50 Monday through Friday. All Early Years students were dismissed at 10:30 for a fifteen-minute morning recess, at 11:45 for a 55-minute lunch break, and at 14:15 for a fifteen-minute afternoon recess. The end of the school day for all students was scheduled for 15:30 (see Appendix B for the Grade 4 timetable).

In Manitoba's English-language programs, the number of school days (194-196 of 365, 10 of which are non-instructional) and the minutes of instruction per week (approximately 1650) are regulated by the provincial government's Department of Education, Citizenship and Youth (no date a). This department also recommends subject area time allotments. For Grades 1 through 6 these allocations are as follows; English Language Arts, 35% of a school year, week, or day; Mathematics, 15%; Science, 10%; Social Studies, 10%; Physical Education and Health, 11%; Arts, 10%; and non-compulsory/optional subjects such as languages and Aboriginal Studies, 9% (Manitoba Education, Citizenship, and Youth, no date b) The 10% for teaching and learning science works out to 33 minutes per school day, 2 hours and forty-five minutes a week, or 3 hours and 18 minutes per six-day cycle.

Science in the Grade 4 classroom where the study was conducted had been scheduled for three 30 minutes sessions each cycle (Appendix B). Like mathematics, English language arts, social studies, health, French, art, and computer lab, instruction in science was the responsibility of the classroom teacher. Specialist teachers were responsible for music, choir, physical education, and library.

As the study was being planned, the decision was made to pilot the light unit in one Grade 4 classroom and to give preference to novice teachers rather than teachers with years of science teaching experience at the Grade 4 level. I made this decision because the study was an intervention study, and the teacher participant would be expected to implement the 'experimental' lessons as designed. By selecting a novice teacher I thought I could reduce, possibly even eliminate, departures from the lessons, particularly the incorporation of activities that teachers with experience thought worked well with former groups of students. As the time for piloting the unit approached, the pool of novice teachers was limited to the teacher candidates in the first year of the two-year after-degree Bachelor of Education program at the University of Manitoba. With hindsight this may not have been the wisest decision. At the time, however, it made sense. These education students had completed the course, Curriculum and Instruction in Early Years Science and Health 1. Having done so, they were aware of children's ideas and strategies for eliciting intuitive beliefs and misconceptions. They had attempted hierarchical concept mapping and used a concept map of the concepts in a science unit to create essential questions. They had created a lesson that addressed a design process outcome and used the 5-E model to plan a lesson incorporating a first-hand investigation. Moreover, during the Year 1 practica candidates are expected to teach with the materials and resources provided by their collaborating teacher. From this pool of approximately seventy Year 1 candidates, I needed one volunteer who would be working with a Grade 4 class studying 'Light' during the Term 2 practicum, and who would have been asked to prepare and teach the light unit. Preference would be given to candidates whose assignments for the C&I science course showed evidence of Early Years children as learners, creative thinking, and knowledge of the science concepts their lessons were designed to help children to begin to understand. Finally, I planned to be present during the teaching of the light unit lessons and envisioned my role as a participant observer. I wasn't to be an inconspicuous researcher sitting quietly, monitoring the effect of the lessons on students' understanding. I also expected that preparatory meetings would be held prior to lessons being taught and that each completed lesson would be discussed. For this reason, priority was given to partner (practicum) schools within twelve kilometers of the university.

A Year 1 candidate, Celia Lantz (pseudonym), expressed immediate interest in the study and agreed in early January to implement the lessons. At the end of the Term 1 practicum, Celia and her collaborating teacher had decided that mathematics (fractions) and science (light) would be the subjects Celia would teach during the six-week practicum in Term 2. Her placement school could be reached by car in approximately fifteen minutes, and both her collaborating teacher and faculty advisor agreed to the study and Celia's participation in it. Celia was older than the majority of teacher candidates admitted to the Early Years program and had worked more than ten years in school divisions outside of Manitoba. Two of her assignments, one using the S.P.I.C.E. Model for developing the design technology lesson (Metropolitan Toronto School Board 1996) and one using the 5-E model were excellent, not only in conception and originality, but also as a consequence of Celia's facility to interest young learners in the science being addressed as well as the design problem or context for learning.

Results and Discussion

Celia began the unit at the beginning of the week with a diagnostic assessment designed as a pre-instruction worksheet (Appendix C). When the students had completed the worksheet, Celia read the picture book, *Musicians of the Sun* (McDermott 2000). It tells the Aztec story of Texcatlipoca, Lord of the Night, who sends Wind to bring the musicians of the Sun to Earth. The story was discussed by the class, and the students talked about how the Sun came to illuminate Earth and what it might be like to live on Earth without Sun's light. Sitting quietly on the floor around Celia, the classroom lights were shut off, and the students experienced the classroom without electric and natural lighting (all window glass had been covered with opaque material). After two minutes, the lights were turned on, and the students talked about the experience before completing a post-instruction worksheet (Appendix C). Before the end of the session, Sun-shaped 'Post-it-Notes' were quickly handed out, and each child was asked to write a question about light that they would to be able to answer before the end of the unit. The Post-it-Notes were displayed on the 'Light' bulletin board and occasionally referred to during the subsequent four weeks.

Tuesday's lesson focused on sources of light, and sources of light like the Sun that are natural and sources of light like the fluorescent bulbs in the classroom that are human-made or artificial. Each child was given a sheet of images (contour drawings) of natural and artificial light sources that every student was asked to sort, using sorting criteria that allowed for categories meaningful to them. The groups of sorted images were to be glued to a sheet of paper and labeled 1, 2, 3... The key to the label for each child's set of categories was to be written on a prepared sheet of paper. At this point, the children were to show their sheet of sorted images to a classmate and ask the classmate to provide a label for each group represented. These labels were written down, along with the name of the classmate and the number of labels that matched the labels on the key. At the end of Tuesday's session for science, the students were either in the process of finalizing their categories or gluing their sorted images on paper. The lesson was picked up on Thursday, and continued. By the end of the session, most children had responses from at least one classmate. Those who worked more quickly and had collected responses from three or more classmates were asked to write about the previous week's field trip to the Manitoba Museum for 'The Amazing Light Workshop'. This was a school program, created by museum employees for the light unit in Grade 4, that addressed most, if not all, 16 knowledge outcomes in 1.75 hours.

All who are familiar with instructional models will have recognized that neither lesson, particularly Lesson 2 that extended over two sessions, moved beyond the engagement and exploration phases. Although there were a number of teaching acts, not one of these could be considered an intellectual act of teaching when Seatter's (2003) criteria are used. All were strategic. That is, they were 'motivating', 'encouraging', and 'guiding' and focused on 'all of the preparation and presentation revolving around the learner' as opposed to the science and scientific ways of knowing the lessons were created to develop (Seatter 2003, pp. 69-70). Research Question 2 that looked for effects of the 'experimental' lessons on student's understanding of science concepts and investigatory methods would not be able to be answered. Rather than terminate the study, the decision was made to focus on aspects that could potentially provide data useful for integrating history of science in lessons designed for the science education of young learners. Thus, the research question for the study became the original Question 1: What observable factors affect the implementation of the historical components incorporated in the science lessons of a Grade 4 light unit? The second research question was eliminated.

The data from the revised study is qualitative and includes the plans for the light unit, field notes of the planning and 'debriefing' sessions, e-mail messages sent from and to Celia, observational notes taken during the teaching of each lesson, digital photographs of five sessions of science, photocopies of the completed diagnostic assessments and sorting sheets from Lessons 1 and 2, photocopies of twenty-one science journals (two journals had been misplaced by students), and digital files of the unit test completed by twenty-three students. The presentation and discussion of the data begins with the factors that constrained or prevented implementation of the science lessons, particularly the historical components. This is followed by a presentation and discussion of the factors that hindered the educative potential of the historical components that were implemented. Hereafter, the stories and narratives from the history of science and technology will be referred to as historical narratives. All names appearing in the text are pseudonyms.

Factors that Inhibited the Implementation of Historical Narratives

TIME

The effect of time and scheduling on teaching and learning is enormous. If this wasn't made clear in the descriptions of Lessons 1 and 2, one only needs to look at the schedule of subjects in Appendix B and imagine preparing and teaching for these six days over a period of 31 cycles. The Early Years classroom teacher is responsible for teaching seven subjects along with technology education and each of these comes with a set of prescribed knowledge, skill, and attitude outcomes. School science instruction, as mentioned in Section 4, is allocated 10% of a school year or 10% of the 5.5 hours of instruction in a school day. If one assumes the minimum 184 days of instruction in a school year, each of the four science units could be addressed over a period of 46 days, which is equivalent to nine school weeks plus a day, or 25 hours and 18 minutes of instruction at 33 minutes per school day. The teacher candidate in this study was given the responsibility of teaching the light unit in fifteen sessions over five weeks. Of these sessions, ten were 45 minutes in duration, two were 60 minutes (as were the sessions for reviewing the unit and writing the unit test), and the construction of periscopes required a 90 minute block of time. The unit was taught in 13 hours. This is just over one half of the time allocated for each Grade 4 science unit. There is no denying that the light unit outcomes can be 'covered' in this period of time. Whether they can be understood by each student and incorporated in conceptual models that are scientifically accurate is unlikely given what we know about teaching for conceptual change (Gilbert & Reiner 2000; Scaife 2000; Taber 2005; among others) and the tenacity by which learners tend to hold on to their own ideas (Harlen, Marco, Reed & Schilling 2003). Observations supporting this view occurred during each science session. Responses to the pre- and post-instruction assessments in Lesson 1 will be used as examples.

Student responses to the first section of the pre-instruction worksheet (Appendix C) illustrated the range of experiential evidence used to support particular beliefs. Three drew a circle around 'I think we can't see in the dark'. Four circled 'I think we can see in the dark'. Fifteen drew a circle around both (Table 1).

PRE-INSTRUCTION	POST-INSTRUCTION		NUMBER OF STUDENTS
Can we see in the dark?	Can we see in the dark?	Do we need light to see?	
Yes	No	Yes	2
Yes	Yes	No	2
No	No	Yes & No	2
No	No	Yes	1
Yes & No	Yes & No	Yes & No	2
Yes & No	Yes	Yes	3
Yes & No	Yes	No	4
Yes & No	No	Yes	5
Yes & No	No	Yes & No	1

Table 1. Student responses to the pre- and post-instruction assessments in Lesson 1

Those who selected the 'can't see' option, wrote about everything being 'black', 'really black', or 'dark and black'. The four who circled 'can see' provided one of two explanations. These were, 'I think we can see in the dark because thar [there] is allway [always] some kind of light at night', and 'If you stay in the dark for a while you can see a little bite [bit] of light so you can see.' The students who selected both 'can see' and can't see' tended to use the explanations provided by the two other groups and to distinguish between darkness (night time with street lights, car lights, Northern lights, and light from the television, Moon, and bright stars) and 'pitch black'. Hannah, for example, wrote: 'Some time you can see because there is a window and the light comes thow [through] a litle [little] but some time it is pich [pitch] dark so you can't see." The intervention, darkening the classroom, caused 17 students to change their initial response, but only seven of the twenty-two responded as a scientist would. Two of the three students who began the lesson believing we couldn't see in the dark maintained this belief but weren't certain, given their yes and no response in the post-instruction assessment, that light was necessary in order to see. One-half of the four students who began the lesson believing it was possible to see in the dark understood that they couldn't see unless light was present. The two other students continued to believe that it was possible to see in the dark and that light was not necessary. Of the fifteen students who circled both possibilities in the diagnostic assessment, thirteen changed their responses on the post-instruction worksheet, but of these, four were the responses expected.

In almost all cases, the intervention did not take into account the evidence students used in the diagnostic to support a specific belief. This is particularly evident in the responses of two students given immediately below.

Matt (Pre-Instruction)

I think we can see in the dark.

How I know. I think this because "if you stay in the dark you wont [won't] be able to see perfectle [perfectly] but you well [will] be abell [able] to see a little bit after about 5-10 minutes."

How can we find out? 'You can turne [turn] of [off] the lights.'

Matt (Post-Instruction)

Can we see in the dark? 'Yes'

Does this mean that we need light in order to see? 'No'

How I know this. I think this because 'you don't need light to see.'

Alex (Pre-Instruction)

I think we can and can't see in the dark because 'sometimes people were [wear] bright T-shirts and you can see them in the dark. If someone is whering [wearing] a dark T-shirt and you turn off the lights you can't see them.'

How can we find out? 'If we try.'

Alex (Post-Instruction)

Can we see in the dark? 'yes'

Does this mean that we need light in order to see? 'no'

How I know this. I think this because: 'people have bright T-shirt[s]. You can see their T-shirts. We don't need light to see because we can see bright T-shirts.'

In other cases, light leaks observed along the ceiling of the darkened classroom were taken as evidence of being able to see even though it was only the light that was seen. It would have been impossible to state the time on the clock, the colour of Celia's socks or hair clip, or even where she was sitting and what she was sitting upon without having observed this beforehand. Few teachers interested in modeling deep approaches to learning science (Chin & Brown 2000), would have left the lesson at the end of 45 minutes and not used the next session scheduled for science to repeat the experience, informed with the students' responses to both worksheets, or enabled children to test the ideas they used to explain and justify particular beliefs. The tight timeline in Celia's case required that she carry on with Lesson 2 and sources of light.

Galili & Hazan (2001) and Seker & Welsh (2006) mention the shortage of time for science instruction as one argument used against including history and philosophy of science. In the context of teaching with the history of science, whether an add-on or integral approach is used, it's not difficult to imagine in mandated curricula with knowledge but no historical outcomes what will be set aside by teachers when the ideas of learners are both erroneous and resistant to change, and time is limited. What happened to the teacher, the learners, and the plan for teaching and learning in Lesson 1 of the light unit was repeated in Lesson 2 and Lesson 3. Then, in Lesson 4 the Archimedes death ray narrative was truncated to allow more time for children's thinking out loud. In Lesson 8, only the first paragraph of the story, 'Newton's Prism', was told so that more time would be available for exploration, discussion, and recording. As well, no mention was made of Newton's colour theory during the role play where each student became a spectral colour and was either reflected by (stayed in front) or absorbed by (moved behind) the colour of poster board held in Celia's hands. In Lesson 10, an assortment of objects including cameras, eye glasses, safely goggles, telescopes, kaleidoscopes, binoculars, a binocular dissecting microscope, a compound light microscope, and a copy of *Micrographia* were set out for children to explore and use, but the historical narratives about the inventors and inventor-scientists were not told.

WEATHER CONDITIONS

In a Grade 4 science unit focused on light and light phenomena, weather conditions can wreak havoc on plans for teaching and learning, particularly the sequence in which new concepts are introduced. During the late winter when the unit was taught, Winnipeg experienced an inordinate number of overcast days, and each day without direct sunshine seemed to fall on a day that science was scheduled. Lesson 4 began the second week of the unit. It focused on the path light travels between the Sun and Earth and from all other natural and human-made light sources on and around Earth. Celia began with two interactive demonstrations conducted in the darkened classroom. In the first she used a flashlight and talcum powder. In the second a laser pointer substituted for the flashlight. Students saw the lit bulb of the flashlight and the patch of light on the wall and talked about how the light got from the flashlight to the wall, a distance of approximately 2.5 meters. Talcum powder was sprinkled in the air between the flashlight and wall and the beam of light 'materialized'. It not only could be observed, but could be observed the entire length of the 2.5 meters as a straight beam. That is, it did not follow an up-and-down, side-to-side, curved, or meandering path. Students were asked to predict, observe, and explain the path traveled by the red beam of the laser pointer.

At this point in the lesson the historical narrative of Archimedes' death ray (using the Sun's light and a parabolic mirror to ignite an enemy ship during battle) was to be told (Appendix D), and students were to watch an online segment of Discovery Channel Canada's 'Daily Planet'. This program aired the reenactments of Archimedes' death ray preformed on the televised program 'Mythbusters' and by Dr. David Wallace, a professor of mechanical engineering at M.I.T. in Cambridge, Massachusetts (Discovery Channel Canada 2005). After discussing what was viewed, the children were to use small, hand-held mirrors to reflect the light from the Sun onto a single target in the schoolyard, and measure the increase in temperature. The fundamental questions to be discussed were: Could you use your mirror to do this if light didn't travel from the source to the target in a straight path? How do you think we make use of this information [property of light] in our daily lives? In addition to the present context, the death ray was also to be referred to in Lesson 11 when light sources that also produce heat were explored.

As mentioned in the previous discussion about the constraint of time, only pieces of the story were told. For reasons to be described later, the online video was not shown. The planned test of Archimedes idea could not be conducted out-of-doors because of the overcast sky, and the classroom simulation merely allowed children to see how a mirror can change the direction of a light beam. With the intention of returning to the schoolyard to conduct the test during the session shadows were to be studied, the lesson quickly moved on to the second activity of the exploration phase. As it turned out, a clear morning on a day with science did not occur until the last lesson before the construction of periscopes. The focus of this lesson happened to be safety, and a welder had been invited to show the materials he used to protect his eyes and vision from the intense light of welding and cutting torches. Re-scheduling was not an option.

TEACHER KNOWLEDGE

Much has been written about the content knowledge, pedagogical knowledge, and pedagogical content knowledge of science teachers, and the concern teacher educators, scientists, historians of science, philosophers of science, and sociologists of science have expressed about the level of teacher knowledge in these areas. Certainly all could be helping to educate and prepare better teachers given the remarkable qualities of the majority of individuals who select teaching as a career. With the exception of courses in teacher education programs that are school-based, the practicum offers the only opportunity for teacher educators to observe the effect of sixteen years (on average) of formal education on individuals' perceptions of what it means to teach and of their capacity to help others learn in a manner that transforms and enriches each life. It is also one of the only opportunities in the career-long education of teachers where teaching acts are expected to be constructively critiqued. The comments that follow are made from this perspective.

In Lesson 8, the engagement phase was designed to guide students' first-hand investigations in the exploration phase. That is, for the practical components of the lesson to be meaningful, the historical narrative, 'Newton's Prism and His Explanation of Colour', had to be told as a story that was to be interrupted in several places by student investigations that simulated Newton's work, albeit at an elementary level. Celia had received the plan for the lesson a week before it was scheduled, and met with the investigator on a Sunday morning to discuss it along with the other science lessons for the week. The three demonstrations were not practiced since it was decided that Celia would tell the story and the investigator, who performed the 'experiments' on previous occasions, would present them as the story was told.

As was the case for all previous science sessions, Lesson 7 had not been completed in the block of time available: the three groups of students explored only two of three stations with dissimilar refraction activities. On the day planned for Lesson 8, Celia sent an email, part of which is excerpted below:

> 'Good morning,
> I really like the lesson today on light and colour.
> The only thing that I am going to change is that I would like to begin with the new lesson and I'd like to complete it from beginning to end. So that means that I am going to move the completion of the final station from yesterday's class on "bending light" to the end of the class if time allows.
> Because there is only 45 mins I don't want to be breaking up the lesson and continuing on Thursday.
> What do you think?'

The light and colour lesson began the session for science. However, Celia ended the historical narrative having read parts of the first paragraph (Section 3.1). Groups of students were assigned a station in the classroom where they were to use the ray boxes and triangular prisms to make Newton's spectrum. They were asked to observe, draw (using coloured pencils), and label the order of spectral colours when the prism was placed on a face and on end. This was followed by the role-play activity described in the previous section. The plan to complete the remaining Lesson 7 station was eliminated and never again addressed. A partial explanation for the spontaneous changes arrived in two of Celia's subsequent email messages. The first included the following passage:

> 'Good morning,
> Sorry I took so long to get back to you…
> I think I'm not quite sure what to do for today's lesson. There seems to be so much to cover.
> I have attached a review plan as well as something on "white light" which I was hoping to get copied on a transparency but was unable to. Can you do that? I know its short notice but it's also not necessary if not. Also are you going to bring your red, green and blue lights so that we can demonstrate how they recombine to produce white light?
> All suggest[ion]s are welcome today...HELP!'

The second announced that Sir Isaac Newton (the investigator in a lab coat) would be visiting the classroom that morning to talk with the students.
Celia gives the appearance of being overwhelmed by the volume of science content in the curriculum and the concepts themselves. She not only asks for help, but

independently makes the decision to turn the lesson over to the investigator. Moreover, the inclusion of 'something on "white light"' and the request for the coloured flood lights to demonstrate the additive properties of light represent ideas that were addressed in the context of Newton's work but never shared with the students. The likely conclusions one is able to draw from this information are: 1) Celia didn't recognize the importance of the historical components for students' understanding of the spectrum and the colours that constitute white light; 2) Celia believed that an understanding of the spectrum and the colours that constitute white light could be better developed using non-historical information and demonstrations; and 3) Celia, herself, did not possess deep understanding of the property of light she was to help students to know and understand. Each of the items, 1, 2, and 3, are embedded in content knowledge. Items 1 and 2 are predicated upon pedagogical content knowledge and how one best brings learners to a scientific understanding of the world that is transformative.

STUDENT KNOWLEDGE AND SKILLS

In several lessons, the background knowledge of the students and the level of development of specific skills proved to be as great an impediment to teaching new science concepts as did time and weather. In Early Years science, children are helped to construct productive questions that lead to first-hand investigations and/or research and to plan such investigations. This begins in Grade 1 with the ability to 'ask questions that lead to explorations of living things, objects, and events in the immediate environment' (SLO 1-0-1a) and to 'follow simple directions while undertaking explorations' (SLO 1-0-4a). It has developed by the end of Grade 3 to an ability to 'ask questions that lead to investigations of living things, objects, and events in the local environment' (SLO 3-0-1a), to 'brainstorm, with the class, one or more methods of finding the answer to a given question' (SLO 3-0-3a), to identify, with the class, variables that have an impact on an investigation (SLO 3-0-3b), to 'create, with the class, a plan to answer a given question' (SLO 3-0-3c), and to 'carry out a plan, and describe the steps followed' (SLO 3-0-4a) (Manitoba Education and Training 1999). In Grade 4 these skills are to be achieved in small groups rather than an entire class. Yet, the children studying the light unit in the second half of their school year had difficulty responding to questions like, 'How could we find out?' In Lesson 1, Elsbeth wrote, 'I think this is a hard question to awser [answer]'. Seven other students suggested testing the ability to see in the dark by turning off the light and closing their eyes. During the lesson on shadows, few students knew how to use a meter stick to

measure the distance of an object from a light source and the distance of the light source from the wall in spite of the fact that standard units are introduced in the 'Shape and Space' strand in Grade 2 mathematics (Manitoba Education and Training 1995).
In the context of the lessons incorporating historical narratives, the discussion of Newton's spectrum and the colours of light began with an elicitation of students' understanding of the subtractive theory of colour. It was believed that mixing of tempra and finger paints in the Kindergarten science unit, 'Colours', and the arts and crafts children should have had in each year of formal schooling would have resulted in a sound understanding of the consequences of mixing colour pigments. The Grade 4 students in Celia's classroom, however, had trouble believing that all of the colours that we see can be represented in paints, dyes, and inks by mixing combinations of the primary subtractive colours along with black and white, where necessary. There was confusion about what primary and secondary colours might be, and around the fact that secondary pigment colours are made by mixing two of the primaries. George, as one example, was adamant that orange could not be made. Like red, blue, and green, it was a pure colour. He also asked about 'periwinkle': What colours would be mixed to make it? Without this knowledge, it was not possible for students to understand the additive theory of light and the idea that objects are the colours they appear to be as a consequence of reflecting some of the colours in white light at the same time the other, non-reflected colours are being absorbed. Although the students when involved in the role play knew when to be reflected (to stay in front of Celia if she was holding up the red poster board and they were the red component of spectrum), and when to be absorbed (to go behind Celia if she was holding up the red poster board and they were the orange, yellow, green, blue, or violet colours of the spectrum), they ended the activity thinking this was true only for the six colours of Newton's spectrum and wondered about all of the other colours in the world.

Factors that Inhibited the Educative Potential of Historical Narratives

THE SOCIAL ENVIRONMENT OF THE CLASSROOM

Programs and professional learning opportunities in the Faculty of Education at the University of Manitoba are guided by five principles. Two of these, collaboration and inquiry-critical reflection, help to define the professional relationship of the collaborating teacher and teacher candidate during the practicum. The collaborating teacher is expected to assume the role of mentor. In this capacity, experiences are to be shared, understandings are to be constructed, agenda are to be negotiated, and 'best practices' are to be developed with guidance (Faculty of Education 2005, p. i). The relationship between Celia and her collaborating teacher was strained. The collaborating teacher assumed more of an evaluative than mentoring position, and this stance had a noticeable impact on the social environment. Celia always felt like a visitor in the classroom. At one point early in the study when she was overly concerned about getting too much talcum powder on the carpeted floor, I asked why she hadn't requested a different placement for Term 2. The response was simply, 'the children'. Celia had established a good rapport with the twenty-three students during the four week practicum in Term 1, and wanted to continue working with them.

With respect to science, however, not being given free access to the second, and unused, computer in the classroom eliminated the educative potential of web-based historical materials. These materials, like the Daily Planet's video of Archimedes' death ray, were important for guiding practical work and introducing or reinforcing new ideas. Plans for using the web had also been developed for the lessons on Newton's spectrum, colour, technological innovations, safety, and the kaleidoscope. Images from books or printed off the web ended up as substandard substitutes.

THE PSYCHOSOCIAL RESILIENCE OF THE CHILD IN THE FACE OF ADVERITY

Four of the five constraints described to this point, weather being the exception, are in the power of teachers and school administrators to eliminate. Resilience, however, is affected more by an individual's social and personal resources than connectedness to classroom and school communities. When these resources ('good relationships with family and friends', 'access to support networks', and 'dispositions such as self-reliance, self-understanding, empathy, altruism, maturity…and a sense of mastery') are limited, individuals can be vulnerable to life's setbacks and challenges and unable to cope with adversity (Moen & Erickson 1995, pp. 170-171). One of the students in Celia's classroom, Henry, started the unit sharing ideas and asking questions with interest and enthusiasm. This waned near the middle of the third week, which also happened to be a time when Henry asked me privately to explain a funeral. Only when George came to school one week later with a model of an 'Archimedes catapult' that he had built was Henry's exuberance once more displayed. They both stayed past dismissal time to determine how well it worked. When the unit test was marked, Henry received 7 of 18 possible points. This was the lowest mark in the class. According to the collaborating teacher, he was in foster care. He had attended the funeral of an uncle, and while at the funeral had seen his mother and met his father for the first time. It's not difficult to appreciate how a child in circumstances such as these could be struggling with issues of more significance than historical narratives and knowledge in and about science.

Implications and Conclusion

As troubling as these six constraints may seem to be in the midst of a study, not one is specific to the incorporation of historical elements in science teaching and learning. These are impediments that exist in all classrooms from Early Years through Senior Years and beyond. Schools have been, and will continue to be, staffed by teachers who delight in literature, language, and the humanities and are baffled by science, technology, and mathematics, and teachers who prefer the closed-door ethic to collegiality, sharing, and cooperation. Both have consequences (outside of government recommendations) for subject area time allotments, the skill, attitude, and knowledge development of students each year they are in school, and the content knowledge and pedagogical content knowledge of teachers. Likewise, there will always be classrooms populated by students who are physically present but have decided for any number of reasons not to learn. So, what's to be done? There is the possibility that it's not reasonable to do more than is currently being done until there are changes to policies and structures that regulate teacher certification, teacher education, and school systems. Progress may be slow, particularly in determining the effect of units of historical presentation on learning, but the studies cited in this paper is moving in the right direction.

The light unit described in this paper is in the process of being redesigned based on feedback from the pilot study. The new unit is interdisciplinary and incorporates outcomes from English language arts, visual arts, mathematics, music, and social studies. It is hoped an interdisciplinary approach will free up time for lessons, assessments, and enrichment activities created to help students learn about light and colour with interest, inquisitiveness, and deep understanding. The unit will be available for implementation by a small number of experienced and novice Grade 4 teachers in January 2007. Interested teachers will be expected to attend two half-day workshops, to keep a reflective journal during their implementation of the unit, and to provide data on students' progress, interest, and understanding of the content as it is presented in the experimental materials. In this way, the questions that initiated the research should be answerable.

Acknowledgements

This publication and the research described herein are supported by the University of Manitoba Centre for Research in Youth, Science Teaching and Learning (CRYSTAL) financed by the Natural Science and Engineering Research Council of Canada (NSERC).

The author wishes to thank the twenty-three children in Room 20, the classroom teacher, and the teacher candidate who implemented the science lessons.

Appendices

APPENDIX A

Knowledge Outcomes for Grade 4, Light

Students will...

01 Use appropriate vocabulary related to their investigations of light.
Include: energy, reflect, absorb, transmit, artificial, light beam, transparent, translucent, opaque, technological development, science, brightness.

02 Give examples of various forms of energy.
Include: light, heat, food, sound.

03 Recognize that energy is an integral part of daily life.

04 Demonstrate that white light can be separated into colours.

05 Distinguish between objects that produce their own light and those that reflect light from another source.

06 Identify a variety of natural and artificial light sources.

07 Observe and describe properties of light.
Include: travels in a straight line, ends as it passes from one medium to another, can be reflected, can be different colours.

08 Explore to determine effects different materials have on a light beam.

09 Recognize that most objects that produce light also give off ehat, and identify objects that produce light but give off little or no heat.

10 Classify materials as transparent, translucent, or opaque.

11 Evaluate the usefulness of a material for a particular task based on its ability to transmit, reflect, or absorb light.

12 Predict location, shape, and size of a shadow based on the position of a light source relative to an object.

13 Identify technological developments that extend our ability to see, and recognize their impact on science.

14 Use the design process to construct a device that transmits and reflects light.

15 Describe practices that help ensure protection of eyes and sight.

16 Identify different uses of light at home, at school, and in the community, and explain how the brightness and colour of the light are appropriate for each use.

Manitoba Education and Training: 1999, *Kindergarten to Grade 4 Science: Manitoba Curriculum Framework of Outcomes*, Winnipeg, pp. 3.50-3.51.

APPENDIX B

Grade 4 Timetable (2005-2006)

<table>
<tr><th>TIME</th><th>DAY 1</th><th>DAY 2</th><th>DAY 3</th><th>DAY 4</th><th>DAY 5</th><th>DAY 6</th></tr>
<tr><td>08:50</td><td colspan="6">REGISTRATION</td></tr>
<tr><td>09:00 – 09:30</td><td>Music</td><td>Physical Education</td><td>Physical Education</td><td>Language Arts</td><td>Physical Education</td><td>Language Arts</td></tr>
<tr><td>09:30 – 10:00</td><td rowspan="2">Math</td><td rowspan="2">Math</td><td rowspan="2">Math</td><td rowspan="2">Physical Education</td><td>Music</td><td>Language Arts</td></tr>
<tr><td>10:00 – 10:30</td><td>Math</td><td>Math</td></tr>
<tr><td>10:30 – 10:45</td><td colspan="6">RECESS</td></tr>
<tr><td>10:45 – 11:15</td><td colspan="5">READING CONFERENCES</td><td>Science</td></tr>
<tr><td>11:15 – 11:45</td><td>Social Studies</td><td>Science</td><td>Social Studies</td><td>Math Games</td><td>Science</td><td>Social Studies</td></tr>
<tr><td>11:45 – 12:40</td><td colspan="6">LUNCH</td></tr>
<tr><td>12-45</td><td colspan="6">REGISTRATION</td></tr>
<tr><td>12:45 – 12:55</td><td>Read Aloud</td><td rowspan="2">Music Portfolio</td><td rowspan="2">Read Aloud</td><td>Read Aloud</td><td>Read Aloud</td><td>Read Aloud</td></tr>
<tr><td>12:55 – 13:15</td><td>Language Arts</td><td>Art</td><td rowspan="3">Art</td><td>Art</td></tr>
<tr><td>13:15 – 13:45</td><td>Choir</td><td rowspan="2">Language Arts</td><td rowspan="2">Language Arts</td><td rowspan="2">Language Arts</td><td rowspan="2">Writer's Workshop</td></tr>
<tr><td>13:45 – 14:15</td><td>Health</td></tr>
<tr><td>14:15 – 14:30</td><td colspan="6">RECESS</td></tr>
<tr><td>14:30 – 14:50</td><td>Agenda</td><td rowspan="2">French</td><td rowspan="2">Music</td><td rowspan="2">French</td><td rowspan="2">French</td><td rowspan="2">Library</td></tr>
<tr><td>14:50 – 15:10</td><td rowspan="2">Computer Lab</td></tr>
<tr><td>15:10 – 15:30</td><td>Agenda</td><td>Agenda</td><td>Agenda</td><td>Agenda</td><td>Agenda</td></tr>
<tr><td>15:30</td><td colspan="6">DISMISSAL</td></tr>
</table>

APPENDIX C

Pre-Instruction Worksheet

Name ___________________

THINKING ABOUT DARKENSS

I think we can see in the dark.

I think we can't see in the dark.

How I know. I think this because ______________________________________

____

____

____

____

How can we find out? ____

____

____

____

____

Post-Instruction Worksheet

Name ___________________

THINKING ABOUT DARKNESS

Can we see in the dark? _______

Does this mean we need light in order to see? _______

How I know this. I think this because ___________________________________

____

____

____

____

____

APPENDIX D

Archimedes' Death Ray

Do you remember hearing the name Isaac Newton when we visited the Planetarium? He was the first scientist to show that white light is made up of all the rainbow's colours, and like Galileo, he was famous for the telescope he invented. Both Newton and Galileo wrote about another amazing scientist, a Greek, who lived in Sicily hundreds of years before they were born. His name was Archimedes. Archimedes was a mathematician, an engineer, and a scientist. He invented a way to pump water for irrigating fields, explained why some things float while other things sink, created pulleys and levers to launch giant ships, constructed catapults, and studied spheres and cones and circles.

Archimedes was one of the world's first great scientist, and one of the subjects he studied was light. He was particularly interested in reflective surfaces and studying what happened to light shining on different kinds of concave and convex mirrors. He used his knowledge of light to set enemy ships on fire. One story that has been told about Archimedes goes like this:

During a war between the Greeks and the Romans, Archimedes had soldiers line up on the shores of the Sea holding small mirrors made from highly polished metal. He then had them hold their mirrors to reflect the Sun's light on one spot, and this one spot was on the side of a Roman ship. The heat created by the light reflected by the mirrors was so strong that it set the ship on fire. Scientists think the idea is possible. Historians don't believe it really happened. What do you think? Is it possible for each of us to hold a mirror so that the light only shines on one spot? How could we find out?

Notes

[1] Using SpringerLink, Taylor & Francis Online Journals, and WileyInterScience, published papers with clear science curriculum, teaching and learning orientations were retrieved from *International Journal of Science Education, Journal of Research in Science Teaching, Science and Education, Science Education,* and *Research in Science Education* using the keyword search, 'history of science', in 'title/abstract', 'fulltext', or 'all fields'. Copies of unpublished conference papers or, in the absence of papers, notes taken during these presentations, and all other cited articles were in the author's possession.

References

Allchin, D.: 2004, 'Psuedohistory and Pseudoscience', *Science & Education* **13**(3), 179-195.

Allchin, D.: 2003, 'Scientific Myth-conceptions', *Science Education* **87**, 329-351.

Allchin, D.: 2002, August, 'Tapping into the Pulse of History of Science with Case Studies', *ActionBioscience* [online]. Retrieved May 10, 2006, from www.actionbioscience.org/education/allchin.html.

Allchin, D.: 2001, 'Kettlewell's Missing Evidence, A Study in Black and White', *Journal of College Science Teaching*, **31**(4), 240-245.

Allchin, D.: 1997, 'Rekindling Phlogiston: From Classroom Case Study to Interdisciplinary Relationships', *Science & Education* **6**(5), 473-509.

American Association for the Advancement of Science: 1993, *Benchmarks for Scientific Literacy*, Oxford University Press, New York.

Asoko, H.: 2002, 'Developing Conceptual Understanding in Primary Science,' *Cambridge Journal of Education* **32**(2), 153-164.

Barth, M.: 2000, 'Electromagnetic Induction Rediscovered Using Original Texts', *Science & Education* **9**(4), 375-387.

Becker, B.J.: 2000, 'Mind Works: Making Scientific Concepts Come Alive', *Science & Education* **9**(3), 269-278.

Begoray, D. & Stinner, A.: 2005, 'Representing Science through Historical Drama: Lord Kelvin and the Age of The Earth Debate', *Science & Education* **14**(3-5), 457-471.

Bellocchi, A.: 2004, 'Designing and Using Historical Vignettes in Science Teaching: A Personal Account', *Teaching Science* **50**(2), 14-17.

Bentley, M.L., 2000, 'Improvisational Drama and The Nature of Science', *Journal of Science Teacher Education* **11**(1), 63-75.

Binnie, A.: 2001, 'Using the History of Electricity and Magnetism to Enhance Teaching', *Science & Education* **10**(4), 379-389.

Carson, R.N.: 2004, 'Teaching Cultural History from Primary Events', *Science & Education* **13**(7), 797-809.

Cartwright, J.: 2005, 'The Discovery of Oxygen: A Resource-based Case Study Approach Using Primary Documents and Role Play', paper presented at the Eighth International History, Philosophy, and Science Teaching Conference, University of Leeds, England, 15-18 July.

Cavicchi, E.M.: 2006, 'Historical Experiments in Students' Hands: Unfragmenting Science through Action and History', *Science & Education*, published online 20 June 2006.

Chan, K.-S.: 2000, *The Impact of Infusing the Interactive Historical Vignettes into 10th Grade High School Science Instruction in Taiwan on Student Understanding of the Nature of Science and Science Achievement*, Unpublished doctoral dissertation, The University of Texas at Austin.

Chin, C. & Brown, D.E.: 2000, 'Learning in Science: A Comparison of Deep and Surface Approaches', *Journal of Research in Science Teaching* **37** (2), 109-138.

Clough, M.P. & Olson, J.K.: 2004, 'The Nature of Science: Always Part of the Science Story', *The Science Teacher* **71**(9), 28-31.

Cotignola, M.I., Bordogna, C., Punte, G. &Cappannini, O.M.: 2002, 'Difficulties in Learning Thermodynamic Concepts: Are They Linked to the Historical Development of this Field?', *Science & Education* **11**(3), 279-291.

Dawes, L.: 2004, 'Talk and Learning in Classroom Science', *International journal of Science Education,* **26**(6), 677-695.

de Berg, K.: 2006a, 'What Happens when Salt Dissolves in Water? An Introduction to Scientific Argument and Counter Argument Drawn from the History of Science', *Teaching Science* **52**(1), 24-27.

de Berg, K.: 2006b, 'The Problem of Determining the Content for the Construction of an Historical Case Study which is Sensitive to Nature of Science Issues and Teaching-Learning Issues', *Science & Education,* published online 13 July 2006.

Degenais, A. 2003, 'Teaching the history of Physics without Lectures', paper presented at the Seventh International History, Philosophy, and Science Teaching Conference, Winnipeg, Manitoba, 30 July – 03 August.

Discovery Channel Canada: October 20, 2005, 'Ancient Weapons of Mass Destruction', Daily Planet Video Archive, www.exn.ca/dailyplanet/view.asp?date=10/20/2005.

Duckworth, E.: 'Critical Exploration in the Classroom', *The New Educator* **1**, 257-272.

Faculty of Education, *University of Manitoba: 2005, Early Years School Experiences Handbook 2005-2006,* Faculty of Education, Winnipeg, MB.

Ford, M.J.: 2003, 'Representing and Meaning in History and in Classrooms: Developing Symbols and Conceptual Organizations of Free-fall Motion', *Science & Education* **12**(1-2), 1-25.

Fowler, P.: 2002, 'Why Teach the History of Science in Science? (Part 2)', *Education Forum: The Newsletter for the Education Section of BSHS* **38**, 1-5.

Harding, P. & Hare, W.: 2000, 'Portraying Science Accurately in Classrooms: Emphasizing Open-mindedness Rather than Relativism', *Journal of Research in Science Teaching* **37**(3), 225-236.

Galili, I.: 2001, 'Weight versus Gravitational Force: Historical and Educational Perspectives', *International Journal of Science Education* **23**(10), 1073-1093.

Galili, I. & Tseitlin, M.: 2003, 'Newton's First Law in Light of a New Study', paper presented at the Seventh International History, Philosophy, and Science Teaching Conference, Winnipeg, Manitoba, 30 July – 03 August.

Galili, I. & Hazan, A.: 2001, '"Experts" Views on Using History and Philosophy of Science in the Practice of Physics Instruction', *Science & Education* **10**(4), 345-367.

Galili, I. & Hazan, A.: 2000, 'The Influence of an Historically Oriented Course on Students' Content Knowledge in Optics Evaluated by Means of Facets-schemes Analysis', *American Journal of Physics* **68** (7), S3-S15.

Gauld, C.: 2004, 'The Treatment of the Motion of a Simple Pendulum in some Early 18th Century Newtonian Textbooks', *Science & Education* **13**(4-5), 321-332.

Gilbert, J.K. & Reiner, M.: 2000, 'Thought Experiments in Science Education: Potential and Current Realization, *International Journal of Science Education* **22**(3), 265-283.

Gorman, M. & Robinson, J.K.: 1998, 'Using History to Teach Invention and Design: The Case of the Telephone', *Science & Education* **7**(2), 173-201.

Hakim, J.: 2002, 'The Story of the Atom', *American Educator* Spring, 12-25.

Harlen, W., Marco, C., Reed, K., & Schilling, M.: 2003, *Making Progress in Primary Science*, RoutledgeFalmer, New York.

Heering, P.: 2000, 'Getting Shocks: Teaching Secondary School Physics through Hsitory', *Science & Education* **9**(4), 363-373.

Hodson, D.: 1999, 'Building a Case for a Sociocultural and Inquiry-oriented View of Science Education', *Journal of Science Education and Technology* **8**(3), 241-249.

Höttecke, D.: 2000, 'How and What Can We Learn from Replicating Historical Experiments? A Case Study', *Science & Education*, **9**(4), 343-362.

Howe, E.M. & Rudge, D.E.: 2005, 'Recapitulating the History of Sickle-Cell Anemia Research', *Science & Education* **14**(3-5), 423-441.

Justi, R. & Gilbert, J.K.: 2001, 'History and Philosophy of Science through Models: Some Challenges in the Case of "The Atom"', *International Journal of Science Education* **22**(9), 939-1009.

Justi, R. & Gilbert, J.K.: 1999, 'History and Philosophy of Science through Models: The Case of Chemical Kinetics', *Science & Education* **8**(3), 287-307.

Iwin, A.R.: 'Historical Case Studies: Teaching the Nature of Science in Context', *Science Education* **84**, 5-26.

Kafai, Y.B. & Gilliland-Swetland, A.J.: 2001, 'The Use of Historical Materials in Elementary Science Classrooms', *Science Education* **85**, 349-367.

Kalman, C.S. & Aulls, M.W.: 2003, 'Can an Analysis of the Contrast between Pre-Galilean and Newtonian Theoretical Frameworks Help Students Develop a Scientific Mindset', *Science & Education* **12**(8), 761-772.

Karplus, R.: 1964, 'The Science Curriculum Improvement Study', *Journal of Research in Science Teaching* **2**(4), 293-303.

Karplus, R. & Their, H.: 1967, *A New Look at Elementary School Science: Science Curriculum Improvement Study*, Rand McNally & Company, Chicago, IL.

Kettle, K.: 2005, 'Science with a Human Touch: Historical Vignettes in the Teaching and Learning of Science', in S. Alsop, L. Bencze & E. Pedretti (eds.), *Analysing Exemplary Science Teaching: Theoretical Lenses and a Spectrum of Possibilities for Practice*, Open University Press, Maidenhead, England, 38-45.

Kettle, K.: 2001, 'Drama in the Science Room: Bringing Eminent Scientists to Life', *OISE Papers in STSE Education* **2**, 91-108.

Kipnis, N.: 2001, 'Scientific Controversies in Teaching Science: The Case of Volta', *Science & Education* **10**(1-2), 33-49.

Kipnis, N.: 1996, 'The 'Historical Investigative' Approach to Teaching Science', *Science & Education* **5**(3), 277-292.

Klassen, S.: 2006, 'The Application of Historical Narrative in Science Learning: The Atlantic Cable Story', *Science & Education,* published online 28 June 2006.

Klassen, S.: 2004, 'Concepts of the Electron', paper presented at the Fifth International Conference for History of Science in Science Education, Keszthely, Hungary, 12-16 July.

Kubli, F.: 1999, 'Historical Aspects in Physics Teaching: Using Galileo's Work in a New Swiss Project', *Science & Education* **8**(2), 137-150.

Lin, H., Hung, J., & Hung S.: 2002, ' Using the History of Science to Promote Students' Problem-solving Ability', *International Journal of Science Education* **24**(5), 453-464.

Lin, H. & Chin, C.: 2002, 'Promoting Preservice Chemistry Teachers' Understanding about the Nature of Science through History', *Journal of Research in Science Teaching* **39**(9), 773-792.

Lonsbury, J.G. & Ellis, J.D.: 2002, 'Science History as a Means to Teach Nature of Science Concepts: Using the Development of Understanding Related to Mechanisms of Inheritance', *Electronic Journal of Science Education* **7**(2). Retrieved from unr.edu/homepage/crowther/ejse/ejsev7n2.html.

Machamer, P. & Hepburn, B.: 2004, 'Galileo and the pendulum: Latching on to Time', *Science & Education* **13**(4-5), 333-347.

Manitoba Education and Training: 1999, *Kindergarten to Grade 4 Science: Manitoba Framework of Outcomes*, Manitoba Education and Training, Winnipeg.

Manitoba Education, Citizenship and Youth: n.d. a, 'Kindergarten to Senior 4: Going to School in Manitoba, School Year'. Retrieved April 5, 2006 from http://www.edu.gov.mb. ca/ k12/ schools/ gts.html.

Manitoba Education, Citizenship and Youth: n.d. b, ' Curriculum Development and Implementation: English Program, Early and Middle Years, Subject Area Time Allotments'. Retrieved April 5, 2006 from http://www.edu.gov.mb .ca/k12/cur /english _pr.html

Matthews, M.R.: 1994, *Science Teaching: The Role of History and Philosophy of Science*, Routledge, New York.

Matthews, M.R., Gauld, C. & Stinner, A.: 2003, 'The Pendulum: Its Place in Science, Culture and Pedagogy', *Science & Education* **13**(4-5), 261-277.

McComas, W.F.: 2005, ' Strategies for Teaching the Nature of Science: What Illustrations and Examples Exist in Popular Books on the Subject', paper presented at the Eighth

International History, Philosophy, and Science Teaching Conference, University of Leeds, England, 15-18 July.

McGinnis, J.R. & Oliver, J.S.: 1998, 'Teaching about Sound: A Select Historical Examination of Research', *Science & Education* 7(4), 381-401.

McKinney, D. & Michalovic, M.: 2004, 'Teaching the Stories of Scientists and Their Discoveries', *The Science Teacher*, **71**(9), 46-51.

McMillan, B.A.: 2004, 'Logbooks and Journals: Using Historical Materials in School Science', paper presented at the Fifth International Conference for History of Science in Science Education, Keszthely, Hungary, 12-16 July.

McMillan, B.A.: 2005, 'Stories of a Nineteenth-century Arctic Whaler and School Science Education', *Canadian Journal of Science, Mathematics and Technology Education* **5**(4), 503-516.

Metropolitan Toronto School Board: 1996, *All Aboard! Cross Curricular Design and Technology Strategies and Activities*, Trifolium Books Incorporated, Toronto.

Metz, D.: 2006, 'William Wales and the 1769 Transit of Venus: Puzzle Solving and the Determination of the Astronomical Unit', paper presented at the Sixth International Conference for the History of Science in Science Education, Carl-von-Ossietzky University, Olderburg, Germany, 10-14 July.

Metz, D.: 2004, 'Using Historical Narratives to Guide Science Experiments', paper presented at the Fifth International Conference for History of Science in Science Education, Keszthely, Hungary, 12-16 July.

Metz, D., Klassen,S., McMillan, B., Clough, M., & Olson, J.: 2006, 'Building a Foundation for the Use of Historical Narratives', *Science & Education*, published online 26 April 2006.

Metz, D. & Smerchanski, G.: 2003, 'Taking a Byte Out of a Historical Experiment: Ole Roemer and the Determination of the Speed of Light', paper presented at the Seventh International History, Philosophy, and Science Teaching Conference, Winnipeg, Manitoba, 30 July – 03 August.

Metz, K.E.: 1997, 'On the Complex Relation Between Cognitive Developmental Research and Children Science Curricula', *Review of Educational Research* **67**(1), 151-163.

Metz, K.E.: 1995, 'Reassessment of Developmental Constraints on Children's Science Instruction', *Review of Educational Research* **65**(2), 93-127.

Mihas, P. & Andreadid, P.: 2005, 'A Historical Approach to the Teaching of the Linear Propagation of Light, Shadows and Pinhole Cameras', *Science & Education* **14**(7/8), 675-697.

Moen, P. & Erickson, M.A.: 1995, 'Linked Lives: A Transgenerational Approach to Resilience', in P. Moen, G.H. Elder & K. Luscher (eds.), *Examining Lives in Context: Perspectives on the Ecology of Human Development*, American Psychological Association, Washington, D.C., 169-207.

Mulholland, J. & Wallace, J.: 2003, 'Crossing Borders: Learning and Teaching Primary Science in the Pre-service to In-service Transition', *International Journal of Science Education* **25**(7), 879-898.

National Research Council: 1996, *National Science Education Standards*, National Academy Press, Washington, D.C.

Niaz, M.: 2001, 'Understanding Nature of Science as Progressive Transitions in Heuristic Principles', *Science Education* **85**(6), 684-690.

Nussbaum, J.: 1998, ' History and Philosophy of Science and the Preparation for Constructivist Teaching', in J. Mintzes, J., Wandersee, and J. Novak (eds.), *Teaching Science for Understanding: A Human Constructivist View*, Academic Press, New York, 165-194.

Olson, J.K., Clough, M.P., Bruxvoort, C.N. & Vanderlinden, D.W.: 2005, 'Improving Students' NOS Understanding through Historical Short Stories in College Level Introductory Geology and Biology Courses', paper presented at the Eighth International History, Philosophy, and Science Teaching Conference, University of Leeds, England, 15-18 July.

Osewold, D.: 2006, 'Students' Conceptions and Historical Ideas about (Mechanical) Waves', paper presented at the Sixth International Conference for the History of Science in Science Education, Carl-von-Ossietzky University, Olderburg, Germany, 10-14 July.

Pohl, W.G.: 2006, Teaching the Atomic Structure of Matter: A History of Ideas, Experiments and Pictures', paper presented at the Sixth International Conference for the History of Science in Science Education, Carl-von-Ossietzky University, Olderburg, Germany, 10-14 July.

Raftopoulos, A., Kalyfommatou, N. & Constantinou, C.P.: 2006, 'The Properties and the Nature of Light: The Study of Newton's Work and the Teaching of Optics', *Science & Education* **14**(7-7), 649-673.

Rice, D.C. & Roychoudhury, A.: 2003, ' Preparing More Confident Preservice Elementary Science Teachers: One Elementary Science Methods Teacher's Self Study', *Journal of Science Teacher Education* **14**(2), 97-126.

Rowell, P.M. & Ebbers, M.: 2004, 'School Science Constrained: Print Experiences in Two Elementary Classrooms', *Teaching and Teacher Education* **20**, 217-230.

Rudge, D.W. & Howe, E.M.: 2004, 'Incorporating History into the Science Classroom', *The Science Teacher* **71**(9), 52-57.

Rutherford, F.J.: 2001, 'Fostering the History of Science in American Science Education', *Science & Education* **10**(6), 569-580.

Scaife, J.: 2000, 'Learning in Science', in J. Wellington (ed.), *Teaching and Learning Secondary Science: Contemporary issues and Practical Approaches*, Routledge, New York, 61-108.

Seatter, C.S.: 2003, 'Constructivist Science Teaching: Intellectual and Strategic Teaching Acts', *Interchange* **34**(1), 63-87.

Seker, H. & Welsh, L.C.: 2006, 'The Use of History of Mechanics in Teaching Motions and Force Units', *Science & Education* **15**(1), 55-89.

Seroglou, F., Koumaras, P., & Tselfes, V.: 1998, 'History of Science and Instructional Design: The Case of Electromagnetism', *Science and Education* **7**(3), 261-280.

Sneider, C.I. & Ohadi, M.M.: 1998, 'Unraveling Students' Misconceptions about the Earth's Shape and Gravity', *Science Education* **82**, 265-284.

Solbes, J. & Traver, M.: 2003, 'Against a Negative Image of Science: History of Science and the Teaching of Physics and Chemistry', *Science & Education* **12**(7), 703-717.

Solomon, J., Duveen, J., Scot, L. & McCarthy.S.: 1992, 'Teaching About the Nature of Science through History: Action Research in the Classroom', *Journal of Research in Science Teaching* **29**(4), 409-421.

Stewart, J. & Rudolph, J.L.: 2001, 'Considering the Nature of Scientific Problems when Designing Science Curricula', *Science Education* **85**, 207-222.

Stinner, A.: 1994, 'The Story of Force: From Aristotle to Einstein', *Physics Education* **29**(2), 77-85.

Stinner, A. & Teichmann, J.: 2003, 'Lord Kelvin and the Age-of-the-Earth Debate: A Dramatization', *Science & Education* **12**(2), 213-228.

Taber, K.S.: 2005, 'Conceptual Development', in S. Alsop, L. Bencze &E. Pedretti (eds.), *Analysing Exemplary Science Teaching: Theoretical Lenses and a Spectrum of Possibilities for Practice*, Open University Press, Maidenhead, England, 127-135.

Tao, P.K.: 2003, 'Eliciting and Developing Junior Secondary Science Students' Understanding of the Nature of Science through a Peer Collaboration Instruction in Science Stories', *International Journal of Science Education* **25**(2), 147-171.

Teichmann, J.: 1999, 'Studying Galileo at Secondary School: A Reconstruction of His "Jumping-hill" Experiment and the Process of Discovery', *Science & Education* **8**(2), 121-136.

Trowbridge, L.W., Bybee, R.W. & Powell, J.C.: 2000, *Teaching Secondary School Science: Strategies for Developing Scientific Literacy*, Merrill-Prentice Hall, Upper Saddle River, N.J., 243-249.

van Tiggelen, B.: 2006, 'Teaching Chemistry through History: Lessons in Polymers and Halogens', paper presented at the Sixth International Conference for the History of Science in Science Education, Carl-von-Ossietzky University, Olderburg, Germany, 10-14 July.

Van Driel, J.H., De Vos, W. & Verloop, N.: 1998, 'Relating Students' Reasoning to the History of Science: The Case of Chemical Equilibrium', *Research in Science Education* **28**(2), 187-198.

Venville, G. and Milne, C.: 1999, 'Three Women Scientists and Their Role in the History of Genetics', *Australian Science Teachers Journal* **45**(3), 9-15.

Wandersee, J.H. & Roach, L.M.: 1998, 'Interactive Historical Vignettes', in J. Mintzes, J. Wandersee and J. Novak (eds.), *Teaching Science for Understanding: A Human Constructivist View*, Academic Press, New York, 1998, 281-306.

Wiggins, G. & McTighe, J.: 2000, *Understanding by Design*, Prentice Hall, Upper Saddle River, N.J.

Winchester, I.: 1989, 'Editorial – History, Science, and Science Teaching', *Interchange* **20**(2), i-vi.

Yip, D.: 2006, 'Using History to Promote Understanding of Nature of Science in Science Teachers', *Teaching Education* **17**(2), 157-166.

William Wales and the 1769 Transit of Venus: Puzzle Solving and the Determination of the Astronomical Unit

DON METZ
University of Winnipeg, Winnipeg, Manitoba, Canada

Abstract: According to Thomas Kuhn, a significant part of "normal science" is the fact gathering, empirical work which is intended to illustrate an existing paradigm. Some of this effort focuses on the determination of physical constants such as the astronomical unit (AU). For Kuhn, normal science is also what prepares students for membership in a particular scientific community and is embodied in some form in our science textbooks. However, neither Kuhn nor the textbook says much about the individuals who practice normal science, especially those who had been relegated to the "hack" duties of long and arduous measurement and calculation.

In this paper, to provide a context for students of astronomy, I will outline the story of the determination of the AU and in particular the contribution of William Wales, an obscure British astronomer. Wales, toiling in the shadow of Halley (of Halley's comet fame), Mason and Dixon (of Mason and Dixon line fame) and the infamous Captain Cook endured a brutal winter in northern Canada for a brief glimpse of the 1769 transit of Venus. In the end, Wales supplied one small piece of the puzzle in the determination of the astronomical unit and he exemplified the human spirit and persistence of a Kuhnian "puzzle solver".

Introduction

In his seminal work, "The Structure of Scientific Revolutions", Thomas Kuhn (1962) formulated a vision of science as having a world view, or "paradigm" guiding scientific activity during a period of "normal science". Kuhn's definition of paradigm "took on a life of their own" (Kuhn 1977, p. xix) and became central to the controversy over scientific progress that fuelled philosophical discussions for many years to come. This debate is not of concern to this paper, however, as Kuhn used the Copernican Revolution to describe a crisis and paradigm shift, I also turn to astronomy to describe the period of "normal science" in the context of Kuhn's description of puzzle solving.

For Kuhn ""“normal science" means research firmly based upon one or more past scientific achievements" (Kuhn 1962, p. 10). This is a period of "mopping up" operations where "the paradigm forces scientists to investigate some part of nature in a detail and depth" (Kuhn 1962, p. 24). According to Kuhn, one of the main occupations of those practicing 'normal science' is that of puzzle solving. A puzzle is a problem, somewhat anomalous, that we believe can be solved with the appropriate innovative, conceptual, instrumental, and mathematical insight. There seems to be no shortage of puzzles in any paradigm and the most classical puzzles have their origins solidly rooted in history. For example, since the dawn of time when humankind first glanced to the starry heavens we've asked ourselves "how big and how far"? From the curiosity of this childlike pondering, the size and expanse of our universe became one of our major puzzles to inhabit the minds of astronomers for thousands of years to come. In the simplest sense, in order to measure the universe, a standard of measurement was needed. That standard became the astronomical unit (AU) or the distance from the earth to the sun, and operating under the paradigm of Newtonian astronomy, the AU became nothing more than a "fact gathering" mission.

Early Attempts to determine the AU

Using the geometry of an eclipse's shadow cone the ancients were able to arrive at rough estimations of the size of the Earth, sun, and moon. Aristarchus devised a simple trigonometric method for determining the distance to the sun. The distance to the sun can be measured by forming a right triangle with the Earth, Moon, and Sun. To find the AU one needs to form a right angle EMS (figure 1), measure the angle MES and know the distance to the moon (EM). When the moon forms a right angle with the sun and the earth we should observe the moon exactly half illuminated and half dark. Judging when the moon is bisected into bright and dark halves is not an exact science and Aristarchus estimated the angle MES to be 87° and that the sun was between 18 and 20 times more distant than the moon. Although he was largely in error (the ratio is closer to 400) his calculations indicated that the sun is much bigger than the Earth.

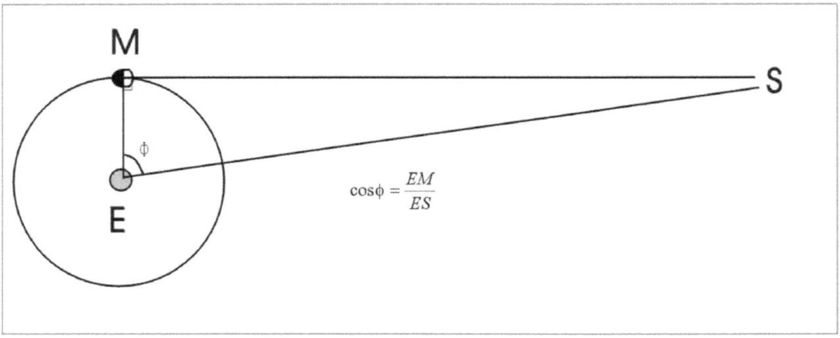

Figure 1. Aristarchus' lunar method

Another leading candidate for determining large distances was parallax. Parallax is the apparent shift of an object against its background as seen by observers in different locations. For example, the observer O in figure 2 sees the star against the background at position B while the observer O' sees the star a position A. The difference between A and B is called the parallax and is usually measured as an angular distance. The star and from the star to the background, by similar triangles we can calculate the distance AB.

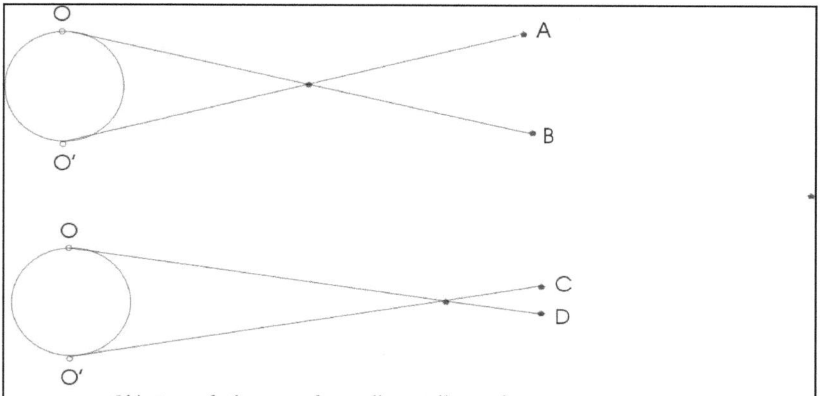

Objects are farther away for smaller parallax angles.

Figure 2. Parallax

We can also see that an object which is farther away will have a smaller angle of parallax. Additionally, if we know the distance OO' and the ratio of the distance from the observer to using methods based on trigonometry and parallax many astronomers estimated the size of the AU mostly in terms of the earth's radius. The acceptance of the Copernican theory lead quickly to the conclusion that the inferior planets will periodically cross the face of the sun (transit). It soon became apparent to astronomers that the transit of a planet could be useful for making astronomical determinations.

Planetary Transits

Kepler, working from his Rudolphine tables, was able to predict in 1627 a transit of Mercury and a month later, the transit of Venus in 1631 and another transit of Venus in 1761[1]. At one time, Kepler reported that he had seen Mercury as a black dot upon the sun (Procter, 1882). However, we know that Mercury is too small to be seen with the naked eye and Kepler himself later admitted that the likely event was sunspots. The first observation of a transit of an inferior planet was made by Gassendi in 1631 when he observed a projection of Mercury on the sun through a small aperture in his window. Gassendi, writing to a Professor Shickhard at the University of Tübingen could claim for the first time in history that "I found him out, and saw him where no one else had hitherto seen him" (c.f. Proctor, p. 8). The transit of Venus in the following month was not seen by Gassendi, or anyone else in Europe, as it occoured during the evening.

Kepler had suggested that in 1639 Venus would pass below the sun not quite close enough for a transit to occour. However, Jeremiah Horrocks (Horrox), a young, unknown astronomer in England, using Lansberg's tables realized that Venus should indeed pass across the face of the sun. Born near Lancashire, England, Horrocks was able to attend the University of Cambridge as a sizar. That is, in return for his tuition, he was required to perform duties for fellow students. Later, as a young amateur astronomer toiling in his "obscure village"of Hoole, Horrocks projected the image of the sun on a small piece of paper in his apartment.

> "I described on a sheet of paper a circle whose diameter was nearly equal to six inches, the narrowness of the apartment not permitting me conveniently to use a larger size ... although it represents the sun's image less, yet it depicts it more clearly and steadily, … the sun's image exactly filled the circle" (Horrocks, p. 45)

Figure 3. Venus on the Sun's Disk (Horrocks)

Horrocks observed the ingress and the motion of Venus across a small part of the sun (figure 3) for about a half hour before the sun set. He recorded his latitude and longitude relative to the Canary Islands and Brahe's Uranibourg observatory. Horrocks had also written to his friend William Crabtree who was able to observe briefly the transit but failed to record any measurements. He also told his brother who, because of weather, was unable to see anything. Horrocks made light of why his friend William Crabtree failed to take measurements during his observations.

> "Rapt in contemplation, he stood for some time motionless, scarcely trusting his own senses, through excess of joy; for we astronomers have as it were a womanish disposition, and are overjoyed with trifles and such small matters as scarcely make an impression upon others" (Horrocks, p. 129).

Horrocks recognized the significance of his discovery and wrote

> "Contemplate, I repeat, this most extraordinary phenomenon, never in our time to be seen again ! the planet venus drawn from her seclusion, modestly delineating on the sun, without disguise, her real magnitude, whilst her disc, at other times SO lovely, is here obscured in melancholy gloom ; in short, constrained to reveal to us those important- truths, which Mercury, on a former occasion, confided to thee" (Horrocks, p. 116).

Horrocks, twenty years of age when he made his most famous observation, also wrote on orbits and the forces governing the solar system. Newton recognized the contributions of Horrocks by citing him in his *Principia.* Destined to greatness, the young Horrocks sadly passed away at the age of twenty-two. In 1874, more than two hundred years after his death, a memorial was placed in a most revered location in Westminster Abbey, directly opposite the great Newton.

Halley's Proposal

Horrocks and later, James Gregory in his *Optica Promota,* suggested that the transit of Venus could be used to investigate the sun's parallax. However, it was left to Edmund Halley to propose the details of the method. Halley was the first person to observe a complete transit of a planet across the face of the sun. He had observed a transit of Mercury from St. Helena in 1677 and noted that the duration of the transit could be measured quite accurately.

Halley began by arguing qualitatively that the distance from the sun to the Earth is much larger than previously computed by Kepler, Riccioli, and Hevelius among others. He first notes that the diameters of Venus and Mercury as observed during a transit are much less than they were thought to be. Arguing that the Earth should subtend an angle between the superior planets Jupiter and Saturn and the inferior Venus and Mercury he concludes that the distance from the Earth to sun is 14 000 earth radii. He further argues that if the sun's parallax is 15 seconds then the diameter of the moon would be larger than the diameter of Mercury and claims that yields " a secondary planet larger than a primary one, which seems repugnant to the regular proportion and symmetry of the mundane system" (Halley, 1716, p. 454). Thus he arrives at a value of 12.5" of parallax such that the moon is smaller than Mercury and the Earth larger than Venus.

Halley's quantitative method is relatively simple to explain but increasingly more complex to compute. Referring to figure 4, let me begin by assuming everything is stationary, in a straight line, that vv' lie on the diameter of the sun, and we have Venus (V) directly between the Earth (E) and the Sun (S).

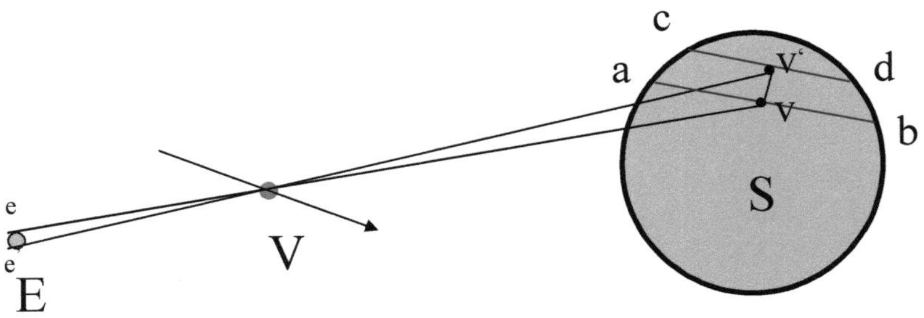

Figure 4: Halley's Method for computing the astronomical unit.

An observer at location *e* on the surface of the earth will observe the shadow of Venus at location v on the sun. A second observer a location e' will observe the shadow of Venus at v'. The lines of sight *e*v and $e'v'$ cross at V and form similar triangles. Although the distance between the Earth and the sun is unknown the proportions of the solar system are well known from Kepler's third law. Thus, we know that the distance *e*V is to the distance Vv in a proportion of about 7 to 18 and the distance vv' is in this same proportion compared to ee'. Knowing the distance ee' (as measured on earth) we can easily calculate the distance vv'. Further, a simple calculation tells us the part of the sun's diameter that vv' forms and thus we know the real diameter of the sun. From the diameter of the sun it is easy to calculate the distance to the earth using a "shadow cone".

Now of course the calculation is complicated by the fact that the Earth rotates, Venus and the Earth move, and the observers are not in simultaneous communication with each other. Halley realized that if the time that Venus took to transit the sun was known that the path distance could also be calculated. Knowing the two path distances ab and cd one could easily calculate the distance vv' and apply the previous method to find the AU. The effects of the rotation of the Earth serve only to lengthen the duration of the transit at one location and shorten it at another location (that is, by choosing locations appropriately). Halley's method then requires an accurate timing of the transit and corrections according to the revolutions of the heavens. Using his method Halley pointed out that an accuracy of one part in five hundred could be attained.

The main disadvantage to Halley's method[2] was that the whole transit had to be observed especially the moments of ingress and egress. The duration of a transit of Venus, can be eight hours, leaves relatively few places on earth where the transit can be suitably viewed. In the eighteenth century such locations were often remote and difficult to travel to and consequently required a special individual capable of making the appropriate measurements.

Halley knowing that he would not live to see the next transit of Venus, exalts future astronomers to apply themselves strenuously to the observation and measurements of the next transit of Venus.

> "I know not if those either of the present or following age shall ever see again; and upon which, the certain and adequate solution of the noblest and otherwise most difficult problem depends"

Halley's confidence that a solution is certain illustrates the puzzle solving nature of the determination of the AU. Who would be the ultimate puzzle solvers? According to Kuhn, no one devotes years of time and energy just to obtain the few facts necessary to achieve an answer. Puzzles test the ingenuity of the scientist and requires only those individuals who can attack the problem with enthusiasm, passion, and devotion. The construction and deployment of innovative apparatus and the persistence to overcome all obstacles would be necessary to bring theory and nature in closer agreement. It seemed that the only things in the way of knowing the AU was a few good men, some time, and money. The lure of the puzzle and the apparent rarity of the transit of Venus was enough to convince many governments to provide funds to mount one of the world's first international scientific expeditions. And, there seemed to be no shortage of "normal scientists" to take on the challenge.

The 1761 Transit of Venus

On June 6, 1761, Venus crossed the face of the sun for the first time in 122 years. Proctor (1882) reports that observers "176 in number ... at no less than 117 stations" observed the international event. Bell (2004, citing Woolf) reported that

> "120 observers were fielded to 62 stations, in expeditions sponsored by Britain, Denmark, France, Germany, Italy, Portugal, Russia, and Sweden. In addition to sites around Europe, astronomers sailed, trekked, or sledged as far afield as Calcutta, Capetown, Constantinople, Newfoundland, Peking (Beijing), and Siberia".

The observers faced huge challenges. Not only did they have to literally carry their observatories half way around the world but the geopolitical rivalries of the time compounded the vagaries of international travel in the 18th century. Many of the astronomers were English and French, combatants at the time in the Seven Year's War between England and France. Neville Maskelyne, Charles Mason, Jeremiah Dixon, and John Winthrop were the notable astronomers from Great Britain who ventured to see the transit while important French observers included Jean-Baptiste Chappe d'Auteroche and Jean-Baptiste Le Gentil.

Maskelyne journeyed by sea to the island of St. Helena in the mid-Atlantic, experienced poor weather and missed the end of the transit. However, Maskelyne would later become the Astronomer Royal from 1765 to 1811 and supervise the investigation of the 1769 transit. Charles Mason and Jeremiah Dixon initially sailed for Sumatra six months before the transit. They quickly retreated after a battle with the French navy where they suffered numerous casualties. The astronomers balked at continuing the mission but under pressure from the Royal Society found their way to Cape Town in South Africa where they experienced good conditions for their observations. The British government would later assign the astronomers to the task of determining the border between Pennsylvania and Maryland now known as the "Mason-Dixon Line".

The 1761 transit saw John Winthrop, a professor of mathematics and philosophy at Cambridge in New England commissioned by the General Assembly of Boston to view the transit of Venus at St. John’s Newfoundland - one of the few places in North America where one could observe the transit.

Many of the astronomers encountered extremed conditions and hardships. For example, the French astronomer Chappe d'Auteroche travelled arduously for five months to Tobolsk in Siberia in 1761 to a hostile welcome. His luck turned worse during the transit of 1769 when he ventured to the Spanish territory of Baja in California. An epidemic took its toll on the members of the expedition with only one person surviving to return the observational data to Paris.

Short (1763) calculated the parallax of the sun from 53 different pairs of locations and reported results that ranged from 5.76” to 10.57”. His mean of the 53 comparisons yielded a value of 8.61”. By rejecting results which were more than 0.5” from the mean (16 rejected) he arrived at a mean of 8.57”. Proctor (1882) reported that in general, the results of the observations from 1761 were less than satisfactory with a solar parallax deduced between 8.5”and 10.5”corresponding to a distance range of about 96 million to 78 million miles. Of course, individual astronomers preferred their measurements over any others.

Proctor points to two significant problems encountered in the transit of 1761, first most of the observations were of the single Delisle form and longitude measurements to the degree of accuracy required could not be made. Additionally, for observations of the ingress and egress it was found that the disc of Venus at that time was distorted by a ‘black drop’ effect.

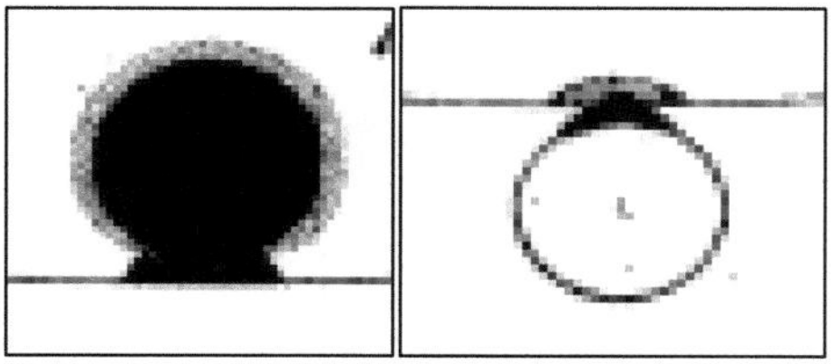

Figure 5: Black drop effect diagrammed by Charles Green
(Cook and Green, (1771)

Observers had also noticed that when Venus is at the point of ingress or egress that the disk of Venus appeared to be connected by a thread between the planet and the edge of the sun resembling at times a drop. The 'black drop' made it difficult to determine the exact moments of contact that were needed to measure the duration of the transit. Consequently, the precision of the transit timings varied considerably due to this effect introducing a significant error in the calculations.

The 1769 Transit of Venus

In 1765 Nevil Maskelyne was appointed the Royal Astronomer and immediately began to attend to the organization of the expeditions to observe the 1769 transit of Venus. Overall, Maor (2000) cites that no fewer than 151 astronomers were sent to 77 sites to observe the transit of June 3, 1769. Of particular interest to this paper are James Cook and William Wales.

In preparation, the Royal Society petitioned the government to finance several expeditions for this rare historical event. Guided by the experiences of the 1761 transit locations around the globe were selected for optimal viewing. We are reminded that in 18th century that navigation was still in its infancy and ocean crossings were hazardous with large parts of the world unexplored and poorly documented. Following a dispute with the Admiralty, the Royal Society accepted the captaincy of James Cook who was instructed to sail his ship the Endeavour to make the necessary astronomical observations and then to continue on to explore the South Seas.

Cook departed in August of 1768 accompanied on board by the botanist Joseph Banks and one of Maskelyne's assistants at Greenwich, Charles Green. Cook arrived seven months later at King Georges Island (Tahiti, Otaheite) in April of 1769 and set up at an observatory at a place they called – and still known as - Point Venus (Figure 6).

Tragically, Charles Green did not live to enjoy the fruits of his labour. He died in Batavia (Djakarta) on the return voyage home. Cook and Maskelyne compiled and edited his data for publication in the Philosophical Transactions.

Figure 6: Cook's observatory. cited from Griffin-Smith.

They were able to observe the transit and record the necessary data. However, Cook commented that "we very distinctly saw an Atmosphere or dusky shade round the body of the Planet which very much disturbed the times of the Contacts".

In May of 1768, a full year before the transit of 1769, William Wales received his instructions from Maskelyne and set sail from Greenwich. Wales left behind his young family and upon successful completion of his duty Wales was promised a pay of 200 pounds. He was accompanied by another assistant, Joseph Dymond, and they hitched a ride with the Hudson Bay Company to make their observations at Prince of Wales Fort near the Churchill River in Hudson's Bay (now the town of Churchill, Manitoba).

Wales kept meticulous records from the time he departed London to the end of his yearlong adventure. He recorded weather conditions daily, continually deduced latitude and longitude of his locations and notable places, observed astronomical phenomena such as lunar eclipses, and computed magnetic variations. Wales was meticulous in his observations and calculations maintaining a keen eye for error and recording variations of the timepieces, noting carefully if a watch was moving too fast or too slow. He was unashamedly honest wishing only to contribute to science that which was indisputable.

Wales acknowledged that, "our situation must be allowed to have been truly dangerous". (Wales, 1770, p. 105). He complained of difficulties with navigation, currents and tides, and poor measurements of longitude because of the motion of the boat. Wales recorded storms and described shrouds of fog and waves beating over the quarterdeck. He took particular interest in the "islands of ice" and detailed how the ship dodged and darted through the flows.

Arriving at Churchill in the month of August, Wales and Dymond found a rather bleak environment waiting for them sleeping on the floor for three weeks until the carpenters could build their beds. After a week unloading the observatory and their instruments, Wales and Dymond began building the observatory by laying the foundation and raising the walls. Griffin-Jones suggests that Wales' observatory was similar to Cook's (figure 6) and Broughton points out that it was a "prefabricated wooden observatory designed by the great British engineer, John Smeaton[3]. To secure the clock – a most critical instrument – they "fixed up a stiff plank of dry English oak to screw the clock to" (Wales, p. 119), and then placed and adjusted equipment including a barometer and two thermometers.

The astronomers daily routine was constantly busy and they worked around the clock the entire year making numerous observations every single day of their stay. They recorded temperature, pressure, wind direction, zenith distances, times, magnetic variations, objects in the night sky, stars, occultations, and innumerable latitude measurements. Included with his quantitative weather observations are personal remarks on the effects of the cold such as the freezing of a half pint glass of British brandy, the bedding frozen to the boards every morning, and the rude awakening by the cracking of the beams in the house which sounded like the retort of guns "rent by the prodigious expansive power of the frost" (Wales, p. 125).

Waiting over a year for the transit of Venus seemed routine to the astronomers. In fact the day seemed to come and go in much the same manner as the hundreds of days past. Measurements were made carefully and regularly during the day and any uncertainties cited. Wales noted what is known as the 'black drop' effect but in contrast to Cook's notes he says

> "We saw nothing like the appearance of an atmosphere round Venus (unless the above-mentioned phenomena may be thought to proceed from thence) either at the beginning, end, or during the time of the transit: nor could we see any thing of a satellite; though we looked for it several times". (Wales and Dymond 1769, p. 482)

Wales' keen eye not only marked scientific phenomenon but the local inhabitants as well. He describes, in contradiction of the prevailing view, the Eskimos not in unflattering terms but as open, generous people having a great genius for the arts. Indeed he must have spent some time with the Eskimos as he reflects on their religious practice and beliefs. Wales also described his winter clothing given to him by the

Hudson's Bay Company, his woollen socks, a beaver skin cap, a white fox's neckcloth, shoes of soft-tanned moose skin and snow shoes. Later, he would lament how these articles were taken from him on his return home by the customs officials.

Wales extended his scientific curiosity well beyond astronomy. He seemed always curious about weather phenomena such sun-dogs and the aurora-borealis which he found somewhat less appealing than in England. He also noted the proliferation of islands of ice (icebergs) and questioned the prevailing theories concerning them noting that they are not ice but frozen snow while speculating on their origins and ultimate demise. Throughout his journal, he returns to this topic several times discounting the current view that they are made entirely of salt water claiming "wiped clean off they are entirely fresh". He also questioned reports that suggest the icebergs last an extremely long time noting that they have disappeared (melted) in the few months of open water.

Although by his own admission he was not enough of a botanist to give a tolerable description of the natural environment and that he lacked the preservation skills for drying plants, he ardently describes, soil, flora, and fauna of the region, including those pesky "moschettos" – too common and too well known". Having weathered a winter and being exposed to "moschettos" a second time he can't help but consider winter as the "most agreeable season of the year".

Wales was well aware of past expeditions to the artic as he cites Crantz's history of Greenland and other times demonstrates a great respect for the practitioners in his discipline. Such a respect surely lead Wales to share his knowledge and experiences as a teacher of navigation once his adventuring days were over. Wales' contribution to the astronomical history in Canada went well beyond his days at Churchill. He tutored the great explorer Samuel Hearne at Churchill and later recommended Philop Turnor to the Hudson Bay Company. Hearne's explorations are legendary and Turnor, an important Canadian surveyor, later became a teacher of the great Canadian geographer David Thompson certainly passing on Wales' methods to several more generations. Griffin-Jones (2003) cites Wales as the inspiration to Coleridge's classic, "The Rime of the Ancient Mariner"[4] and while his contributions might remain obscure his legend lives on.

Following his posting to Churchill, Wales later sailed with Cook on two other voyages. Eventually, he settled down as a teacher of navigation at the Mathematical School in Christs Hospital, London. In 1765 Wales had married Mary Green, Charles Greens' sister and in a tribute to Cook's astronomer he would later write Green's biography in 1788. In a testament to his scientific accomplishments Wales was named a fellow of the Royal Society in 1776 (sponsored by Cook) and in 1795 was appointed to prestigious Board of Longitude.

Concluding remarks

The transit of Venus of 1769 yielded more acceptable results for the AU. Hornsby (1771) calculated a value of 8.78" and Euler (1772) a value of 8.82" which began to converge on our present day value of 8.79". The arduous calculations were in no small way the culmination of the long preparations and meticulous measurements of hundreds of astronomers around the world like William Wales.

The story of William Wales and his journey to northern Canada illustrates the sheer persistence and utter devotion of people like Wales, his brother in law Charles Green, James Cook and countless others. Such a tale helps students connect easily to Kuhn's meaning of normal science. At the same time challenges to Kuhn can be raised. If Wales was open to new interpretations – such as his theories on icebergs – was he merely preoccupied with puzzle solving or was his intention to reveal new ideas about nature which could exist even if they fell outside of the prevailing paradigm? Additional questions such as "What is the state of astronomy today?" also challenge Kuhn's view. As today's astronomers toil outside of the boundaries of visible light and formulate new theories of the cosmos are they merely puzzle solving or are there better ways to portray the nature of science than Kuhn's view?

Notes

[1] Kepler predicted that in 1639 Venus would be too far below the sun for a transit, he was later proven wrong on this account.

[2] An alternative method was devised by the Frenchman Delisle who recognized that a single observation at ingress or egress could be used if each observer marked the exact local time and knowing their longitude is able to convert this time to Greenwich time.

[3] Cited in Broughton from the minutes of the Council of the Royal Society in March and April, 1768 and in the Churchill post journal.

[4] McGoogan (2004) concludes that it was Hearne who was the inspiration for the Ancient Mariner. In any case Wales' influence remains intact.

References

Babbage, Charles (1830). *Reflections On The Decline Of Science In England And On Some Of Its Causes.* Project Gutenberg Release #1216 (February 1998).

Bell, Trudy E. (2004). Quest for the Astronomical Unit, *THE BENT OF TAU BETA PI* SUMMER 2004.

Brosse, Jacques (1885). *Great Voyages of Discovery, Circumnavigators and Scientists, 1764-1843*, Facts on File, Inc., 1985.

Broughton, Peter (2003). Historic Transits Of Venus From A Canadian Perspective, *Royal Astronomical Society of Canada Historical Committee*, November, 2003, http://www.rasc.ca/historical

Cook, James. & Green, Charles. (1771). Observations Made, by Appointment of the Royal Society, at King George's Island in the South Sea; *Philosophical Transactions*, Vol. 61, pp. 397-421.

Euler, Leonhard. (1772). A Deduction of the Quantity of the Sun's Parallax from the Comparison of the Several Observations of the Late Transit of Venus, Made in Europe, with Those Made in George Island in the South Seas: Communicated by Mr. Euler, Jun. Secretary of the Imperial Academy of Petersburg; In a Letter to Charles Morton, M.D. Sec. R.S. and Acad. Imper. Petroburg, *Philosophical Transactions*, Vol. 61, pp. 397-421.

Halley, E.(1716). "A new Method of determining the Parallax of theSun, or his Distance from the Earth," *Philosophical Transactions of the Royal Society of London*, No. 348, p. 454, 1716. Translated from the Latin in *Philosophical Transactions of the Royal Society* of London. Abridged Transactions of the Royal Society, Volume VI, pp.243-249, published in 1809.

Hornsby, T. (1771). The Quantity of the Sun's Parallax, as Deduced from the Observations of the Transit of Venus, on June 3, 1769: *Philosophical Transactions*, Vol. 61, pp.574-579.

Horrox, Jeremiah (1639). *The Transit of Venus Across the Sun,* translated by Arundell Blount Wratton (1859), William Macintosh, London.

Kuhn, Thomas S. (1970). *The Structure of Scientific Revolutions,* 2nd edition, The University of Chicago Press, 1962, 1970.

Kuhn, Thomas S. (1977). *The Essential Tension : Selected Studies in Scientific Tradition and Change*, The University of Chicago Press.

Maor, E. (2000). *June 8, 2004: Venus in Transit,* Princeton University Press.

McGoogan, Ken (2004). *Ancient Mariner: The Arctic Adventures of Samuel Hearne, the Sailor Who Inspired Coleridge's Masterpiece*, Carroll & Graf Publishers.

Proctor, R. A. (1882). *Transits of Venus, A Popular Account of Past and Coming Transits*, London: Longmans, Green, and Co.

VanHelden, Albert (1985). *Measuring the Universe.* Chicago: University of Chicago Press

Wales, William (1770). Journal of a Voyage, Made by Order of the Royal society, to Churchill River, on the North-West coast of Hudson's Bay; Of Thirteen Months Residence in That country; and of the Boyage Back to England; In the Years 1768 and 1769. *Philosophical Transactions*, Vol. 60, p. 100 - 136.

Wales, William & Dymond, Joseph (1769). Astronomical Observations Made by Order of the Royal Society, at Prince of Wale's Fort, on the North-West Coast of Hudson's Bay. *Philosophical Transactions*, Vol. 59, p. 467 – 488.

Whatton, A. (1859). *Memoir of the Life and Labours of Jeremiah Horrocks, To Which Is Appended A Translation of His Celebrated Discourse The Transit of Venus across the Sun*, (London 1859). William Macintosh, London.

Woolf, H. (1959). *The Transits of Venus: A Study in Eighteenth-Century Science*. Princeton, New Jersey: Princeton University Press.

Short history of the use of historical experiments in German physics lessons

FALK RIESS
University of Oldenburg, Germany

When we look at science education we can observe two poles lying far apart from each other, they are experimentation and history of science. Almost everybody will admit that both of them are relevant and necessary, but it seems to be a hard task to bring them together. I will give you a short history of physics education in Germany in respect to the topic of the conference[1].

Let me begin with a remark of Friedrich Paulsen, a well-known German historian of pedagogy from the 19th century with which he tried to characterise the use of experiments in high school teaching during the 1860s:

> 'In (contemporary) physics teaching the experiment did not stand at the beginning ... but came afterwards as quite unnecessary confirmation of a theory: every four or six weeks the class gathered together in the hall presumably to watch experiments (that didn't work properly in most cases), in fact the students got up to mischief in that big and unfamiliar room (Paulsen 1865).

Experiments are the key

During 19th century science education became an important subject of school policy and of pedagogy in Germany. Text books for use in schools of different levels and for teacher training were written, and lot of journals on this topic as well as national associations were founded in the last quarter of the century. The main principle, the credo of science education was formulated by one of the great old men of German physics pedagogy, Ernst Grimsehl: 'The method of scientific research must be the model for the teaching method on any level of physics education' (Grimsehl 1911, p. 2).

While the first practical laboratory work of students was performed in the private homes of the university professors (Jolly in Heidelberg 1846, Magnus in Berlin in the 1850s) around the 1850s the first practical laboratory studies for the students were introduced first at the Polytechnical Schools (Eisenlohr in Karlsruhe), then at the universities (Weber in Göttingen).

Soon science educators as well as state regulations for science teaching regarded experimenting as crucial for the learning process: 'The centre of physics education is formed by the experiment'. (Kiessling after Pahl 1913, p. 310). 'In the first physics

course the experiment in a form as simple as possible has to serve as a basis'. (Curriculum for the Prussian high schools 1892 and 1901, after Grimsehl 1911).
Consequently, some of the more prosperous high schools were equipped with measuring instruments, demonstration apparatus and laboratories during the 1890s. Bernhard Schwalbe, the founder of one of the most famous school laboratories in Berlin shaped his equipment after British (Armstrong) and American models.
Perhaps not astonishing in the pre World War I German Reich the method of the laboratory courses was borrowed from military practice: "zerstreuter Angriff = scattered attack" where the groups carry out different experiments (and cannot be controlled completely by the teacher), and working "in gleicher Front = in equal front" where all the students are occupied with the same experiment at the same time and the measurements taken can be made the basis of the following lessons (from Hahn 1912, p. 183).
Thus the introduction of laboratory practice at schools was due to the development of experimental physics at the universities and to the nationalist and pre-military education in German high schools at the beginning of the 20th century (from König 1962).

Short history of the historical approach in science teaching in Germany

The first science educator of German tongue to have stressed on the importance of historical considerations in modern physics teaching whom I have found is one of the founders of neo-positivist thinking, Ernst Mach. He wrote in a book entitled "Der relative Bildungswert der philologischen und der mathematisch-naturwissenschaftlichen Unterrichtsfächer der höheren Schulen" (The relative educational value of mathematics and physics in high schools) (Leipzig und Prag 1886, p. 22): 'I would be satisfied if every young man would have re-experienced some mathematical or physical discoveries and pursued their consequences ... Science education would follow perfectly and naturally reading the great scientific classics' (cited after Göllnitz 1920, p. 131). In a textbook meant for the first courses of secondary schools he stated in 1887: 'Whenever possible we use the very naive, simple and classical observations and thoughts from which the great researchers have built physics' (E. Mach, J. Odstreil: Grundriß der Naturlehre, Prag 1887).
History of science was not en vogue during the Weimar culture nor was it in the 3rd Reich of the so-called "national socialists" unless it could be used to show the differences between German physics and Jewish physics and the superiority of the first over the latter (Philip Lenard, Deutsche Physik).
After World War II the restoration of a conservative school policy in Western Germany that was oriented towards an elitist and hierarchical system prevented a discus-

sion about content and methods of science education. Martin Wagenschein (1896-1989), a science teacher and teacher educator, who is hardly known internationally and only verbally recognised in Germany tried to establish a science pedagogy that is focussed on the genesis of knowledge in history and in the individual. His method which he called "historio-genetic" and "exemplary" was based on frequent conversations with young children and on historical studies. He laid emphasis on phenomena and their explanations starting from every day experience. In addition he intended finally to make the (historical) process of physical inquiry the subject of philosophical consideration (Wagenschein 1980, p. 188)

His criticism concerning the practice of university science teaching is summarised in the sentence: 'Today's science education teaches us to register as finished and settled what we better should remember as questions which come from the dark and because of which more capable persons than we have failed'. (Wagenschein 1980, p. 310).

During the students' movement in the late 1960s and early 1970s reading and studying the leading Marxist theorists of the 19th and 20th century lead to a general criticism of society, science and technology, and education. These ideas of an intellectual rebellion were not only introduced into the educational discussion but lead to some practical attempts to change the content of science lessons. The late Jens Pukies taught classical mechanics and thermodynamics following a modified historio-genetic method with a strong emphasis on philosophical and historical topics: 'The philosophical and economical implications of science which were constitutive for the development of science must also be constitutive for the teaching of science. Science teaching will thus contribute to the responsibility and autonomy of the student' (Pukies 1979, p. 7).

Nevertheless, it must be stressed that it was a very small group of people who tried to change the practice of science education. School politicians and school administration were especially suspicious of those radical propositions. In fact there was very little effect on the reality of science education not to be compared with the STS and HPS movement in Britain and the US.

What would deserve a more detailed analysis but can only be mentioned shortly is the use of history in the German Democratic Republic, a state that - as long as it existed - claimed to be based on socialist principles. We read in an early textbook for teachers on physics education:

> Students should recognize that under the rule of exploiting classes the scientific and technological development only is supported to certain level by the social conditions which impeded the progress from that stage on. Only with the con-

> struction of the socialist order of society these obstacles are removed once and for all (Hamann 1961, p. 6).

To transform these findings into relevant teaching material seemed to be a challenging task, but the difficulties to do so turned out to be considerable. In 1971 the next official book publication on physics education had to admit:

> Our current curriculum still is too rigid, does not reflect sufficiently our philosophical knowledge and does not show science as a process ... The difficulties in forming a curriculum and its transformation into teaching methods lie in the fact that logical and historical structure often seem to be contradictory (Haspas 1970, p. 114ff).

These problems seem to have been overcome when eight years later a joint publication of Soviet and German (democratic) science educators came out where the subject of history in physics education was treated under the title "The Historical Principle" on five pages. A lot of propositions were made how to integrate historical subjects into the systematic course. Also it was recommended to read original texts of famous physicists. In fact the effect must have been very small because the physics textbooks which were standardized and mandatory throughout the entire country did not contain corresponding material nor was it produced by teachers or universities. Only occasional attempts were made in teacher training seminars (see Wünschmann et al. 1978, pp. 40-45).

Introducing historical experiments into physics teaching

Besides the more general comments on the use of historical material for science teaching which have been cited above one of the first authors who dealt with historical experiments for teaching purposes specifically was Carl Ramsauer, a professor of experimental physics in Berlin who was born in Oldenburg in 1879. After his retirement he began to work on historical experiments and published - without actual historical ambition as he confessed - a book called "Fundamental experiments in physics represented historically". This book is a very useful collection of descriptions and analysis of the most famous experiments in physics. Figures and data were taken from the original publications, and Ramsauer as a gifted and competent experimentalist points out the crucial difficulties of every set-up and measurement. Besides this he gives a realistic view of the role of experiments in his foreword:

> By the way, do not think that these fundamental experiments are easy reading because they are only experiments. What was very hard work for the great researchers to plan and perform requires an even amount of work for the reader ...

> But this is rewarding. For my person I have to confess that I only feel as a proper physicist after having come to close contact with the whole of the sources of experimental physics - unfortunately only in my seventh decade. I would appreciate it very much if this book could impress the historical basis of experimental physics on teachers and students of high schools (Ramsauer 1953, p. VII).

The most severe fault of his book is its complete lack of historical considerations besides technical or scientific ones and the fact that he never tried to repeat one the experiments he is reporting on. Beginning in the late 1960s, Samuel Devons has run his history-of-physics laboratory at Columbia University for more than twenty years. His addressees were mostly science teacher students and science teachers on the job.

> The aim was not to present the history of physics as something distinct from, or additional to, the science itself, but rather to exploit the rich historical material and background as one vehicle for the better understanding of physics itself ... so the challenge remains for the student to comprehend all that is actually observed ... whether in the context of the prevailing knowledge and understanding, current at the time of the original discovery, or in the, possibly contrasting, framework of the student's own comprehension of contemporary physics (Devons 1987, p. 315-326, Devons, Hartmann 1970, p. 44-49).

Jürgen Teichmann from the Deutsches Museum in Munich worked on replications and their use for science teaching during the late 1970s and the early 1980s. He recommended to use history as a "quarry" for didactic ideas. 'History can teach us that scientific knowledge is not necessarily definite, but also how difficult it is to put it in question from nowadays. Using particularly simple examples history can make clear that science, technology and society cannot be seen as independent from one another' (Teichmann 1978, p. 185-212).

Almost 20 years ago two books were published with a lot of material on historical experiments in physics. They are meant specifically for use in schools and universities. Unfortunately they do not treat the problem of replication seriously as they leave historical authenticity behind which is sacrificed to easy reproduction. On the other hand they spread well known ideologies. 'Particularly meaningful is the simplicity and clarity of the historical set-ups. The most important reasons for this often were the material limitations but also the conscious orientation on the most important in order to delete disturbing influences as far as possible' (Wilke 1987). 'During re-experimenting I did not emphasize to reconstruct a set-up in the proper materials but to reconstruct the original function' (Achilles 1989).

Within the last 20 year six textbooks published in Germany on physics education. Three of them do not contain any mentioning of history of science as a means for better comprehension of science, two of the other three only dedicate two sentences to the use of history in science teaching; only the book by Jörn Willer (2003) has a chapter about quantum mechanics and its historical development.

Common misunderstandings concerning the use of historical experiments in teaching

Much of the historical material meant to be used in teaching - even Harvard Project Physics - tend to write a history of science ex post as a mere history of winners. But in order to give a realistic image of the development the defeated ways of thinking must also be known. Not 'wrong' theories are overthrown by the 'right' ones but from investigating the process of decision making in science for or against certain models or theories one can learn that the criteria of success depend on the purposes and intentions which might be either technical, institutional or political ones; in any case they are influenced by society. Once scientists and lay persons are aware of this they are in principle able to decide about the future direction of research in science and technology in a democratic process.

Usually a cluster of ideologies is connected with historical experimentation. The set-ups are claimed to be simple and the measurements easy to perform, none of which is true if you try to repeat a single experiment. It is crucially dependent on what concept you have, what variable you observe and what result you expect. The admiration of "great physicists" because of their outstanding experimental abilities is a product of writing some kind of "whig" history. In most cases they did not work alone, and their success relied to a great extent upon some hardly known instrument maker or helping person, not to mention the women who cared for the geniuses. So history of science is reduced to the life and deeds of heroes whose battlefield was nature and whose weapons were measuring instruments. Examples of how **not** to deal with historical experiments in physics education can be found in the series of "classic physics experiments" which are offered by PASCO Scientific (http://www.pasco.com/products/groups/55-372-1.html, 11.11.2006).

The conference has the aim to show alternatives which are historically correct *and* inspiring for science comprehension of young people.

Notes

[1] The translations of the quotations originally in German were made by the author; he is responsible for any misunderstandings resulting from this.

References

Achilles, M.: 1989, *Historische Versuche der Physik. Funktionsfähig nachgebaut*, Springer, Berlin.

Devons, S.: 1987, 'The Art of Experiment: Teaching Science Teachers', *Hyperfine Interactions* **33**, 315-326

Devons, S.; Hartmann, L.: 1970, 'A History-of-Physics Laboratory', *Physics Today* **23**, 44-49

Göllnitz, H.: 1920, *Beiträge zur Geschichte des physikalisch-chemischen Unterrichts an den höheren Schulen Deutschlands*, Leipzig.

Grimsehl, E.: 1911, *Didaktik und Methodik der Physik*, München.

Hahn H.: 1912, 'Die Entwicklung des physikalischen Arbeitsunterrichts', *Jahrbuch der Pädagogischen Zentrale des Deutschen Lehrervereins* **2**, 181 - 189

Hamann, G. (ed): 1961, *Physikunterricht*, Berlin/GDR.

Haspas, K. (ed): 1970, *Methodik des Physikunterrichts*, Berlin/GDR.

König, H.: 1962, *Imperialistische und militaristische Erziehung in den Hörsälen und Schulstuben Deutschlands 1870-1960*, Berlin/West 1962

Mach, E.; Odstreil, J.: 1887, *Grundriß der Naturlehre*, Prag.

Pahl, F.: 1913, *Geschichte des naturwissenschaftlichen und mathematischen Unterrichts*, Leipzig.

Paulsen, F.: 1865, 'Physikunterricht am Christianeum in Hamburg-Altona 1863-66', in F. Paulsen, *Geschichte des gelehrten Unterrichts*, Leipzig

Pukies, J.: 1979, *Das Verstehen der Naturwissenschaften*, Braunschweig .

Ramsauer, C.: 1953, *Grundversuche der Physik in historischer Darstellung, Erster Band*, Berlin, Göttingen, Heidelberg.

Teichmann, J.: 1978, 'Das historische Experiment im Physikunterricht - wissenschaftstheoretische Betrachtungen und praktische Beispiele', in: M. Ewers (ed), *Wissenschaftsgeschichte und naturwissenschaftlicher Unterricht*, Bad Salzdetfurth, 185-212.

Wagenschein, M.: 1980, *Naturphänomene sehen und verstehen,* Stuttgart.

Wilke, H.J. (ed): 1987, *Historische physikalische Versuche*, Berlin/GDR.

Wünschmann, M. et al: 1978, *Methodik des Physikunterrichts in der DDR und in der UdSSR*, Berlin/GDR, Moskau.

Willer, J.: 2003, *Didaktik des Physikunterrichts*, Frankfurt.

History of Science in the Service of Middle School Science Teacher Preparation[1]

DAVID W. RUDGE

The Mallinson Institute for Science Education, Western Michigan University, Kalamazoo, USA.

Abstract: *Enlist, Equip and Empower (E^3): A Program for the Preparation of Middle School Science Teachers* is an NSF sponsored project aimed at recruiting qualified and caring individuals into the profession, providing training specific to the needs of middle school science teachers, and supporting teachers during the first crucial years of their careers. In this paper I discuss three ways history of biology is being used to promote understanding of the nature of science. These include recapitulating the discovery of blood circulation (William Harvey), discovering why dark moths became more common in soot-darkened areas after the industrial revolution (Bernard Kettlewell), and exploring a mystery disease (sickle-cell anemia) from the standpoint of six subdisciplines in biology (Anthony Allison). The essay concludes that this instrumental use of history has enormous potential for teacher preparation programs.

Introduction

The United States faces a critical shortage of qualified middle school science teachers that is expected to worsen as retirements add to attrition. In the state of Michigan alone, approximately 25% of school districts report great difficulty recruiting qualified middle school science teachers, and another 60% report moderate difficulty (East Lansing Collegiate Employment Research Institute (2000)). The causes of this shortage reflect longstanding problems associated with *enlisting* people into the profession, *equipping* them with a deeper understanding of science (content and process), and *empowering* them to continue teaching at the middle school level. The following essay discusses how history of science is being used as part of a coordinated effort to improve the overall preparation and support of middle school science teachers. It begins by briefly reviewing the goals and objectives of *Enlist, Equip And Empower* (E^3), an NSF sponsored project to promote middle school science teaching in Western Michigan. After briefly discussing the context of several course revisions being done on this project, the essay turns to a discussion of three innovative ways history of science is being used instrumentally to promote the project goals. The essay discusses the overall philosophy behind this use of history and concludes by summarizing the enormous potential using history in this way has for teacher preparation programs.

E^3: A BRIEF OVERVIEW

Enlist, Equip And Empower (E^3): A Program for the Preparation of Middle School Science Teachers is an integrated approach to improving the quality of middle school science teaching. Developed at Western Michigan University (WMU) and funded by the National Science Foundation, this innovative approach to teacher preparation includes activities aimed at recruiting caring and qualified individuals into the profession, the development of a new program of instruction at WMU specific to the needs of middle school science teachers, and meetings and workshops that empower in-service teachers during the crucial first years of their careers.

Enlist Objective: Recruiting caring, capable individuals into the teaching profession is increasingly problematic. Many students entering education programs at WMU, currently the fourth largest producer of teachers in the United States (WMU Office of University Relations, 2001), do so from a widespread perception that a teaching credential can serve as job insurance (National Commission on Teaching and America's Future, 1996). Most are not interested or prepared to teach at the middle level. E^3 recruitment efforts have included the creation of a brochure and project web site, a series of presentations at introductory courses attended by over 650 elementary education students, presentations to area high school students, the creation of a data base of over 300 self identified prospective middle school science teachers, and a series of informal meetings developed to foster the development of a professional learning community.

Equip Objective: Preparing individuals to teach science is also problematic. To teach effectively, middle school science teachers must have a thorough grasp of content, understand how children learn, and be able to use inquiry-based methods (National Commission on Teaching and America's Future, 1996; American Association for the Advancement of Science, 1993). At the start of the grant in Spring 2002, WMU, like other institutions throughout the United States, had two teacher preparation programs specific to the needs of elementary (grades K-8) and secondary teachers (grades 7-12). The former are required as part of their preparation program to take six science content courses representing the life, earth and physical sciences. The latter major in a particular science. Neither program is sufficient to train future middle school science teachers, who, like elementary teachers, are responsible for teaching science content in all three content areas. Students with elementary certification generally have insufficient science depth (Dolster, Jackson & Smith, 1997). Students with secondary certification only have sufficient science content in one of the three areas they will need to succeed at the middle school level. (Young adolescents who attend middle school are also demographically distinct as learners from those in elementary and secondary settings.)

To address these needs, the core of the E^3 suggested program for the preparation for future middle school science teachers is a six science content course sequence specifically developed for the needs of elementary school teachers. These courses (two each in the earth, physical and life sciences) have been developed with reference to state and national benchmarks for scientific literacy and are taught in the form of 24 student lab sections. While not methods courses, the student-centered, inquiry-based pedagogy of these courses is specifically intended to role model how future teachers of elementary classrooms should one day teach. To provide future middle school science teachers with the science content they will need to succeed, participants are encouraged to take a series of additional science content courses that are taught, for a variety of logistic constraints, in the form of traditional large lecture classes. Additional specific courses were chosen to complement the elementary science content courses mentioned above with reference to state and national benchmarks for science literacy at the middle school level. The E^3 project staff approached the instructors of these courses about the possibility of revising pedagogy in ways that would make the courses more student-centered and inquiry-based for the benefit of not only future middle school science teachers (who would always be a distinct minority), but also all other students taking these courses.[2] Project staff also assisted a faculty member responsible for a methods course taken by elementary education majors in the creation of a special section of the course devoted specifically to the needs of future middle school teachers.[3]

Empower objective: Supporting qualified middle school science teachers is also a challenge, since many teachers change careers within the first five years (Hope, 1999). The National Science Education Standards (National Academy of Sciences, 1996) emphasizes that the ongoing professional development of science teachers must provide teachers with opportunities to engage in scientific inquiry, reflect on the nature of science, and enhance and refine their pedagogy (National Research Council 1999, 2000; Loucks-Horsley, Hewson, Love, Stiles, 1998). Beyond this, professional development programs must foster support and collegiality among teachers. Without ongoing support, it is not surprising that many middle school science teachers leave the profession (National Commission on Teaching and America's Future, 1993). To better address the needs of inservice teachers, E^3 project staff have worked with area school districts to promote the professional development of area inservice middle school science teachers at three participating school districts (Kalamazoo Public Schools, Portage Public Schools and Vicksburg Community Schools), through a series of meetings offered during the school year and week long science content workshops during the summer. Much of the success of the project has been due to the full time assistance of a full time middle school science teacher in residence, who has been intimately associated with the planning and execution of each of these activities.[4]

Three Ways History is Used

As noted above, a central concern for the E^3 project has been to help future middle school science teachers develop a deeper understanding of science. This includes considerations of both science content and issues associated with the nature of science as a process.

CONTEXT

BIOS 1700 and BIOS 2700, *Life Science for Elementary Educators I and II*, form an integrated biology sequence for elementary education majors developed with reference to several professional guidelines regarding scientific literacy (e.g. American Association for the Advancement of Science 1993; Michigan State Board of Education 2000 [1991]). Both courses are required as part of a science/math minor taken by all elementary education students, and, as noted above, are strongly recommended for those pursuing a career as a middle school science teacher.

With reference to science content, BIOS 1700 consists of three units that provide an overview of the biological subfields of human anatomy and physiology (emphasizing the cardiovascular system), ecology, and evolution. BIOS 2700 includes units devoted to genetics, molecular and cell biology, and a capstone unit that engages students in an exploration of a single biological phenomenon from all of these perspectives. The two courses have also been developed to help students develop a more sophisticated view of the nature of science than is normally presented in textbooks, emphasizing the importance of thinking about biology as a process rather than simply as a body of knowledge. They explicitly invite students to reflect on how people and children learn science, and consider the ramifications of their reflections for how they should teach. Like other courses taught through The Mallinson Institute for Science Education, these two courses are taught primarily by means of open ended problems, in which instructors invite students to recognize patterns, identify a problem they will pursue, consider how to design an investigation to solve the problem, collect and interpret results, and then persuade others of their interpretation.

History of biology is used extensively to promote learning of both science content and also an understanding of issues associated with the nature of science in both of these courses. Some of these uses of history are rather traditional, in that they explicitly invite students to reflect on the thinking of past scientists. In BIOS 2700, for instance, students read a brief chapter that summarizes Lynn Margulis's research on endosymbiosis, a theory for the evolution of cell organelles in eukaryotes (Hagen, 1996). As they read the chapter and also during discussions in class, students are invited to reflect on questions that direct their attention to nature of science issues, such as how and why the same set of facts might be accounted for by two quite different theories. Students also regularly view a 1987 film available through Films for Humanities entitled "Life Story", based upon James Watson's autobiographical account of his and Francis Crick's discovery of the structure of DNA. As they watch the film, students are invited to reflect upon a series of thought provoking questions associated with the nature of science, such as whether science is a creative endeavor, the role of women in science during the 1950s, and who discovered the structure of DNA.

Each of these examples is quite effective in humanizing science, drawing attention to the institutional structure of science and how it affects the process of scientific discovery, and are both interesting and entertaining in their own right. There is also value in having students gain some understanding of the history of biology for its own sake. While experience with the course suggests these uses of history are quite effective for some students, it's clear that they are less effective for others. Part of the problem is that many students who choose to pursue teaching as an avocation are not particularly interested in learning or teaching science, let alone history of science. For some, history of science *per se* represents a complicating distraction, one that neither furthers their own understanding of the science nor deepens their interest in the subject. For others, the act of reflecting upon the actions and decisions of others is itself inherently difficult, particularly if it involves evaluating the views of past scientists in a context quite divorced from their own.

Beyond the foregoing, it seems quite clear that from a constructivist theory of learning (Driver & Oldham 1985), effective use of history of science in the classroom must take into account prior (alternative) conceptions students have about both science content and process, as well as issues associated with the nature of science. It is widely recognized that these prior or alternative conceptions are often quite resistant to change. Meaningful learning can only take place in instructional environments where students are explicitly invited not only to share their prior conceptions, but also to reflect upon the limitations of their alternative conceptions as well as the relative advantages of goal conceptions. The National Science Education Standards (1996) also emphasizes the importance of guiding students in active, extended scientific inquiry, which contrasts with traditional approaches to the use of history in which it is often used as an "add on" to humanize the science.

These considerations suggest a fundamentally different approach to the use of history in the science classroom. Instead of having students learn about the work of past scientists, the following three examples show how history of science can be used instrumentally to construct a learning environment that invites students to share and critique their own alternative conceptions of science content, the process of science, and issues associated with the nature of science. In each case, history is used *instrumentally* to construct a learning environment in which students are invited to reflect on their own deliberations, rather than those of past scientists, as they actively investigate a biological phenomenon over the course of several days.

WILLIAM HARVEY'S DISCOVERY OF THE CIRCULATION OF BLOOD[5]

In BIOS 1700, a unit developed with reference to William Harvey's discovery of the circulation of blood illustrates how and why one may sometimes need to "corrupt" history of science in the service of other learning objectives (c.f. Allchin 1993).

The unit begins by eliciting student ideas about the cardiovascular system, an organ system most antecedently claim to know a great deal about. It begins by having students observe what happens to their bodies after they climb up and down two sets of stairs. Students notice that they are breathing more heavily, and their pulse and heart rate have increased. Students are also invited to reflect on why these changes are occurring, with all ideas being accepted without comment or criticism by the instructor. The instructor then provides a brief lecture about the history of research on the human body from the time of the ancient Greeks up until the time of William Harvey, with special attention on Galen's account of the arterial, venous and nervous systems. In the next class, students examine histology slides and dissect sheep hearts. While students interpret these activities as aimed primarily at the identification of tissues and the names of parts, the instructor repeatedly invites students (both orally and in writing) to reflect on several curious features, e.g. the fact that arteries are generally more muscular than veins, veins have valves whereas arteries do not, the heart is asymmetrical, etc.

In the final class, a group discussion of these same questions invites students, *unbeknownst to them*, to recapitulate the sort of reasoning that led William Harvey to discover the circulation of blood. The instructor never tells them this, it is a discovery they come to on their own as a result of their own deliberations and careful scaffolding on the part of the instructor. To facilitate student insights in light of the fact that they are working in a different context than Harvey did, during this discussion the instructor provides additional evidence that either Harvey did *not* use or invites students to deliberate in a way that Harvey is not believed to have reasoned. For example, students are asked to reflect upon why sheep and humans have four-chambered hearts, as opposed to three or five chambers. Attention is directed to the fact that whereas mammals (e.g. humans and sheep) and birds have four-chambered hearts; reptiles and amphibians have three-chambered hearts, and fish only have two chambered hearts. Students are then invited to consider what mammals and birds have in common in contrast to fish, amphibia and reptiles (mammals and birds are referred to as "warm-blooded" because they maintain a constant body temperature; whereas the others are referred to as "cold-blooded" because their internal temperature is set by the ambient environment). This, coupled with students' antecedent understanding that maintaining body temperature is due to cell respiration, a process that requires oxygen and the elimination of carbon dioxide, leads students to recognize that warm-blooded animals have a greater need for oxygen and the elimination of carbon dioxide. This clarifies why it is particularly important for birds and mammals to separate oxygenated from and de-oxygenated blood, and therefore why it would be advantageous for both birds and mammals to have four chambers (two that direct deoxygenated blood from the body to the lungs; two that direct oxygenated blood from the lungs to the body) rather than having these two types of blood mixed, as is the case in two- and three-chambered hearts.

After reaching an understanding of how the heart works that conforms with our modern understanding, it is important that students learn how our use of history has departed from what Harvey is believed to actually have done. In a concluding segment, students view a film that presents a historical reenactment of Harvey's vivisections, an important line of evidence Harvey used that for ethical and practical reasons we do not. Students are asked to comment on this method of instruction, the use of history of biology to teach biology, as well as nature of science issues, such as the possibility we can have good evidence to believe in the existence of structures one cannot directly observe (e.g. capillaries). The unit concludes by inviting students to use the Biology Explorer™: Cardiovascular System™ to study other aspects of the cardiovascular system not revealed by this historical overview, including a study of how blood pressure is controlled by the cardiovascular system and the causes and treatment of several associated diseases.

H.B.D. KETTLEWELL'S RESEARCH ON INDUSTRIAL MELANISM[6]

A second example of how history of science can be used instrumentally to promote learning of science and issues associated with the nature of science occurs during the evolution unit of BIOS 1700. Students are presented with a "mystery phenomenon" (the phenomenon of industrial melanism) and invited to reason about it in a manner that parallels that of past scientists who worked on the phenomenon.

This unit begins by introducing students to some observations made by naturalists and amateur entomologists of chance findings of a rare dark form of a heretofore common moth that was known only by its common pale form. Using a Powerpoint presentation, the instructor shares evidence of dark moth sightings with students, inviting them to notice a pattern in the distribution of the ever more common dark moth over time. Students learn that over the courses of only about one hundred years (1850-1950) the dark form was not only found in an increasing number of locations throughout Britain, but also that the frequency of the dark from was becoming more common. Indeed, in some locations the dark form went from being entirely unknown to constituting 99% of the whole population. The presentation invites students to recognize that the dark form is becoming more common in only certain regions of the British countryside. Once some of the students notice this, the instructor invites them to consider what these areas have in common (they are all near cities), and further what was happening in Britain during this time period (the industrial revolution). Additional supplied observations reveal that while the dark form has become more common in soot darkened woods near manufacturing centers, the pale form continues to dominate in unpolluted rural areas. This leads to the mystery: Why is the dark form becoming more common in the vicinity of manufacturing centers?

It has long been recognized that students often have preconceived ideas about scientific phenomena that strongly parallel those of past scientists. In the contest of this unit, it is entirely predictable that some students will come up with theories that strongly resemble what are now regarded as erroneous theories (acquired inheritance, mutation), and further that even those who recognize the possibility that it can be accounted for by natural selection will still belabor under misconceptions about how natural selection works. It would be tempting at this juncture for the instructor to simply publicly label student alternative conceptions as mistaken and present the goal conception. Instead, the instructor invites students to share the reasons why they think their respective theories could indeed account for the mystery phenomenon. In the next class the instructor shares brief summaries of scientists who had theories similar to those of students (acquired inheritance (Jean Baptiste De Lamarck), mutation (Hugo De Vries), and natural selection (Charles Darwin). The instructor then shares the fact that several scientists in the past actually used these theories to account for the mystery phenomenon (Nicholas Cooke, James W. Heslop Harrison, and Edmund Briscoe Ford). This has the effect of distancing the student from the theories they proposed, such that a critique of the theory need not be taken as a personal attack upon the views of the student. With these examples in mind, students are invited to discuss what theories are in more general terms, with the instructor inviting students to use examples from other contexts (e.g. the theory of plate tectonics).

The instructor next invites students to consider how they might test these theories that purport to account for the mystery phenomenon. This includes considering what further observations one might make, and also whether one could conduct an experiment. As before, student responses at this point are entirely predictable and can be used to motivate a discussion of the results of actual experiments were conducted along the lines of what the students suggested. Students are specifically invited to consider how proponents of the three theories might respond to these results, and during the discussion that follows, it becomes clear that the results are ambiguous. The class concludes with a more general discussion of what experiments are in science, and further whether they are necessary for progress in science, again using examples from other contexts.

In the final class, students view a film created by H.B.D. Kettlewell that discusses his famous experiments on natural selection and, in particular, demonstrates that birds differentially prey upon the more conspicuous form of the moth in the polluted and unpolluted settings as predicted by natural selection. The students at this point are all in agreement that the reason that the dark form has become more common in the vicinity of industrial sites is because of natural selection: birds in these areas are less likely to capture them than the pale forms which in these settings are quite conspicuous. This is the goal conception, and as such, it would be tempting to leave it at that. Instead the instructor shares the rest of the story, namely a series of interpretive problems that have recently been raised about Kettlewell's work and our understanding of the phenomenon of industrial melanism. The point of doing this is not to raise questions about whether this is indeed an example of natural selection (it is), rather, we do so to invite these future teachers to consider how very simplistic textbook accounts of biological phenomena like industrial melanism are, and what they can and should do with their future students to give them some insights into the actual more complicated practice of science. A concluding discussion once more focuses on how this use of history has departed from the actual history of events, again emphasizing how and why other pedagogical goals (such as addressing student misconceptions) have trumped our need to faithfully present the history of this episode.

ANTHONY ALLISON'S WORK ON SICKLE CELL ANEMIA[7]

A third example of how history of science can be used instrumentally to promote learning of science and issues associated with the nature of science occurs during the eight class long capstone unit of BIOS 2700. A primary goal of this unit is to have students bring to bear their previous understanding of multiple subdisciplines (anatomy and physiology, cell biology, ecology, evolution, genetics, and molecular biology) on a single biological phenomenon. To this end, students are presented with a "mystery disease" (sickle-cell anemia) and invited to reason about it in a manner that parallels that of past scientists who worked on the phenomenon. Each class invites students to reason about the mystery disease from the perspective of one subdiscpline in biology in an open-ended problem solving environment. Tentative conclusions reached one day feed into the activities of the next, and sometimes lead students to reevaluate their previous understandings. In this way, students can directly experience how the findings of one subdiscipline might affect their understanding of the phenomenon from the perspective of another subdiscipline.

Students are initially presented with a mystery patient who suffers from a series of symptoms (anatomy and physiology). Recognition that at least some of the symptoms might be related to a problem with the cardiovascular system leads to a study of the patient's blood, which reveals that some red blood cells have a curious narrow shape compared to normal "donut-shaped" blood cells (cell biology). Students predictably raise questions about whether patient's family has a history of this illness (genetics), which in turn raises questions about whether the illness might have some environmental cause (ecology). Students learn that the patient's family is from Uganda, where curiously a high incidence of the gene that causes the disease occurs in areas with a high incidence of another deadly disease (malaria). Students consider and reject the hypothesis that the mystery disease (which has a genetic basis) might be malaria (which is caused by a parasite transmitted by mosquitoes). Resolution of this altogether curious coincidence depends upon recognizing that in addition to its harmful side effects, the gene responsible for the mystery disease confers partial immunity against malaria (evolution). The class then investigates how the gene that causes the disease might also prevent malaria, and ultimately concludes that the gene responsible for the mystery disease is a point mutation in the gene that codes for normal hemoglobin. In malaria free environments, this mutation is disadvantageous, because whenever blood oxygen is low (as for instance when the patient engages in strenuous activity), the blood cell takes on a characteristic sickle cell appearance that can easily block the flow of blood in small blood vessels known as capillaries. In malaria-infested environments, having blood cells adopt a sickle cell shape is actually beneficial in preventing spread of the parasite. In addition to science content and learning about the process of science through their own deliberations and attempts to solve each day's problem, students are also encouraged to explicitly discuss and reflect on issues associated with the nature of science, such as whether scientific knowledge is tentative and how scientists having different theoretical perspectives might disagree when trying to make sense of the "same" empirical evidence.

As with the previous two examples, history of science is used instrumentally to promote learning science (content and process) and also issues associated with the nature of science. The unit invites students to discuss and reflect on the findings of multiple scientists, only some of whom are identified by name, in an order dictated by pedagogical considerations rather than the actual chronology of events. While there was indeed a mystery patient, the pedigree students use to study the genetics of the disease is artificially contrived to promote student thinking of the disease as due to a dominant, rather than a recessive gene, in parallel with how scientists initially thought about the disease. Like the previous two examples, the unit ends with an explicit discussion of how and why the unit has departed from the actual history of events.

Conclusions

In a provocative article, Monk & Osborne (1997) lament the relative lack of progress history of science has made into the (secondary) science classroom, despite the "continued special pleading" by Michael Matthews (1994) and others regarding the enormous potential of history of science for promoting interest in science, scientific literacy and critical thinking skills, to name a few. According to Monk & Osborne, the problem is essentially two-fold. First, most science teachers operate from a relatively impoverished understanding of the nature of science, one that leads them to focus on the *products* of science, rather than the *process* of scientific discovery. Teachers often regard methods used by scientists as unproblematic, requiring only a minimal justification. Indeed, if they use it at all, history is used as an "add on" meant chiefly to humanize the science (Gallagher 1991). Second, even when teachers have a relatively sophisticated understanding of issues associated with the nature of science, their decisions on what to teach and how to teach are often dictated by the imperatives of the classroom, such as classroom management issues, considerations of concept acquisition and development, and the necessity to teach specific content to prepare students for standardized tests (Hodson 1993). When one considers further that science teachers are neither trained as historians nor particularly interested in history of science, it becomes hard to fault their reluctance to include history of science as an integral part of the science classroom.

At least part of the solution to the above problem lies in the development of more sophisticated teacher training programs that incorporate history of science as a central, rather than a peripheral, feature. Teacher training programs can and often are developed by individuals who have relatively sophisticated understandings about nature of science issues. Moreover, the curriculum for teacher training programs is not (at least currently) so restrictively set by state and national standards as to preclude extended, open-ended problem solving approaches to history such as those discussed above. In short, teacher trainers have the luxury to go into more depth, thereby aiding future educators in developing sophisticated understandings about nature of science necessary for them to make informed, reflective decisions about what nature of science issues to teach and how best to help their students learn.

Part of the problem also lies in the mistaken presumption that a science teacher must be completely faithful to a historian's account of the history of an episode when using it in her classroom. This is the tacit presumption in Monk & Osborne's advocated approach to history using historical vignettes that provide students with a rich context upon which to reflect upon the deliberations and actions of past scientists. There are certainly some educational objectives for which this is important and indeed essential, as for instance when a teacher decides that she actually wants student to learn some history of science. This being said, there are many educational objectives (e.g. deeper understanding of scientific concepts, addressing student misconceptions) that do not necessitate slavish attention to historical detail for its own right. Rather than insisting students reflect on nature of science issues in contexts they are unfamiliar with (and moreover may not engage them), it seems clear that there is much to be gained from students reflecting on nature of science issues with regard to the context of their science classroom. The latter is a rich context in its own right, one that stems from prior conceptions brought to the classroom by the students as well as the instructors' careful decisions regarding what issues to raise, what evidence to share, and how to scaffold learning as it occurs. The teaching of an introductory class by its nature always involves a series of simplifications on the part of the instructor- careful deliberate decisions regarding what information to share and what to withhold, mindful of the insights (and alternative conceptions) students bring into the classroom and also common pitfalls in their reasonings. It is nothing short of ridiculous to fault teachers (who have no particular training in history) for departing from the actual history of events when their objectives depart from those of historians of science. And, as illustrated above, there are ways history of science can instrumentally be used to further objectives shared and valued by teachers that can indeed promote a greater understanding of science content, the process of science, and even nature of science issues held dear by historians and philosophers of science.[8]

Acknowledgements

This material is based on work supported by the National Science Foundation Grant No. 0202923. Any opinions, findings and conclusions or recommendations expressed in this material are those of the authors and do not necessarily reflect the views of the National Science Foundation.

Notes

[1] This is a revised version of a paper presented at the Sixth International Conference for History of Science in Science Education, Oldenburg, Germany, July 10-14, 2006, organized by Peter Heering (Carl-von-Ossietzky Universität Oldenburg, University of Augsburg) and Daniel Osewold (Carl-von-Ossietzky Universität Oldenburg).

[2] For a comprehensive review of the literature on techniques used to promote student centered and inquiry based learning in the context of large lecture science classrooms, see Geer & Rudge (2002).

[3] For specific information about the suggested program of middle school teacher preparation and other activities related to the grant, please visit the project website: www.wmich.edu/science/e3/index.html .

[4] For a more comprehensive overview of the project as a whole, see Rudge & Burke (2004); for a summary of results associated with one of the large lecture science content courses revised as part of this project, see Cutrim, et. al. (2006).

[5] The approach taken in the following example is adapted from Allchin (1993).

[6] For further information about this example, including a lesson plan that describes how the unit is conducted in more detail, see Rudge (2004).

[7] This unit was developed by Eric Howe (Howe 2004). For more information about this example, see Howe & Rudge (2005), Howe & Rudge (2006).

[8] For advice on how to incorporate history of science into a science classroom, see Rudge & Howe (2004).

References

Allchin, D.: 1993, 'Of Squid Hearts and William Harvey', *The Science Teacher* **60**(7), 26-33.

American Association for the Advancement of Science: 1993, *Benchmarks for Science Literacy*, Oxford University Press, New York.

Cutrim, E., Rudge, D., Kits, K., Mitchell, J. & Nogueira, R.: 2006, 'Changing Teaching Techniques and Adapting New Technologies to Improve Student Learning in an Introductory Meteorology and Climate Course'. In E. Cutrim, M. Ramamurthy, S. Nativi, & L. Miller (eds.) *Advances in Geosciences (ADGEO) Special Issue: Earth System Science Data Access, Distribution and Use for Education and Research*, Volume 8: 11-18. (www.copernicus.org/EGU/adgeo/8/contents1.html)

Doster, E. C., Jackson, D. F., & Smith, D. W.: 1997, 'Modeling Pedagogical Content Knowledge in Physical Science for Prospective Middle School Teachers: Problems and Possibilities', *Teacher Education Quarterly* **24** (4) 51-65.

Driver, R. & Oldham V.: 1985, 'A Constructivist Approach to Curriculum Development', *Studies in Science Education* **13**:105-122.

East Lansing Collegiate Employment Research Institute: 2000, *Finding Teachers: How Hard is it for Michigan Public School Districts?*, Michigan State University, East Lansing, MI (ERIC Document Reproduction Service No. ED 450 097).

Gallagher, J.J.: 1991, 'Prospective and Practicing Secondary School Science Teachers' Knowledge and Beliefs about Philosophy of Science', *Science Education* **75**: 121-133.

Geer, U.C. & Rudge, D.W.: 2002, A Review of Research on Constructivist-Based Strategies for Large Lecture Science Classes *The Electronic Journal of Science Education* **7**(2) http://unr.edu/homepage/crowther/ejse/ejsev7n2.html

Hagen, J.: 1996, 'Chapter 3 Lynn Margulis & the Question of How Cells Evolved'. In J. Hagen, D. Allchin & F. Singer (eds.), *Doing Biology*, Harper Collins Publishers, New York, pp. 23-36.

Hodson, D.: 1993, 'Philosophical Stance of Secondary School Science Teachers, Curriculum Experiences, and Childrens' Understandings of Science', *Interchange* **24**: 41-52.

Howe, E.M.: 2004, *Using the History of Research on Sickle-Cell Anemia to Affect Preservice Teachers' Conceptions of the Nature Of Science*. Unpublished doctoral dissertation, Western Michigan University, Kalamazoo, Michigan.

Howe, E. M. & Rudge, D.W.: 2005, 'Recapitulating the History of Sickle-Cell Anemia Research: Improving Students' NOS Views Explicitly and Reflectively', *Science & Education* **14**(3-5): 423-441.

Howe, E.M. & Rudge, D.W.: 2006, 'Using Explicit and Reflective Pedagogy and the History of Science to Affect Students' Nature of Science Conceptions: The Results of an Empirical Study', *Proceedings of the Annual International Conference of the National Association for Research in Science Teaching (NARST) San Francisco, CA, April 3-6, 2006.*

Hope, W. C.: 1999, 'Principals' Orientation and Induction Activities as Factors in Teacher Retention', *Clearing House*, **73**:1, 54-56.

Loucks-Horsley, S., Hewson, P. W., Love, N. & Stiles, K. E.: 1998, *Designing Professional Development for Teachers of Science and Mathematics.* Corwin Press, Inc., Thousand Oaks, CA.

Matthews, M.: 1994, *Science Teaching: The Role of History and Philosophy of Science*, Routledge, London.

Michigan State Board of Education: 2000 [1991], *The Michigan Essential Goals and Objectives for Science Education (K-12),* Michigan Department of Education: Lansing, MI. (http://www.michigan.gov/documents/MichiganCurriculumFramework_8172_7.pdf)

Monk, M. & Osborne, J.: 1997, 'Placing History and Philosophy of Science on the Curriculum: A Model for the Development of Pedagogy', *Science Education* **81**, 405-24.

National Academy of Sciences: 1996, *National Science Education Standards*, National Academy Press, Washington, DC

National Commission on Teaching and America's Future: 1996, *What Matters Most: Teaching for America's Future*.

National Research Council: 1999, *How People Learn: Bridging Research and Practice*, National Academy Press, Washington, DC.

National Research Council: 2000, *Inquiry and the National Science Education Standards: A Guide for Teaching and Learning*, National Academy Press, Washington, DC.

Rudge, D.W.: 2004, 'Using the History of Research on Industrial Melanism to Help Students Better Appreciate the Nature of Science' and 'The Mystery Phenomenon: Lesson Plans'. In D. Metz (ed.) *Proceedings of the Seventh International History, Philosophy and Science Teaching Group Meeting* (Winnipeg, Canada), pp. 761-772 and 773-811.

Rudge, D.W. & Burke, M.: 2004, 'Lessons Learned from the E^3 STEMTP Grant Project to Improve Middle School Science Teaching', *Proceedings of the Teacher Preparation PI Conference Arlington, VA, March 14-15, 2004.* (http://stemtec.org/piconf/Proceedings/)

Rudge, D.W. & Howe, E.M.: 2004, 'Incorporating History into the Science Classroom', *The Science Teacher* **71**(9): 52-57.

WMU Office of University Relations: 2001, *WMU Points of Pride.* http://www.wmich.edu/sub/points-of-pride.html

Beyond the Textbook: Formative Traditions, Objects, and the Science Museum of the Future

CHRISTIAN SICHAU

Deutsches Museum, München, Germany

Abstract: Science education in museums has often and for a long time followed closely the model of traditional science education in schools: The focus has been on teaching elementary principles with the help of simple demonstrations; a few 'heroic' discoveries have been interspersed which served as a foundation of a tradition. Only few historical objects came in as an add-on. In this paper I will sketch out a new approach of science education in museums in which historical objects figure prominently as the keystones of the stories told. Several examples will illustrate the potential of such an approach.

In March 2006 the Deutsche Physikalische Gesellschaft (German Physical Society, DPG) published some „Theses for a modern university curriculum for teacher students".[1] History is not present within this document. Interestingly, and consequently, under the heading "Advanced Training" programs offered by "Science Centers", "Schülerlabore" (i.e. laboratory courses for pupils by and within major research institutes), "Saturday Morning Physics" (i.e. lectures at Universities for pupils) are explicitly mentioned. Museums do not appear among the institutions listed here. This document by the DPG cannot, of course, be seen as setting decisively the official policies on science education in Germany; nevertheless, it reflects some arguments present in the discussions about science education within this country. Further, it provides evidence that at least with regard to one major actor in these discussions museums (and the history of science) are not seen as an important element of an overall strategy for science education. The tendencies present within this document get added weight if they are seen within the context of the various activities within the "Einsteinjahr" in 2005 in Germany.

Thus, the 'traditional' science museum like the Deutsches Museum has come – for two reasons – under pressure: First, the approach taken in the past is simply outdated and fails to engage the public today in any meaningful way; second, given the (heated and political) discussions about science education – its aims, its methods, its 'content' – in present society museums have to find a new position in a new and open environment which is still 'under construction'. Hence, we need to ask: What was and what is, what should and what could be the role of a museum like the Deutsches Museum in science education?

In order to re-define the possible aims and purposes of the Deutsches Museum with regard to science education it is – given its long tradition – necessary to start with an analysis of the approaches taken in the past. This will form the first section of this paper. Within such an analysis, the institution itself – its structure and character as well as the forces that act upon it and shape its course – has to be taken into account.
In the second part of my paper I will then introduce a new approach which might provide an opportunity for museums to find a very specific and unique position among all those institutions and organisations which, in one way or another, take part in science education programmes.

The Deutsches Museum as a Science Center

The Deutsches Museum was founded more than one hundred years ago.[2] Its foundation was an attempt by engineers and scientists to portray and explain their work and its importance for society to the wider public. It should re-configure the sets of cultural values attributed to certain forms of knowledge and expertise: Science and engineering was to be raised to a higher status, on level with the canon defined by the "Bildungsbürgertum". At the same time, organising and presenting technical and scientific knowledge within the museum was not only directed 'outward' to the public, but also had implications for the self-perception of engineers, scientists and their disciplines. Thus, from the outset, the museum was both: a museum <u>for</u> science and technology as well as a museum <u>of</u> scientists and engineers. They acted as advisers, reviewed lists of objects to be collected and exhibited, made further suggestions, supported the museum in its struggle to obtain the desired objects etc (see figure 1). In total, the museum should provide an encyclopaedia of accepted knowledge. The various parts and sections in the museum served to define disciplines and their relation to each other.
With a reverential look to the 'great achievements' of the past and by taking up a notion used hitherto mainly in the arts some objects were raised to the status of "Meisterwerke" ("masterpieces"). These few masterpieces were the highlights of exhibits organised along the lines of disciplines (see figure 2).

Gruppenverzeichnis und Referentenliste.

Heft	Gruppe	Referenten die Herren
I	Mathematik.	Prof. Dr. W. v. Dyck, z. Z. Rektor Magn. der Techn. Hochschule München.
"	Uhren.	Prof. Dr. Goepel, Vorstand der Württembergischen Fachschule für Feinmechanik, Schwenningen a. N. Geh. Kommerzienrat A. Junghans, Schramberg.
"	Meſswesen.	Prof. Dr. Gerland, Bergakademie, Clausthal. Prof. Dr. E. Voit, Techn. Hochschule München.
"	Geodäsie und Kartographie.	Prof. Dr. Max Schmidt, Techn. Hochschule München.
"	Astronomie.	Prof. Dr. H. C. Vogel, Geh. Oberregierungsrat, Direktor des Astrophysikal. Observatoriums Potsdam.
II	Physikalische und technische Mechanik.	Prof. Dr. Prandtl, Universität Göttingen. Prof. Dr. W. Wien, Universität Würzburg.
"	Physikalische und technische Optik.	Dr. S. Czapski, Bevollmächtigter der C. Zeiß-Stiftung, Jena. Prof. Dr. E. Wiedemann, Universität Würzburg.
"	Wärme, einschlieſslich mechanische Wärmetheorie.	Prof. Dr. W. C. Röntgen, Kgl. Geheimrat, Universität München.
"	Physikalische und tech- Akustik.	Prof. Dr. H. Ebert, Techn. Hochschule München. Prof. Dr. O. Fleischer, Universität Berlin.
III	Magnetismus und Elektrizitätslehre.	Prof. Dr. L. Grätz, Universität München.
"	Telegraphie und Telephonie einschlieſslich Funkentelegraphie.	Kgl. Oberpostrat Emil Bieringer, München. Dr. H. Scholl, Fabrikbesitzer, München.
"	Elektrotechnik.	Prof. G. Ossana, Techn. Hochschule München.
IV	Chemie, Elektrochemie.	Prof. Dr. W. Nernst, Geheimrat, Universität Berlin. Prof. Dr. W. Ostwald, Geh. Hofrat, Leipzig.
"	Chemische Groſsindustrie.	Prof. Dr. H. Bunte, Geh. Hofrat, Techn. Hochschule Karlsruhe. Prof. Dr. O. N. Witt, Geh. Regierungsrat, Techn. Hochschule Charlottenburg.
"	Zuckerfabrikation und Gärungsgewerbe.	Prof. Dr. C. J. Lintner, Techn. Hochschule München.
"	Gaserzeugung.	Prof. Dr. H. Bunte, Geh. Hofrat, Techn. Hochschule Karlsruhe. Dr. E. Schilling, München.
"	Beleuchtungswesen.	Prof. Dr. W. Wedding, Techn. Hochschule Charlottenburg.
V	Heizung und Lüftung.	Prof. H. Rietschel, Geh. Regierungsrat, Techn. Hochschule Berlin-Grunewald.
"	Kälteindustrie.	Prof. Theodor Ganzenmüller, Weihenstephan.
"	Städtehygiene.	Dr.-Ing. Stübben, Geh. Oberbaurat, Berlin-Grunewald.

Figure 1: List of external advisers to the Deutsches Museum – among them eminent scientists like Nernst, Ostwald, Prandtl, Röntgen, Wien or Wiedemann, 1905.
Photo: Deutsches Museum.

Figure 2: Special showcase with Roentgen's apparatuses, c. 1906. Photo: Deutsches Museum.

The other objects formed, again within a discipline-based perspective, the basis of genealogies ("Entwicklungsreihen") of particular instruments or technological devices.[3] By doing so, the progressive nature of technology and science was clearly marked out. Further, past and present became directly linked: Any individual working already within, entering or planning to enter the profession was given a precisely defined place in history and a sense of participating in a tradition.

However, the museum wasn't meant just to portray the history of the various disciplines and to gain more recognition in society for the achievements of scientists and engineers, it was a very important mission of the museum to explain natural phenomena and laws as well as to show how things work.[4] A set of different techniques, like working models and push-button-experiments, were used in order to achieve this aim. Thus, it comes as no surprise, that the Deutsches Museum has also been described as being "the first science center of the world" (see figure 3).[5]

Figure 3: Simple electric and electro-magnetic demonstrations in the Deutsches Museum, c. 1930s. Photo: Deutsches Museum.

Due to the impact of its foundation on museums all over the world, to its sheer size and to its exceptional position the Deutsches Museum has influenced greatly our notion what a museum on science and technology should be.[6] Building upon that image, laying great emphasis on its 'traditions', and referring to the enormous visitor numbers the museum seems to be firmly rooted, having found its position within the large field of 'science education'. As sketched out in the introduction, this position is now for various reasons increasingly questioned. Change seems immediate, but tradition and long time success make change difficult.

Thus, it might be helpful to look in more detail into the history of the museum. By doing so, we can discern that changes have already taken place in the past. They can be seen as changes of emphasis within the complex mixture of motifs and interests described above. Nevertheless, these changes were real, and they could be 'felt' – by the staff, by the scientific advisors and by the visitors. A major change within the physics galleries came with their renewal in the late 1950s and 1960s. The focus on science education became stronger and, as a consequence, many historical objects were removed in favour of more simple demonstration apparatuses.[7] The galleries became even more "science centre-like" than they had been before. "Dead Objects / Historical Objects" – this was one of two categories which appeared in internal planning documents for the new galleries. The opposing category was called: "Experiments" or "Demonstrations" (the German word used was: "Versuche"). The term 'dead object' used by one of my predecessors makes drastically clear a fundamental tension within a museum for science and technology such as the Deutsches Museum between an emphasis on basic concepts of science explained by simple demonstrations and a historical perspective on the development of science, its practice and its concepts. We can find similar confrontations today, where everything has to be 'interactive'.

The aim of the new exhibit was to explain fundamental physical principles; as the curator at the time explicitly stated, an elementary physics-textbook served as the basis of the exhibition-concept. The curator presented several arguments in order to show that history was only of secondary importance. He introduced the "man of the street" who was supposedly not interested in historical details but only in the main principles of physics. History was seen as complicating matters and therefore banned to a few sidelines for "höhere Ansprüche" ("higher demands"). [8]Another way to express this sentiment was to draw a line between "the main principles of physics" and further "details". The principles to be displayed and explained were supposedly "essentially timeless" ("ihrem Wesen nach zeitlos").[9] Linked with this approach a "Leitmotif" was formulated by the curator according to which "all complex measurement devices should be banned since they cannot help the understanding of the laws of nature, even if they are important with regard to technology."[10] This was directed not only against historical apparatuses and instruments but also modern ones: Hardly any object of contemporary research in physics managed to find a place in the new exhibit.

We can describe these changes in the conception of the physics exhibit as a form of 'purification': The messiness of physics with its increasingly blurred boundaries within a context of the emerging "techno-science" was – in the eyes of the curatorial staff – to be stopped in order to enhance the comprehensiveness of the 'basic principles'.[11] Applications like television and communication technologies which had been an integral part of the physics galleries were separated from them. Today, you will find in the physics galleries of the Deutsches Museums the consequences of this change; some parts have remained more or less still the same, some have changed its outlook but not the underlying conceptual approach.

Object-stories

With this exhibit-rationale objects were used only as "illustrations" or "representations" of seemingly 'timeless physical principles'. The physics galleries have started to resemble more a three-dimensional elementary text-book (see figure 4).

Regardless of the time of their construction and use objects were not understood as historical objects, but rather taken out of history into the realm of the abstract. Many were simply taken out of the galleries altogether, because they wouldn't function within such a concept. Furthermore, collecting was not continued on a level adequate for a national museum.

Figure 4: Entry to the section on mechanics in the new physics' galleries, c. 1959. Photo: Deutsches Museum.

In the following I will sketch a different approach in which historical objects are at centre-stage.[12] This will allow us to relate more directly with each other the various tasks of a museum: as a teaching institution, as an archive of cultural heritage, and as a research centre. Further, it will make a visit to the museum a unique experience – visitors will encounter things which are only present within a museum and not anywhere else. Additionally, as I will argue in the next part of my paper, the stories attached to those objects open up new possibilities within science education: Museums can focus more closely on their specific part within the general efforts made in science education by various institutions and organisations. The key question will be: Can we use the objects in thoughtful and provocative ways? In order to be able to answer this question we need to look more closely at the objects at hand, looking to them for evidence of points to be made, and starting to write our stories around the objects.[13] I will now discuss three examples of object-stories,[14] all related to the Einstein-exhibitions we had at Munich and at Berlin in 2005.[15]

When I started to look into our store rooms for instruments and experiments which might be useful for an exhibition on Einstein and his scientific work I came across two remarkable experimental set-ups (or at least some remains of these set-ups): One apparatus by Walter Kaufmann (1871-1947) and one apparatus by Alfred Bucherer (1863-1927). Both are roughly from the same time, 1906 and 1908 respectively; both came into the museum at nearly the same time, in the early 1920s; with both similar experiments were performed and they were historically very strongly linked (see figure 5 and 6).[16]

Figure 5: Apparatus used by Walter Kaufmann in his experiments on the determination of the change of mass of electrons with velocity (as presented to the Deutsches Museum in 1922). Photo: Deutsches Museum.

Figure 6: Apparatus used by Alfred Bucherer Kaufmann in his experiments on the determination of the change of mass of electrons with velocity (as presented to the Deutsches Museum in 1921). Photo: Deutsches Museum.

The two instruments were used to examine the deflection of very fast electrons (obtained from radioactive β-decay) in magnetic and electrostatic fields in order to measure the charge to mass ratio of these electrons. The aim of the experiments was to determine the change of mass of electrons with speed. Kaufmann had started his research before Albert Einstein had published his theory of Special Relativity. His research was firmly rooted within the discussions of an "electromagnetic worldview" (i.e. the attempt to make electrodynamics the true, all encompassing basis of physics).
However, when Kaufmann published his results in 1906 his results did not confirm the predictions based upon Einstein's theory. A discussion started about the discrepancies between theoretical prediction and experimental data. Whereas Kaufmann saw his data as clearly proving that the theory was incorrect, Einstein wanted to retain the possibility that his new theory was right. He tried to present Kaufmann's experimental data as being in fairly good agreement with his predictions – given the experimental difficulties. Without new experiments the conflict could not be solved. Thus, Alfred Bucherer stepped in. He had also for some time worked on his own electron theory and was therefore an expert in this field. After doing some experiments he switched to Einstein's theory and published his results under the title: "An Experimental Confirmation of the Principle of Relativity".
So both instruments are important for our understanding of the early history of relativity. Both instruments are among the very first which showed that particles change their mass with speed. However, when the two objects came into the Deutsches Museum in the 1920s, they were presented instead in a long line ("Entwicklungs-reihe") of electric discharge tubes. Within this particular "Entwicklungsreihe" instruments and apparatuses were brought together to establish a unity of the research on the electron. The different contexts of the various research projects were ignored; instead the contemporary understanding of the electron and its properties formed the main criterion for selecting the objects. This "Entwicklungsreihe" disappeared with the renewal of the physics galleries in the 1960s. Some discharge tubes remained in the galleries on electricity, but the 'discovery of the electron' became part of the section on atomic physics. The apparatus by Bucherer and Kaufmann had for some time found a place in the new 'historical' section on atomic physics, later they were removed from the gallery and put into the store rooms of the museum.
Thus, a rich story – rich in regard to the historical context as well as in terms of physics – has been left untold and out of sight. With the Einstein-exhibition at Berlin there came the possibility to change this situation. However, new problems emerged. As noted above, exhibitions are often based on stories which have been written (by

academics) without taking into account the material heritage. Hence, within many traditional accounts of early relativity these two objects are not mentioned or – at least – do not figure prominently. The Berlin exhibition, at least in this part and according to my opinion, did follow the traditional account. The story came first, and as the objects didn't quite fit into this prefabricated story they were presented only somewhere in the background. There was no real attempt to put the objects more to the centre of the story; thus, an opportunity was lost to go beyond textbook-accounts by making use of the objects for shedding a different light on a historical episode.
However, maybe, this was part of the problem: taking the stories about those two objects seriously would mean that we have to re-position the theory of relativity as well. Suddenly, there appears a rival theoretical concept, the "electromagnetic worldview", which has not been integrated into our currently accepted understanding of physics. It does not fit easily into the story of the progressive development of our understanding of the world.
There is a second problem here: The stories we use as the basis for new exhibits are often stories written by historians of science outside the museums. Objects do not play a significant role in many of these stories. It is the curator in the museum who has to bring in the objects – and the stories attached to them. This makes exhibition-making a very complex task for any curator involved.
I will discuss my second example only very shortly. It is a gyrocompass made by the company of Hermann Anschütz-Kaempfe (see figure 7).[17]

Figure 7: Gyrocompass made by Anschuetz-Kaempfe, developed with the help of Albert Einstein, 1926. Photo: Deutsches Museum.

With the construction of modern war ships and submarines as well as with the development of airplanes, the gyrocompass became in the early 20th century a crucial instrument for navigation, substituting the magnetic compass. Einstein got involved with it, first as an expert in a patent law suit, later as a paid consultant to the Anschütz-company. He had a holding in a patent for some improvements of the gyrocompass. The gyrocompass itself and its development would be already a good example to discuss the relation between science and technology, scientists and engineers.[18] Taking Einstein into this story, it is even more interesting as it might be used to challenge the popular image of Einstein as "theoretician". Yet, I think, this potential has not been fully exploited in the Einstein exhibitions. It has been used more in the form of an anecdote and not as a challenge to our perceived understanding of the relation between science and technology, between academia and commerce. It might be possible to discuss the gyrocompass within the context of the emergence of industrial research laboratories, of technical universities ("Technische Hochschule"), and of a class of "technical physicists" ("Technische Physiker") in the early 1920s.

With my third example, however, we move back to relativity theory. It is an interferometer used by the German physicist Georg Joos (1894-1959) in 1930 to examine the question whether an ether(-drift) exists (see figure 8).

Figure 8: Interferometer used by Georg Joos in his ether-drift-experiments, 1930.
Photo: Deutsches Museum.

The experiment is one of the many in the tradition of the famous Michelson-Morley experiments. Joos did his experiments in order to demonstrate that relativity theory is correct with regard to the claim that the speed of light does not depend on its direction. Without going into detail here, I will only draw attention to a few aspects which are important for my overall argument.[19]

First, as in the case of Kaufmann and Bucherer, Joos is also a scientist who played for some time an important role in the scientific community but is today largely unknown. His experiment does not belong to the great icons – or "masterpieces" – in the history of physics like, say the famous discoveries by Ohm, Hertz, or Röntgen. Kaufmann, Bucherer, Joos do not represent any great 'heroic' event, but they stand for the 'normal' course of science.[20] 'Heroic history' might be more suitable for the purposes of marketing, but it does distort the historical development. It is constructed out of singular events with long periods in-between which have no significance. Thus, science is construed as nothing but a series of outstanding 'discoveries'. We need the Kaufmanns, Bucherers and Joos's to correct that misleading view.

It is part of 'normal' science that here the seemingly 'great icons' are put under scrutiny: 25 years after Einstein had introduced his Special Theory of Relativity, there were still experiments going on – like the one by Joos – to '(dis-)prove' it. In order to explain this fact we have to go into the nitty-gritty details of experimental work. According to simplified textbook-accounts, Michelson and Morley had already shown in the 1880s that no ether(-drift) can be found experimentally. This story was taken up by the Deutsches Museum which in the 1920s made a model of the Michelson-Morley experiments (see figure 9).

Figure 9: Model of the Michelson-Morley-Interferometer made by the Deutsches Museum in 1922. Photo: Deutsches Museum.

The model – like many textbooks – shows how the experiment functions "in principle", and "in principle" every experiment seems to be easy and straightforward: A lamp produces a beam of light which is then split into two travelling at right angles to each other. The two beams are reflected at the end of the path by mirrors and are then superimposed. The resulting image showing interference fringes helps to clarify the question whether the speed of light was different for any of the two beams. Arguing on such a 'in-principle' level the outcome of an experiment can be nothing but decisive, there must be a clear 'Yes!' or 'No!'. We are then at a loss when we try to explain the work of Geog Joos to our visitors. We can only exploit the story of this object if we start, instead, clarifying the huge difference between an 'in-principle' description and the functioning of a real apparatus. We have to emphasize the limited and 'provisional' character of experimental work. Michelson and Morley did not show once and for all that there is no ether drift. They could only show that – given their apparatus, their experimental procedures, their theoretical assumptions and bias – any etherdrift was much smaller than they had expected. Not everyone was convinced that this was the end of debate – regardless of Einstein's theory. Too many experimental details were involved which usually are left out of the 'in-principle' description: Neither temperature nor magnetic effects nor the medium through which the light passes (air, vacuum, or water) – are part of the "in principle"-explanation but they were all very important for the construction of the instrument and the evaluation of the experiment. Thus, the Joos interferometer in its materiality makes evident the complexity of the experiment. It is not at all necessary – or even useful – to attempt explaining all technicalities of the experiment in the museum; however, we should do our best to create an awareness that our real world is a more complicated and complex thing than many textbooks want us to believe. Further, we should strive to give a better account of the nature of experimental work and science in general.

We will stay for a while with the Joos-interferometer because it provides a good example to explore the wider background of an experiment. When Joos started his project in 1925 he was right from the start aware of the fact that he had to take the political aspects of his scientific enterprise into account. Writing to his friend and colleague Arnold Sommerfeld in December 1925 about the first planning stages, Joos remarked: "We want to handle the affair very discretely, because otherwise obstructionism might occur. Among non-physicists here Relativity theory is considered to be a subversive activity."[21] The experiment clearly had an important political dimension. Hence, it comes as no surprise that in Joos' publication in the "Annalen der Physik" in 1930 one is looking in vain for the name 'Einstein' or the word ‘Relativity theory’. Joos watched carefully his wording in the various accounts he gave about his work. This political side of the experiment was also present when, in 1935, the interferometer was offered to the Deutsches Museum by the Zeiss company. The instrument travelled from Jena to Munich with the meaning already established at Jena in its baggage and both were erected together in the museum’s gallery. The positioning by Zeiss and Joos of the instrument provided the basis for the presentation in the museum, in the gallery as well as in guide-books. According to its own conceptual basis, the museum defined the interferometer as a “hervorragendes Meisterwerk neuzeitlicher Präzisions-mechanik und -optik” (“outstanding masterpiece of modern precision mechanics and optics”). By using this terminology the apparatus was presented as an exemplar of a strong specific German tradition: precision mechanics and optics. The politically potentially problematic scientific dimensions of the experiment – the association with Einstein’s theory of relativity – were thus further moved to the background. Only years after the end of the second-world-war, the linkage between the experiments and their relevance for the theory of Relativity was re-installed in the museum.

So, there is another very interesting story to be told. But it is a story which raises important questions and the answers may seem to be disturbing in various aspects. We are not used to discuss the political dimensions of science, doing so would mean to change the very idea of science traditionally held, or: the idea of science as presented in the ‘purified’ museum-version. Further, we are also confronted with the issue of the museum’s authority.

However, although this part of the story was not told, the Joos interferometer could fulfil for some time an important function which had been attached to it during its early career in the museum: It served as a signal to all visitors and interested parties that the museum did not only show antiquated objects but also kept up with the development of modern experimental physics.[22] According to the curator at the time, the interferometer belonged to the group of ‘modern apparatuses’ which stood opposite the ‘historical instruments’.[23] For this reason the interferometer (supposedly) became ‘a main point of attraction in the physics gallery’.[24] However, with the renewal of the physics galleries in the late 1950s and 1960s Joos’ apparatus was finally dismantled. Not only had the apparatus grown older and the excitement about Relativity had abated, but, even more importantly, it couldn’t find a place within the new concept. The new curator chose as a “Leitmotif” for the new exhibit the proclamation, already quoted above, that all complex measurement devices should be banned from the galleries. Hence, the Joos-interferometer was put into the store rooms of the museum in 1957.

Conclusion

The decision to remove the interferometer shows clearly the shift of emphasis that came with the new physics exhibit in the late 1950s and 1960s. As I have indicated, this shift has proved very influential; it has dominated the physics galleries up till today. It fits well with 'modern' tendencies in exhibit-making. The various activities in the Einstein year in Germany can be characterised in this respect by the following elements. First, it is especially young kids which are targeted by most activities. Especially with regard to the discussion of the PISA-results in German society children are now seen as the primary group which has to be addressed.[25] Second, the dominant aim of the activities is to raise 'excitement', in most cases no further educational aims are specified. In order to achieve this 'interactives' form the focus of many exhibits. The term 'interactive' is used to make an explicit contrast to historical documents, ('static') objects, photographs, and texts. 'Interactives' are seen as the key to make science more interesting for the general public; the argument of the past that no one would be interested in historical 'details' is now repeated in a slightly different form.[26] Thus, with these 'interactives' science is presented in the light of simple natural phenomena. A second method to induce 'excitement' consists in introducing the public to modern scientific research.

Historical objects and the stories attached to them do not fit into such a programme which tends to make science (again) appear 'timeless'. They need a different environment; one which is friendlier to the many facets of scientific practice (and its historical development). Hence, a discussion is very much needed on the aims of the various educational activities; however, within our present-day 'event-culture' and the widespread talk on 'making science exciting' this discussion will prove very difficult.

The few examples given here may indicate that object-stories do have a huge potential for science education: Elementary physical problems are mixed with experimental difficulties, fundamental beliefs with pragmatic political concerns, curiosity with commercial interests and so on. The stories attached to the historical objects invite us to explore more deeply the nature of science and its historical development. However, in order to be able to follow this path there is a central precondition: We need more and better object-related research. We need a new history of science which from the very beginning integrates instruments and apparatuses into the stories. Further, we need a willingness to create more room for manoeuvre for science education in the museum. Thus, there is a second precondition: We need a more differentiated and nuanced view on science education which attributes different roles to the various actors involved.

Notes

[1] The document is published under the title: "Thesen für ein modernes Lehramtsstudium im Fach Physik – Eine Studie der Deutschen Physikalischen Gesellschaft e.V.", March 2006. It is available for download under: www.dpg-physik.de (19.10.2006).

[2] For the history of the museum see, for example: (Duffy 2002), (Füßl & Trischler 2003), (Füßl 2003), (Mayr 2003). For the development within the department of physics: (Fuchs 1957). Fuchs was the first curator in physics.

[3] The relation between the masterpieces and the 'Entwicklungsreihen' was described by Fuchs in the following form: "Die Meisterwerke dieser Forscher … bilden die Marksteine der Entwicklung. (…) Hier, vor diesen Werken, die wie ein Denkmal aus der verwirrenden Fülle aufragen, soll der Besucher sich sammeln und in Ehrfurcht und Dankbarkeit der Großen seines Volkes gedenken. Die Begeisterung beim Anblick der Leistungen bedeutender Männer wandelt sich in ernstes Streben, wenn der Besucher sich den Entwicklungsreihen physikalischer Erkenntnisse und Apparate zuwendet …" (Fuchs 1925, p. 104).

[4] Fuchs already pointed out the great importance of demonstrations in his publication: "Neben den Originalapparaten und Nachbildungen berühmter historischer Apparate sollten die Physiksäle auch mit Demonstrationsapparaten ausgestattet werden, an denen den Besuchern die fundamentale Bedeutung der ausgestellten Objekte gezeigt werden soll." However, as this citation indicates, most demonstrations had a specific purpose and a close relation to the historical objects – at least: this was the intention. Fuchs stated: "Es erschien v. Miller wichtig, die demonstrative Erklärung eines Apparates in unmittelbarer Nähe des Originals zu stellen, und so legte er die Führungslinie logisch durch die historischen Reihen und Demonstrationen hindurch" (Fuchs 1957, p. 17). This changed in the 1950s and 1960s, see below.

[5] See, for example: (Fehlhammer 2005).

[6] For a detailed discussion of the impact of the Deutsches Museum on similar institutions see. (Finn 2003).

[7] How controversial this shift of emphasis was can be seen in various documents (correspondence, minutes of meetings etc.) held in the Verwaltungsarchiv (VA) of the Deutsches Museum. In a letter from a member of the "Vorstand" to an external adviser the rationale of the new exhibition was explained: "Zunächst wollen wir das didaktische Prinzip wesentlich stärker hervorheben, als das früher der Fall war. Dem Schüler oder dem 'Mann von der Straße' sollen vor allem die physikalischen Grundbegriffe und ihre Bedeutung für die Technik in einfachster und leichtest durchschaubarer Form klar gemacht werden. (…) Der zweite Leitgedanke ist der, dass wir um des besseren Überblicks willen – wie das auch andere

Museen gemacht haben – aus der großen Fülle der ausgestellten Objekte vieles in das Depot verweisen müssen" (Letter from W.O. Schumann an Prof. Bergmann, 30.01.1959, DM VA).

[8] In 1956 Eduard Maurer, successor of Fuchs as curator of Physics, explained in a planning meeting on the mechanic-section, "daß er ganz zweckbewusst auf die Hervorhebung der elementaren Volksschulphysik hinziele. Sie biete nach seiner Ansicht die beste Grundlage zur Darstellung einer Physik für den 'Mann von der Straße' " (DM VA). In the same meeting W.O. Schumann (see note 7) gave support to Maurer by stating: "[M]an wolle vor allem Physik für die Schuljugend bringen. Die Volksschulphysik sei nicht so schlecht, wenn sie richtig gebracht werde. Sie solle das Skelett bilden, das auch der flüchtige Besucher zu sehen bekomme. Die Nebenlinien könnten dann den höheren Ansprüchen dienen."

[9] This phrase was used in an argument about the new introductory room for mechanics. In the minutes of a planning meeting from 02.07.1956 it is stated: "Diese zur Aufstellung vorgesehenen Gegenstände seien ihrem Wesen nach zeitlos" (DM VA).

[10] Undated note in the Verwaltungsarchiv of the Deutsches Museum, the German wording was: "Zu verbannen sind aber alle komplizierten Meßinstrumente, deren nähere Kenntnis zum Verständnis der Naturgesetze doch nichts beiträgt, mögen sie auch für die physikalische Technik, speziell die physikalische Meßkunst, von höchster Bedeutung sein." According to the note, this 'Leitmotif' seems to have been attributed to John Tyndall. However, whereas another part of the note can indeed be traced back to a German translation of Tyndall's "Lectures on Electricity", I have not been able to locate the source of the quotation given above.

[11] Eduard Maurer repeatedly expressed in various meetings and letters a concern, that "die Physik zu ausführlich dargestellt und der Besucher verwirrt und überfordert werde", „dass das historische Institut zuviel Wissen in die Abteilung hineinstopfen möchte", or that „eine Überforderung des Besuchers und ein ‚Rumpelkammereindruck' zu vermeiden, die Abteilung nach Möglichkeit zu ‚lichten' [sei]" (DM VA, see note 7).

[12] There already exists a body of interesting literature on object-based history of science and object-based exhibition-making, see for example: (Kopytoff 1986), (Hörning 1990), (Butler 1992), (Durant 1992), (Pearce 1992), (Silverstone 1992), (Lubar & Kingery 1993), (Hochreiter 1994), (Arnold 1996), (Kingery 1996), (Pearce 1996), (Sichau 2000), (Bennett 2005), (Alberti 2005).

[13] This question has been asked by B. Finn, curator at the Smithsonian Institution in Washington. The full quotation is as follows: "Can we use the objects in thoughtful and provocative ways? (…) Stories need to be written around the objects, looking to them for evidence of points to be made. Much more often the story comes first (from the academics) and the objects are hung on as ornaments. We need to force ourselves to do otherwise." (Finn 2000, p. 157)

[14] A full analysis would require a more detailed discussion on techniques of museum display, design, the use of media, the role of accompanying texts, guided tours etc. Thus, the examples

given here serve only to show on the one hand the potential of object-stories and on the other hand some conceptual difficulties.

[15] Both exhibitions were accompanied by catalogues: (Bracher; Hartl; Sichau 2005) and (Renn 2005).

[16] For a more detailed account of the story and for further references see: (Sichau 2005a).

[17] A full historical analysis on the development of the gyrocompass is given in: (Broelmann 2002).

[18] According to Fuchs, gyroscopes and gyrocompasses were originally presented together in the section on mechanics in the physics department where the technical application of the gyroscope figured prominently in the exhibit; see (Fuchs 1957, S. 16).

[19] For further information on the Joos-interferometer, its story and for further references see: (Sichau 2005b; Sichau 2005c; Sichau 2007).

[20] At the 2006-meeeting of the "Artefacts"-conference series the idea of the "icon" was discussed from various perspectives ("Constructing and Deconstructing Icons of Achievement in Science and Technology", Stockholm, Nobel Museum, 14.-16.09.2006). By introducing the term 'normal science' I am not referring to Thomas Kuhn's notion of 'normal' and 'revolutionary' science; instead, the term 'normal' is used in contrast to the term 'icon', i.e. singular events in the history of science described as 'discovery' or 'breakthrough'. The collective construction of 'discoveries' – that is the process how experiments or publications etc. are transformed and turned into acts of 'discovery' – and their role within scientific disciplines is discussed in: (Caneva 2005). Caneva's analysis is much in line with and sheds some more light on the concepts used in the Deutsches Museum in the early 20-century.

[21] The German original text is: "Wir wollen aber hier die Sache ziemlich diskret behandeln, da sonst Quertreibereien möglich sind. Unter Nichtphysikern hier gilt nämlich die Relativtätstheorie zum Teil als staatsgefährdend" (Letter from Georg Joos to Arnold Sommerfeld, 30.12.1925, Archive of the Deutsches Museum).

[22] To keep the galleries 'up-to-date' and to incorporate modern apparatuses was an important issue for the Deutsches Museum, especially for Jonathan Zenneck who was head of the Deutsches Museum from 1933 to 1953. In 1938, for example, Zenneck visited the Conservatoire des Arts et Métiers in Paris. He was not impressed, to the contrary. According to his opinion the Conservatoire had become like a 'dead tree' because it had stopped growing. It had not taken in the latest developments of science. Thus, for Zenneck it was nothing but a collection of antiquated objects. In order to avoid such a development the Deutsches Museum should constantly try to bring in new objects within its galleries (Zenneck 1942).

[23] Fuchs gave the characterisation of the Joos-interferometer quoted above for example in: (Fuchs 1942) and (Fuchs 1950).

[24] On the occasion of Joos' 60th birthday and later in various obituaries after his death Walter Meißner used the phrase, that the interferometric apparatuses used by Joos "gehören im

Deutschen Museum zu den Glanzstücken der physikalischen Sammlung"; see, for example: (Meißner 1954).

[25] Bundesministerin Edelgard Bulmahn programmatcially announced: "Viele Veranstaltungen richten sich ganz besonders an junge Menschen, an die 'Erben' Einsteins. Gerade sie wollen wir für Forschung und wissenschaftliche Themen begeistern. Wettbewerbe, Kinder-Universitäten, Schülercamps, Ausstellungen, Ausprobieren und Staunen sind dabei" (Rede der Bundesministerin für Bildung und Forschung, Edelgard Bulmahn, anlässlich der Eröffnung des Einsteinjahres 2005, am 19.01.2005 in Berlin). This special focus was justified by referring to the results of the PISA-study for Germany. Thus, in a concept-paper for the Einstein-year it was explicitly stated: "Nicht erst seit PISA wird beklagt, dass die Kenntnisse von Schülern in den Naturwissenschaften außerordentlich lückenhaft sind. Es ist daher allgemeiner Konsens, dass Schülerinnen und Schüler eine herausgehobene Zielgruppe im Einsteinjahr sind" ("Einsteinjahr 2005. Eine gemeinsame Initiative von Bundesregierung, Wissenschaft, Wirtschaft und Kultur – Konzept", veröffentlichlicht vom Büro Einsteinjahr 2005 Berlin, 17.09.2004).

[26] Thus, for example, we can read in a museum-publication: "Zu viele Dokumente und Fotos, noch unterstützt durch lange wissenschaftliche Texte, können heute kaum noch mit dem Interesse eines großen Publikums rechnen. Deshalb muss ein wirkungsvoller, interaktiver Gegensatz dazu aufgebaut werden." In this museum broschure it is further stated, that instead of explaining all known historical facts about Einstein's life the focus would be on "einleuchtende Demonstrationen und einfach, d.h. verständlich aufgebaute physikalische Experimente" ("Einstein begreifen : Begleitheft zur großen Sonderausstellung des Landes Baden-Württemberg", Landesmuseum für Technik und Arbeit Mannheim, 2005). According to this statement, there exist two opposite techniques in exhibit-design: The 'old-fashioned' visual and textual (object-based?) approach and the new emphasis on 'interactives'. Scientific explanations are per se declared to be uninteresting and too long. Similar statements can be found in many more publications by various museums and organisations, especially in the context of the Einstein-year.

References

Alberti, S.: 2005, 'Objects and the Museum', *ISIS* 96, 559-571.

Arnold, K.: 1996, 'Presenting Science as Product or as Process: Museums and the Making of Science'. In: (Pearce 1996, pp. 57-78).

Bennett, J.: 2005, 'Museums and the History of Science – Practitioner's Postscript', *ISIS* 95, 602-608.

Brachner, A. & Hartl, G. & Sichau, C.: 2005: *Abenteuer der Erkenntnis. Albert Einstein und die Physik des 20. Jahrhunderts*, Deutsches Museum, München.

Broelmann, J.: 2002, *Intuition und Wissenschaft in der Kreiseltechnik 1750 bis 1930*, Abhandlungen und Berichte des Deutschen Museums, Neue Folge, 17, München.

Butler, S.: 1992, *Science and Technology Museums*, Leicester University Press, Leicester.

Caneva, K.: 2005, 'Discovery as a site for the collective construction of scientific knowledge', *Historical studies in the physical and biological sciences* 35 (2), pp. 175-291.

Duffy, E.: 2002, *Representing science and technology: Politics and display in the Deutsches Museum*, Chapel Hill, Univ. of North Carolina, Diss.

Durant, J. (ed.): 1992, *Museums and the Public Understanding of Science*, Science Museum, London.

Fehlhammer, W.: 2005, 'Science Center – Irrläufer der Edutainment-Welle oder Herausforderung für die Industrie- und Technikmuseen?'. In: H. John & I. Mazzoni (eds.), Industrie- und Technikmuseen im Wandel. Standortbestimmung und Perspektiven, Transcript Verlag, Bielefeld, pp. 163-172.

Finn, B.: 2003, 'Der Einfluss des Deutschen Museums auf die internationale Landschaft der Wissenschafts- und Technikmuseen'. In: (Füßl & Trischler 2003, pp. 397-405)

Fuchs, F.: 1925, 'Die Physik und die Astronomie im Deutschen Museum', *Kosmos – Handweiser für Naturfreunde* 4, 104-113.

Fuchs, F.: 1942, 'Neuzugänge in Physik und Astronomie', *Industria – Blätter der Frankfurter Zeitung für Technik und Wirtschaft*, 16.07.1942.

Fuchs, F.: 1950, 'Die naturwissenschaftlichen Gruppen im Deutschen Museum', *Naturwissenschaftliche Rundschau* 10, 468-469.

Fuchs, F.: 1957, *Der Aufbau der Physik im Deutschen Museum 1905 – 1933*, Abhandlungen und Berichte des Deutschen Museums 25 (3), München.

Füßl, W.: 2003, 'Konstruktion technischer Kultur. Sammlungspolitik des Deutschen Museums in den Aufbaujahren 1903 – 1909'. In: U. Hashagen et al. (eds.), *Circa 1903. Artefakte ind er Gründungszeit des Deutschen Museums*, Abhandlungen und Berichte des Deutschen Museums, Neue Folge, 19, München, pp. 32-53.

Füßl, W. & Trischler H.: 2003, *Geschichte des Deutschen Museums. Akteure, Artefakte, Ausstellungen*, Prestel, München.

Hochreiter, W.: 1994, *Vom Musentempel zum Lernort*, WBG, Darmstadt.

Hörning, K.: 1990, 'Vom Umgang mit den Dingen. Eine techniksoziologische Zuspitzung'. In: P. Weingart (ed.), *Technik als sozialer Prozeß*, Suhrkamp, Frankfurt, pp. 90-127.

Kingery, W.D. (ed.): 1996, *Learning from Things: Method and Theory of Material Culture Studies*, Smithsonian Institution Press, Washington.

Kopytoff, I: 1986, 'The Cultural Biography of Things'. In: A. Appadurai (ed.), *The Social Life of Things: Commodities in Cultural Perspective*, ed. Cambridge University Press, Cambridge (UK), pp. 64-91.

Lubar, S. & Kingery, W.D. (eds.): 1993, *History from Things: Essays on Material Culture*, Smithsonian Institution Press, Washington.

Mayr, O.: 2003, Wiederaufbau: Das Deutsche Museum 1945 – 1970, Abhandlungen und Berichte des Deutschen Museums, Neue Folge, 18, München.

Meißner, W.: 1954, 'Zum 60. Geburtstag von Georg Joos', *Zeitschrift für angewandte Physik* 6 (5), 193-194.

Pearce, S.: 1992, *Museums, Objects, and Collections: A Cultural Study*, Leicester University Press, Leicester.

Pearce, S. (ed.): 1996, *Exploring Science in Museums*, Athlone, London.

Renn, J. (ed.): 2005, *Albert Einstein – Chief Engineer of the Universe. Einstein's Life and Work in Context*, Wiley-VCH, Weinheim.

Sichau, C.: 2000, 'Die Joule-Thomson-Experimente: Anmerkungen zur Materialität von Experimenten', *NTM* 8, 222-243.

Sichau, C.: 2005a: '6qm Wandfläche und zwei deplatzierte Artefakte: Relativitätstheorie im Deutschen Museum'. In: J. Renn (ed.): *Albert Einstein – Ingenieur des Universums. Hundert Autoren für Einstein*, Wiley-VCH, Weinheim, pp. 170-173.

Sichau, C.: 2005b, 'Das allmähliche Verschwinden von Einstein: Die Experimente von G. Joos zur Relativitätstheorie'. In: J.Renn (ed.): *Albert Einstein – Ingenieur des Universums. Hundert Autoren für Einstein*, Wiley-VCH, Weinheim, pp 194-199.

Sichau, C.: 2005c, 'Eine Äthersuchmaschine. Die Experimente von Georg Joos zur Relativitätstheorie'. In (Brachner et al. 2005, 78-85).

Sichau, C.: 2007, 'Hard Times – The Difficult Lives of Three Instruments in the Museum'.In: *Artefacts – Studies in the History of Science and Technology*, London, accepted.

Silverstone, R.: 1992, 'The Medium is the Museum: on Objects and Logics in Times and Spaces'. In: (Durant 1992, pp. 34-42.)

Zenneck, J.: 1942, 'Die Anpassung von naturwissenschaftlichen und technischen Museen', *Technikgeschichte – Beiträge zur Geschichte der Technik und Industrie* 30, 143-148.

From Theory to Practice: Placing contextual science in the classroom

ARTHUR STINNER
University of Manitoba, Canada

Abstract: In this article I will present a brief background discussion of the need for contextual teaching in science (physics) with emphasis on the large context problem (LCP) approach that was developed at the University of Manitoba. In the Appendix I will place a detailed outline of a book/DVD that I am now preparing, entitled: *From Intuitive Physics to Star Trek: Large Context Problems (LCP) to enrich the teaching of physics.*

The contextual approach to the teaching of physics

In developing a program of contextual science my aims were, first, to liberate teachers from the tyranny of textbooks, and secondly, to humanize the teaching of science by placing it in rich contexts, informed by the nature and history of science. Research in science education shows that students often leave school with a knowledge of isolated, disconnected facts, based on memorization and problem solving ability by the use of algorithmic procedures. Textbook-centered teaching also promotes a poor understanding of the nature of science and a spotty knowledge of the history of science (Rigden, 1991). All this, however, does not suggest the need for a total replacement of textbooks and the abandonment of teaching algorithmic procedures. It only recommends the displacing of textbooks from their conventional central position to one of a good source of reference. Algorithmic procedures should still be taught, not just as an aid to memorizing but rather as an economic way to solve certain important class of problems. Using textbooks as reference only and teaching algorithmic procedures should be more along the lines suggested by Kuhn in providing the understanding of the central problems of physics that he first called exemplars, or model solutions (Kuhn, 1962).

Working with exemplars involves working with instruments in the laboratory and practising problem solving. The well-known exemplars of elementary physics are connected with *the inclined plane*, *the pendulum*, *Atwood's machine*, *the ballistic pendulum*, *the wave-tank*, and more recently, *the electronic air table*. In his *The Structure of Scientific Revolutions*, Kuhn argues that a physics student becomes acquainted with and discovers the fundamental notions of Newtonian dynamics only through the application of these concepts to both problem solutions in the laboratory and on paper. He argues that this is the way students should make "contact with nature", Students do not learn physics by reading the definitions of *force, mass, instantaneous velocity,*etc. and the memorization of algorithms which are then applied to "type" problems in the textbooks.

According to contextual learning theory, learning occurs only when students (learners) process new information or knowledge in such a way that it makes sense to them in their own frames of reference. This approach to learning and teaching assumes that the mind naturally seeks meaning in context, and that it does so by searching for relationships that make sense and appear useful (Driver, 1989).

Research also shows that presenting major concepts and ideas in science imbedded in context by way of a coherent story is essential in retaining these concepts and ideas. Indeed, if science is taught as a 'fact' without context or a coherent story line, students will try to invent their own. It is arguably a truism that appropriately designed contexts which attract young students' interests often create great motivation to learn science (Kenealy, 1989).

Setting appropriate contexts, however, go beyond serving as a source of motivation. There is strong evidence that we must connect cognitive activity to context, that learning methods imbedded in context are not merely useful; they are essential. In spite of a general agreement about the importance of contexts our textbook-centered teaching in science seems to stubbornly ignore it (Roth, 1993).

The Large Context Problem (LCP)

The superiority of a contextual approach over the conventional textbook-centered teaching in physics became clear to me after designing and successfully using my first contextual setting for a senior physics class in a Canadian (Toronto) high school. What I later came to call the "large context problem" (LCP) approach was originally developed as a response to the discovery that *learning could be well motivated by a context with one unifying central idea capable of capturing the imagination of the students.* The second aim for developing contexts was that the topics, the concepts, and the general content of the textbooks should be covered by the context. What I found was that the questions and problems generated by the context were intrinsically more interesting and less contrived than the corresponding ones found in textbooks.

Early experiences using LCPs

I have discussed the rationale and the design of LCPs, embedded in a rich theoretical Background provided by the *contexts of inquiry* in detail elsewhere (Stinner, 1989a, 1989b, 1994); Stinner & Williams, 1993). LCPs are contextual settings that are designed by the teacher in collaboration with students. Each LCP should be so designed that most of the physics for a particular topic would have to be used for the successful completion of the problems suggested by the context. What is so attractive about this kind of setting is that the questions and problems are generated *naturally* by the context and will *include* problems that are given *out of context* (in a contrived way) in a textbook for a given topic. Designing contexts on this scale gives the instructor the status of researcher and the student the feeling of participating in an on-going research program.

Indeed, many of the questions and problems generated do not have obvious answers for the student or the instructor. The ability to answer questions and solve problems that do not have textbook answers, using elementary physics only, is very rewarding for both students and teachers. A contextual approach to the teaching of physics may be more time-consuming then the conventional textbook approach. However, the understanding of the student as well as the quality of interaction between the student and the teacher is lifted from an ordinary to a high-grade level. Indeed, solving problems that are naturally generated by a context that attracts the imagination of the student are more likely to make contact with nature than solving contrived problems in text books.

Examples of LCPs that we have developed over the years are: "Physics and the Bionic Man", "The Physics of Star Trek", "Physics and the Dam Busters", "Hitchhiking on an Asteroid", "Calculating the Age of the Earth and the Sun", "Pursuing the Ubiquitous Pendulum", and "Sudden Impact: The Physics of Asteroid/Earth Collisions"(See references).An example of how one can use the contexts of inquiry and the history of science in a major topic is : 'The Story of Force: From Aristotle to Einstein" (Stinner, 1994 b).

Even a cursory survey of journals like *The Physics Teacher*, will provide the physics instructor with plenty of examples of such contextual settings: "The Physics of the Play Ground", "The physics of Toys", "Physics and Skiing", and historical surveys, like "Is Maxwell's displacement current a current?", "Newton's Thermometry: The role of Radiation" and "The search electromagnetic induction". These could be easily adapted and transformed into LCPs or investigations using *the concepts of inquiry*. Later I presented general guidelines for the planning and the development of large context problems, as can be seen below in the Appendix.

In my science education classes at the University of Manitoba I have had students design LCPs for science in general and physics in particular with notable success and enthusiastic cooperation. Many of these LCPs were later used in physics classes when these students began teaching physics. Can we place LCPs in a central position in existing curricula and teaching practices?

Originally the LCP was placed peripherally to the textbook and the curriculum. I used LCPs mainly to reinforce core material to the extent that time was available. The high school Physics curricula in Canada (in most provinces) at that time generally consisted of a core Cntent surrounded by "options". The core and the sequencing of topics was conventional.

The options were contexts such as "Solar Energy" and "Motion: Earth and Sky" that I realized could be developed into good LCPs. Unfortunately, the options were generally considered "interesting supplementary material", and if used at all, discussed only in a hurried manner. Both teachers and students are primarily interested in "covering the material" in preparation for the next level of physics or science course. The mandate to "cover" the material in the curriculum is still with us (Stinner, 1994).

Examples of innovative physics teaching

Innovations in science (physics) teaching all seem to rest on one simple premise:
a better learning experience results from an active engagement of the student. Many of these innovations can be placed in the following categories: (1) microcomputer–based laboratories, (2) active engagement in lectures, (3) collaborative learning, and (4) structured problem solving. Priscilla Laws and her group at Dickinson College replaces the standard calculus-based physics course; and the virtual physics experimental site. Physics 2000, are good examples of the first approach. Harvard professor of physics , Eric Mazur's *Peer Instructor*, and the physics education research group at the University of Minnesota have developed "rich context problems" for collaborative learning (Stinner, 1994).

I would also like to mention the work of my colleague Wytze Brouwer at the University of Alberta in improving physics education. The detailed account of the collaborative approach, replacing the conventional lecture-centered teaching of large classes I first year physics, is described in his article (Brower, 1995).

The work of Paul Hewitt is well known. His book *Conceptual Physics* is a successful attempt to present the qualitative aspect of the concepts of elementary physics. This is done through visuals, demonstrations, hands-on, minds-on activities, verbal explanations, and dialogues. There is also a quantitative aspects to this approach but it is kept to a minimum (Hewitt, 1990).

Finally, James Trefil and Robert Hazin's *The Sciences: An Integrated Approach*, also tries to present the concepts of physics of physics, but uses much more quantitative support. Laws and definitions are given verbally, graphically and pictorially first, but and only then expressed symbolically. It is a nice attempt to balance the quantitative

and qualitative aspects of physics, somewhere between the conventional textbook-centered approach and Hewitt's conceptual approach.
Even a cursory survey of journals like *The Physics Teacher*, will provide the physics instructor with plenty of examples of such contextual settings: "The Physics of the Play Ground", "The physics of Toys", "Physics and Skiing", and historical surveys, like "Is Maxwell's displacement current a current?", "Newton's Thermometry: The role of Radiation" and "The search for electromagnetic induction". These papers could be easily adapted and transformed into LCPs or investigations using the approach described here.

Local efforts

In the recent physics curriculum reforms in the province of Manitoba contextual teaching and history of science has been incorporated to provide a more humanistic approach to learning physics. In *senior one* students investigate historical models of charge and build and perform experiments with historical apparatus such as the electrophorous. In grade 11 the historical development of the model of light is used as a backdrop to study models in general, as well as laws, and theories explicitly. Several historical case studies (such as the life of Madam Curie (Radioactivity), the history of rock and roll (electromagnetism), and the Doppler effect) have been developed by our teachers which can be implemented in the classroom.
There is also considerable anecdotal support for the success of contextual teaching by my former students who are now classroom teachers. We have one study that measured both academic and attitudinal in a senior high school chemistry class where interactive historical vignettes were introduced. Our conclusion was that academically there was no significant improvement but the attitudes of students toward chemistry measurably improved. We are about to publish these findings.
I am hopeful that the book and the DVD I am now developing will be available by the late spring of 2007. What remains is a systematic testing of contextual using the LCP approach in a high school class. I am looking forward to a large scale study that uses the contextual approach as outlined here, where textbooks are consulted, but not centralized.

Concluding remarks

The foregoing discussion about the role of imagination in science, however, suggests that a fairly radical change from the conventional text book-centered physics teaching is required if we want to teach an authentic science (physics) to future physicists or scientific (physics) literacy to the general public. Granted, the main task of physics educators has always been to prepare young people for physics research and the professions that require a good basic understanding of physics. Today, however, we must go further and ensure that all students leave school with a basic scientific literacy that includes a knowledge of elementary physics. In order to achieve these goals, we have to make our learning contexts richer and more challenging for both the university-bound student who is required to study physics and the student who is looking for general physics literacy. However, providing rich contexts is a necessary but not sufficient condition for successful teaching of authentic science. We must also educate and train young science (physics) teachers to have a good understanding of the nature and history of science and of how students learn science.

Cutting the umbilical chord with textbook-centered teaching will be successful only when teachers of science (physics) have a deep understanding of the *contexts of inquiry* and have a more than cursory acquaintance with the history of science. We should try to convince textbook writers and publishers to write and publish post-Kuhnian text books that pay attention to the requirements of good pedagogy, the importance of contextualizing the teaching of physics, the nature and the history of science.

Idealistically, we want to help students to traverse, what Alfred N. Whitehead referred to in his "The aims of education", as "the path from romance to precision to generalization" in our teaching of physics (Whitehead, 1929).

APPENDIX 1

From Intuitive Physics to Star Trek:

Large Context Problems (LCP) to enrich the teaching of physics.

I am now preparing a book and a DVD of Large Context Problems (LCP) to enrich the teaching of physics. The book will contain the chapter topics listed below, and the DVD the LCP, also listed below:

Outline of the book

CHAPTER 1: THE IMPORTANCE OF CONTEXT

- The contextual approach to the teaching of science.
- The senior years: toward a comprehension of organized scientific knowledge
- Contemporary contextual approaches.

CHAPTER 2: THE PHYSICS CURRICULUM

- A brief history of physics teaching
- The science curriculum
- Physics and scientific literacy

CHAPTER 8: THE IMPLEMENTATION OF THE LARGE CONTEXT PROBLEM.

A list of interactive contexts, large context problems, historical contexts, and case studies are given.

A suggested classification of the above into the traditional physics curriculum,

According to: Kinematics, Dynamics, Wave Motion, Optics, Electricity and Magnetism is given.

CHAPTER 9: THOUGHT EXPERIMENTS AND CONTEXTUAL TEACHING.

The function and place of thought experiments in the teaching of physics.

These thought experiments will include those of Galileo, Newton and Einstein.

The content of the DVD:

LCP 1: INTUITIVE PHYSICS AND MOTION

In this LCP we will test and challenge students' conceptual understanding of motion.

History of science: Ideas of Aristotle, Galileo and Newton

LCP 2: MOTION AND THE PENDULUM

This context is based largely on "The Pendulum: Scientific, Historical, Philosophical, & Educational Perspectives", and on the work of the author and Don Metz (The Ubiquitous Pendulum, *The Physics Teacher*, (2003). Almost all elementary kinematics and dynamics can be taught and/or reinforced.

LCP 3: GALILEO, NEWTON, AND ROBOTICS

This LCP is based on four sources: Galileo's "Two New Sciences", G.B.S. Haldane's celebrated article "On Being the Right Size", published in 1928,

Mel Siegel's, a robotics research professor, "When Physics Rules Robotics";

published in 2005, and, finally, the author's updated version of the article.

"Physics and the Bionic Man" published 25 years ago in *The Physics Teacher* and *New Scientist.*

LCP 4: MACROROBOTS FOR ENERGY PRODUCTION

The macrorobots we will discuss are the Giant Wind Turbines (GWT) and the giant solar furnace in Souther France (GSF). Discussion of electricity, magnetism, mechanical energy, radiation, optics, wave motion, thermodynamics, solar energy, thermonuclear reactions, and BB radiation.

LCP 5: A ROTATING SPACE STATION (RSS).

The design and the physics of a RSS is presented, based on *2001: Space Odyssey.*

A three tear-level system for training astronauts to go to the Moon and Mars will be investigated. The physics that is part of this context is dynamics and gravitational theory. The historical context discussed is Newton's description of satellite motion in the *Principia.*

LCP 6: PHYSICS ON THE MOON:

The physics of low gravity environment is investigated and the physics of living on the Moon discussed. Structures, mobility, astronomic observations, Olympic games, etc are topics investigated. The physics involved is elementary kinematics and dynamics, the strength of material discussed in chapter 1.

LCP 7: JOURNEY TO MARS: THE PHYSICS OF TRAVELLING TO THE RED PLANET

(This LCP is based on my article of the same title, published in *Physics Education*, 2005).

The history of the importance of Mars to the understanding of the solar system is presented. Several scenarios to travel to the red planet will be described and the physics of the journey explained. An interactive computer program that we developed will allow students to plan their own journey.

LCP 8: THE AGE OF THE EARTH AND THE SUN

(This LCP is based on my article of the same title, published in *Physics Education* in 2002)

Review of main attempts made to calculate the age of the earth and the sun, from Newton's thought experiment to Bethe's thermonuclear model of the sun's energy

Special attention is paid to the debate about the age of the earth in the second half of the nineteenth century that involved Kelvin and Helmholtz.

LCP 9: ASTEROID/EARTH COLLISIONS

(This LCP is based on the article of the same title, written with Don Metz, that was published by *The Physics Teacher*, in 2002).The physics of asteroid/Earth collisions will be discussed.

The physics of a simple computer model for "impact scenarios" will be developed. Several famous collisions (Tunguska, Yucatan) presented in detail.

The energy of the recent Tsunami will be compared to an asteroid collision of average size.

LCP 10: THE PHYSICS OF STAR TREK

Based on an article of the same title, written with Ian Winchester. Published in *New Scientist* 1981). The study of motion in the three regions of physics is discussed:

1. Speed of less than 10% the speed of light (Newtonian),
2. Speeds greater than 10% but less than the speed of light (Einsteinian),
3. Speeds greater than the speed of light (superluminal, or tachyon-like).

Note: This LCP is based on the old STAR TREK series.

APPENDIX 2

Guidelines for Writing LCPs

1. Map out a context with one unifying central idea that is deemed important in science *and* is likely to capture the imagination of the student.

2. Provide the student with experiences that can be related to his/her everyday world as well as being simply and effectively explained by scientists' science but at a level that "makes sense" to the student.
3. Invent a "story line" (may be historical) that will dramatize and highlight the main idea. Identify an important event associated with a person or persons and find binary opposites, or conflicting characters or events (Egan, 1986) that may be appropriate to include in the story.
4. Ensure that the major ideas, concepts and problems of the topic are generated by the context naturally; that it will include those the student would learn piece-meal in a conventional textbook approach.
5. Secure the path from romance to precision to generalization (Whitehead, 1929). This is best accomplished by showing the student that
   a. problem situations come out of the context and are intrinsically interesting;
   b. that concepts are diversely connected, within the setting of the story as well as with present-day science and technology;
   c. there is room for individual extension and generalization of ideas, problems and conclusions.
6. Map out and design the context, ideally in cooperation with students, where you as the teacher assume the role of the research-leader and the student becomes part of an on-going research program.

7. Resolve the conflict that was generated by the context and find connections between the ideas and concepts discussed with the corresponding ones of today.

References

Brouwer,W. (1995).Toward a more conceptual way of teaching physics. *Physics in Canad.* 55-61. pp. 335-346.

Driver, R. (1989). Students' conceptions and the learning of science. *Int. J. Sci. Educ.*, vol.11, No.5, 481-490.

Dyson, F. (1958). Innovation in physics. *Scientific American,* **129**(3) 74-82.Egan, K. (1986). *Teaching as Story Telling.* The Althouse Press: University of Western Ontario.

Harré, R. (1970). *The principles of scientific thinking.* Chicago: University of Chicago Press.

Hestenes, D. et al. (1992). Force concept inventory. *The Physics Teacher,* **30**, 141-151.

Hewitt, P. G. (1990, May). Conceptually speaking. *The Science Teacher,* 35-57.

Holton, G. (1978). *The scientific imagination.* New York: Cambridge University Press.

Kenealy, P. (1989). "Telling a Coherent 'Story': A Role for the History and Philosophy of Science in a Physical Science Course". HPSST, *Proc. of the First Int. Conference*, Don Emil Herget (ed).

Kuhn, T. (1962). *The structure of scientific revolutions.* Chicago: The University of Chicago Press.

McCloskey, M. (1983, April). Intuitive physics. *Scientific American*, 122-130.

Merzyn, G. (1987). The language of school science. *International Journal of Science Education,* **9**(4), 483-489.

Miller, A.I. (1986). *Imagery in scientific thought.* Cambridge, MA: The MIT Press.

Nersessian, N. (1995). Should physicists preach what they practice? *Science & Education,* **4**(3), 203-226.

Nersessian, N. (1989). Conceptual change in science and in science education. *Synthese,* **80**, 163-183

Pearson, K. (1936). *The grammar of science.* London: J.M. Dent & Sons Ltd. (Original work published (1892).

Popper, K. (1968). *The logic of scientific discovery.* New York: Harper and Row, Publishers.

Rigden, J. S. *et al.* (1991 Jan.) Tune in, Turn off, Drop out. *The Sciences.* p. 16-20.

Roth, M. and Roychoudbury R. (1993). The conceptual map as a tool for the collaborative construction of knowledge: a microanalysis of high school physics students / J. Res. Scie. Teaching 30, 503-34.

Stinner, A. (1973). An Enrichment Program for Grade 13 Physics, *The Crucible* (The Journal of the Science Teachers' Association of Ontario) *4* (6), 22-30.

Stinner, A. (1978). A Solar House for Northern Latitudes, *The Physics Teacher, 16* (1), 25-30.

Stinner, A. (1980). Physics and the Bionic Man, *The Physics Teacher, 18* (5), 352-362.

Stinner, A., & Winchester I. (1981). The Physics of Star Trek, *New Scientist, 92,* 1285/1286, 884-888.

Stinner, A. (1989 a). Physics and the Dambusters, *Physics Education, 64*(8), 260-271.

Stinner, A. O. (1989b). The Teaching of Physics and the Contexts of Inquiry: From Aristotle to Einstein. *Science Education, 73* (5), 591-605.

Stinner, A. (1990). Philosophy, Thought Experiments, and Large Context Problems in the Secondary Physics Course, *International Journal of Science Education* 12/3: 244- 157.

Stinner, A. (1992). Science textbooks and science teaching: From logic to evidence, *Science Education,* 76, 1-16.

Stinner, A. (2003). Scientific Method, Imagination, and the Teaching of Science. *Physics in Canada.* Vol. 59, No 6, 335-346.

Stinner, A. (1994). The story of force: From Aristotle to Einstein. *Physics Education, 29,* 77-86.

Stinner, A. (1995). Contextual Settings, Science Stories, and Large Context Problems: Toward a More Humanistic Science Education. *Science Education, 79* (5), 555-581.

Stinner, A. (1997), From Intuitive Physics to Star Trek. *Physics in Canada, 53* (1), 23-33.

Whitehead, A. N. (1929). *The aims of education and other essays.* New York: McMillan.

From Babylon to the Big Bang – Are there Revolutions in Astronomy?

JÜRGEN TEICHMANN

Deutsches Museum and Ludwig Maximilians University, Munich, Germany

Abstract: More than 40 years ago, Thomas S. Kuhn presented the notion that revolutions highlight the development of science. However, Kuhn does not give an account of the long lasting and rich cultural background to these revolutions and pays little attention to technological developments. Without these, we cannot fully understand the history of science. Using examples from astronomy, I will illustrate these enduring rich cultural backgrounds and long range changes.

It may be better for this discussion to choose the term "evolutions" rather than "revolution". I think there are 3 periods that we must discuss: the discovery of the possibility of astronomy as an exact science in Greek culture, the scientific revolution of the sixteenth and seventeenth centuries, and the development of modern astrophysics in the nineteenth and twentieth century. These periods can be used in the class room to motivate students, assigning them to three different groups at the same time, to start exciting debates about the methodology of science, about the Kuhnian concept of radical changes in science, about the relation of man and the cosmos, and about the role of society in the interpretation of nature.

Introduction: The birth of observational astronomy

Within all cultures, the observation of the sun, moon and the stars played an important role. Three main reasons were responsible for the "birth" of an astronomical science in most cultures:

1. Although the sky was considered an immaterial counterpart of our daily world, it deeply influenced the earth by heat and light from the sun, provided a very exact clock and influenced other natural and man made-phenomena. For most cultures the sky became the residence of gods. (I will not discuss this further on, since we would have to look at the different mythologies of all cultures. This would take us too far afield.)
2. The movement of celestial bodies defined time measurement and chronological order. This was most important for practical life in a highly developed culture and fundamental for the self-determination of a society between a distant past and a near future. Therefore, calendars often became symbols of power.
3. The celestial bodies, mainly the stars and the sun, aided in navigation and provided orientation, indispensable when travelling in deserts or on the open sea.

Egyptian calendars date from about 3000 B.C. and our division of the year by 365 days was first introduced in Egypt. But this culture was not interested in continuous observation of complex celestial movements. It was different in Mesopotamia (what today is Iraq). As early as 2000 B.C. an astonishing "science" of astronomy developed in this culture by the second century B.C..

Unfortunately, it was an inquiry based on arithmetical manipulations resulting in rules and algorithms. We know of no geometrical conceptions, depicting the daily or other periodic movements in the sky. It was a science of long columns of numbers, observed and recorded and then calculated using very refined interpolation formulas.

Mesopotamia was also strongly interested in using a lot of periodic celestial phenomena to predict the future, the future of celestial movements themselves, for example, of eclipses. They used these predictions for founding a system of astrology: the first exact horoscope that been found dates from about 410 B.C.. In fact, moon eclipses could be predicted to with an accuracy of about 1 hour.

We find those interests also in other cultures, like the Chinese. There is evidence to believe that they became acquainted with them from Mesopotamia. But China never developed, or adapted a very good mathematical astronomy. The Chinese were more interested in irregular phenomena – like supernovae.

India was more influenced by Mesopotamia. Both cultures liked mathematical recording of long-time periods. For example, because it was considered a significant bad sign when all 7 planets (sun, moon, Mercury, Venus, Mars, Jupiter, Saturn) would meet within the last degree of the star constellation cancer, it was worthwhile to calculate this extremely rare event.

I only want to make a few remarks about the third reason for the development of astronomy. To use the sky for orientation and navigation was not so important for travelling in most cultures because landmarks, like mountains, served as signs. But for Chinese or Mesopotamian caravans in flat deserts it was necessary. And it was also essential for the primitive Polynesian cultures on the Pacific Islands. Unfortunately, we do not know much about them and their concepts that allowed them to navigate and orient themselves. Not before the big travels of discovery in the fourteenth and fifteenth centuries on the great oceans did navigation become a very important challenge to improve the science of astronomy.

The invention of the celestial constellations, which helped orientation and navigation, is deeply connected with the concept of the sky as an immaterial counterpart of our daily world and therefore deeply embedded in the different cultural developments.

It is interesting to note that most cultures used figures and images rather than stories to make the 'abstract' sphere of shining bodies and (mainly) points come alive: gods, mythical personalities, animals, objects. Very few cultures used stories, but some South American cultures did. An example is connected to our star constellation Orion. In South America the central star was imagined as a dangerous thief, who was captured by two policemen (to the right and left) and for punishment thrown before four vultures (around the two policemen and 1 thief) who would proceed to devour him. In Greek mythology, instead of being in a story, these stars became a mythological figure: the big hunter (with 3 belt stars and 4 stars round them: 2 at the shoulders, 2 at the legs).

Picture 1: The star constellation Orion, from a copper engraving 1733

The three fundamental evolutions of astronomy

I adopt the Kuhnian model of Scientific Revolutions (the changing of paradigms or his better defined changing of disciplinary matrixes). However, I see Kuhn's ideas in a wider relationship with long range developments in society, including technology, experimental style, philosophy of quantification, and these as partly independent factors. Therefore, the term "evolution" may be a better word for those changes. In this sense we may find to date three fundamental evolutions in astronomy:

Fourth century, B.C. to the second century A.D.

The discovery of the possibility of astronomy as an exact geometrical science started with the Greeks, the first exact science in history at all, developed in close relationship with Euclidian geometry. Of course, we must give credit to the Mesopotamians with their arithmetical calculations.

From the beginning this new science was combined with arguments which we today relate to physics: about matter found on earth in contrast to that of the sky and about the causes of the eternal celestial movements. Celestial "physics" was considered totally different from the physics accepted as valid on earth, namely terrestrial physics. So there was a clear separation between terrestrial and celestial physics. All this was bound into a convincing system of experience and arguments. Before and outside the Greek culture such reflections, if any, were expressed mainly in the form of myths.

On the other hand, the geometry-based Greek science excluded irregular phenomena like novae, sunspots, and star colours. For example, Chinese astronomers observed sunspots already in the period of B.C., whereas in Europe these phenomena were not studied before the invention of the telescope in 1609.

The sixteenth and seventeenth centuries:

Mankind has been displaced from a central position in the cosmos. The sky is no longer different from our material world. For example, "dirty" sunspots are seen as turning around with the sun's surface. There is an invisible reality (for the naked eye), which can only be mediated by instruments with the just invented telescope (1609). Moreover, the same natural laws apply to the whole of the cosmos. Finally, all useful knowledge comes from mathematically formulated arguments, based on observation and experiment. Europeans can conquer nature using instruments, the way they conquered the vast oceans and foreign countries by superior technology.

The nineteenth and twentieth centuries:

The sky now becomes a laboratory for physics and chemistry, beginning with the application of spectroscopy around 1860. Later, in the early twentieth century, our common sense conceptions of space, time and action are fundamentally changed, by relativity theory and quantum physics.

Examples of the three revolutions

Let us start with two examples of the first evolution in Greek culture. For the first time, the Greeks had the concept of complete circular movement, including parts of the movements, "under" the earth, which were not directly visible. And they combined the more complicated movements using more and more circles. E.g., the motion of the sun representing a daily circle was combined with a yearly one; or the planetary retrograde movements, using at least 3 circles. All these circles should really exist in the form of hollow spheres of a celestial element which was considered transparent, invisible and without friction.

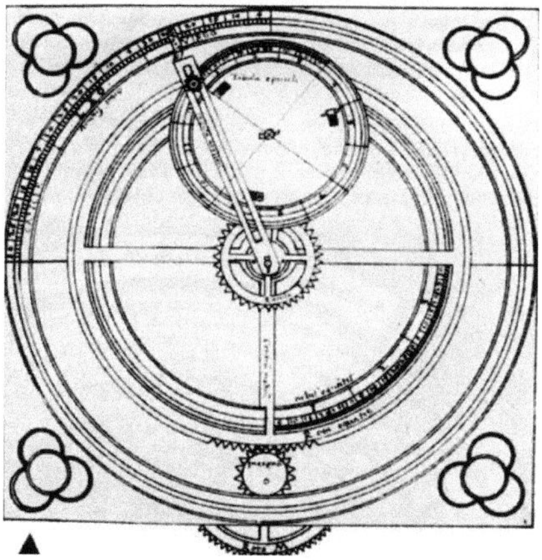

Picture 2: The Greek geometrical cosmology of circles on circles. Here we see the movement of Venus in a planetarium taken from the year 1397: the big epicycle is turning on a circle round the earth in the centre

For 200 years this model of the sky was not very exact – but at the end it became fruitful. When the Greeks added the exact dates of Mesopotamia to their models, in the 2nd century B.C., they had a very good mathematical model of astronomy, the synthesis of all this was achieved by Ptolemy in the 2nd c. A.D. The model was useful, especially for exact calculations of the difficult paths of the 7 "planets" (meaning "wandering" in between the other stars in Greek). Even in the 16th century this system was used and considered reliable.

My second example is a very astonishing materialization of such a geometrical theory from about 80 B.C.: It is, in modern interpretation, an analogue calendar computer adding the sun and moon movements measured in respect to the fixed stars and transforming them into the phases of the moon. Recent interpretations even suggest that the complete instrument originally was a whole planetary machine.

This is by far the most complex instrument we know from that time in the world. Not until 1500 years later (when the mechanical clocks in Europe – invented around 1300 – had been refined for 100 years) will we find similar complex mechanisms. How was this possible to construct in the Greek workshops of the time? Why don't we find any comparable mechanisms from this time in Greece or Roman Italy and not anything written about such precision technology? Why did such standards have no influence on other technologies?

Picture 3: The Antikythera mechanism, interior gears - an incredibly complicated astronomical "computer" by the Greeks, from about 80 B.C. Reconstruction, in the Deutsches Museum Munich, according to D.de Solla Price 1975

I think this had to do with the sharp division between handicraft (which was considered dirty and not very valuable) on the one hand and philosophy (including mathematics) on the other, which was part of the higher educational level (e.g. in the thinking found in Plato's Academy). At the same time it is true that the Greeks were able to assimilate different influences from different cultures, even if they seemed contradictory. This presumably had to do with the fact that there were very different "nations" at the beginning which grew up together, and then settled further out at different places around the Mediterranean Sea, but still remaining in cultural contact , and outside a central religion. In any case, the seemingly contradictory cases of

mathematical astronomy and mechanical technology fitted together, superbly illustrated with this example. Greek engineering in other branches also produced first rank technology. It may also be that Greek handicraft that developed turning wheels, and later tooth wheels, influenced, at least in the beginning, the concept of circulation in the sky.
This, of course, is merely a hypothesis. We only know that Greek aesthetic philosophy influenced astronomy: e.g. the sphere as the most perfect body in Plato's philosophy, with constant velocities of circular movement, as the most perfect arrangement.

The second fundamental evolution, in the sixteenth and seventeenth centuries

Now we encounter such famous names as Copernicus, Kepler, Galileo, and Newton, often called the scientific revolution. The scientific revolution first began with big changes in the Christian societies of middle Europe. Without those changes it would not have been possible to study Greek astronomy and then later critically question this astronomy.
It should be noted, that before the thirteenth century the Islamic-Arab culture had preserved and to some extent improved, but not significantly changed this knowledge.
In middle Europe, towns grew up into self-governing, and soon powerful, communities, as for example Venice or Florence. Christianity developed as a dynamical, albeit often brutal but in the beginning was innovative. The rediscovery of Greek culture, later on called "Renaissance", or rebirth, was in reality the birth of a new understanding of life and the cosmos. Here we have the central emergent new attitude to the "discovery of the natural world" as an object which can be conquered by rational thinking and using technical instrumentation. The discovery of America by Columbus in 1492 and the discovery of mountains on the moon by Galileo's telescope in 1609 are examples illustrating this new attitude for discovering nature. In addition, we encounter an emergent new attitude to the "discovery of the_individual", that is, the single personality, the genius, whose work was evaluated not based on noble birth, but by innovative abilities.
When the German painter Albrecht Dürer (who also was innovative in technology) around 1500 visited Italy, at this time the most modern part of Europe, he said that in Italy he is treated like a noble man (a "Herr"), but in Germany he is only an artisan.
But in spite of all that, progress was needed for a further 100 years until the idea of Copernicus of changing the place of sun and earth found able defenders. And we all know about the struggle between Galileo and very reactionary Catholic Church around 1630. It was indeed a revolutionary idea for traditional belief, when the scientific revolution postulated that there is no place for God in the visible sky anymore, that no

celestial element exists. Celestial matter now became dirty matter like that of our earth, and in such an enormous quantity that nobody could have imagined earlier. The cosmos has not been created for mankind, we are only a moving dust atom in a – possibly – infinite space.

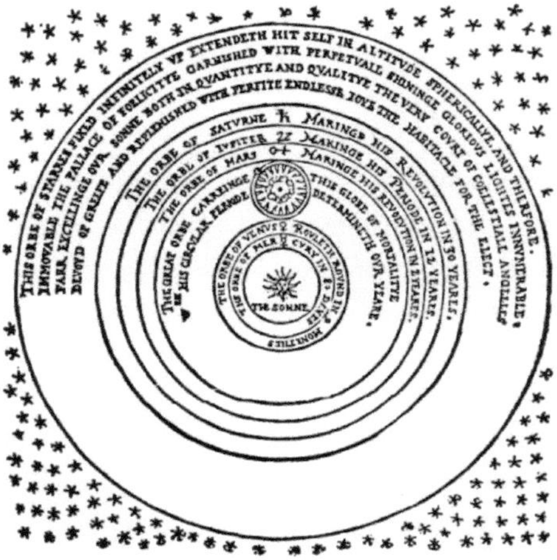

Picture 4: The first drawing in history with an "infinitely up extended" sphere of the fixed stars – by Thomas Digges from 1576 - a consequence of Copernican axioms showing no visible parallax of those stars.

In the year 1686 a lady and a 'marquis' were able to discuss the Copernican system with the following reflections: If there are totally different civilizations all over the cosmos we have to be principally tolerant of them, therefore we must learn to practice tolerance on earth. And in another scene the Marquis says: The moon looks like a perfect shining body because it is so far from us. But it is made of similar matter as the earth. We cannot imagine the earth as shining because we do not see it from a large distance. And now the lady answers: This is the same when we look at the higher nobility which is glittering for us so strangely, but in principle this nobility is extremely similar to us.

Of course, we still do not have the principle of political equality of the later French Revolution, but the new astronomy, together with the related celestial mechanics, was one of the starting points of the famous Enlightenment in France and in Europe in general, which prepared the field for the first democratic revolutions in America and later, in 1789, in France.

Picture 5: Marquis and Marquise, discussing 1686 about the Copernican system - from a booklet of the secretary of the French Academy of Sciences, Bernard le Bovier de Fontenelle from 1686: "Discussion about the plurality of worlds" – which became, after the Holy Bible, the most read book in the 18th century.

The third fundamental evolution in the nineteenth century

The final evolution started in an age of uncritical "technicism" and scientific positivism in Europe. There seemed to be no frontiers for quantitative science and technology. Mechanical analogies along mechanical, as well as early electrical technology, often served as models for all knowledge. One important example is the French philosopher Auguste Comte and his arguments. Around 1850 he declared that all knowledge in history develops along 3 different phases: first the *theological* one, then the *metaphysical* one and then the *scientific positive* phase. He considered the most positive science the science of astronomy which is based on mechanical, that is, on physical principles. He further postulated for the future a development of a "social physics" as an exact knowledge about society, today called sociology. He argued that astronomy had one big problem that can never be solved, that is the problem that it was not possible to develop a physics and chemistry of stars because we cannot experiment with them.

Comte did not know that a German optician, Joseph Fraunhofer, in 1814 had discovered dark lines in the rainbow spectrum of the sun, which he could not explain. But shortly after Comte's philosophical reflections, these dark lines were explained in 1859 as a kind of telegram by the sun, indicating which chemical elements were part of its lower photosphere. From this date on, in concert with other technologies like photography, photometry and new theoretical branches of physics like thermodynamics, from the 20th century on, relativity theory, , quantum physics, the profession of an astronomer changed fundamentally. Step by step he became a physicist who no longer has a direct cognitive or emotional contact with "glancing star points".

Today, we still call everything that changed in this direction, astrophysics. But this separation is not as clear and useful, as it was around 1900. At that time the sky literally became a laboratory. Of course it remained impossible to make direct experiments with celestial bodies but we see in front of our instruments billions of different mixtures of chemical elements with different temperatures, under different pressures, showing more variations than physics and chemistry alone could provide on earth. This went on to produce the big "laboratories" for particle astronomy in our days.

At the end of the 19th century, this quickly growing astrophysics also demanded a new methodology of research. Now, scientifically speaking, young United States often progressed faster than the strongly traditional astronomers in Europe (mainly in Germany).

Seen from the theoretical side, the frontier between astronomy and physics opened wider and wider: The theory of general relativity, developed by 1916, changed our concepts of space and time radically: Space was no longer an empty vessel filled with matter. Matter itself defined and formed space! This led to new definitions of finiteness and infinity, and a new cosmology was born.

But it was not before the nineteen twenties, at the same time when quantum physics was being developed, did the last part of the old Greek thesis that the earth was the center of the world, fall down. In addition, it was found that the Milky Way is not unique in the universe. Because the other galaxies are so far away, we have to

conclude that they are of the same size, and not only small companions near our home galaxy. And the most surprising result was that all galaxies fly away from each other.
It is interesting to note that the founder of psychoanalyses, Sigmund Freud, declared that there are 3 decisive violations of human self-understanding (from the 16th c. to the beginning of the 20th c.). These are:

- Human beings are not central in the cosmos (since Copernicus)
- Human beings in principle are not exceptional among all living beings (since Darwin)
- Mankind has also lost its belief in the omnipotence and central place of the consciousness within the human soul (since Freud himself).

We should now add, that perhaps in the 21st century we will also loose the belief that this universe with these physical laws is the only one that is possible - hand in hand with new concepts of subjective and objective reality.
After World War II, astronomy in the non visible regions of electromagnetic waves developed and became very important. This was mainly:

1. Radio astronomy, which was partly a child of war.(In world war II, it was discovered that radiation coming from the sun disturbed telecommunication.) Important discoveries were made from the sixties on, the discovery of pulsars, the micro-wave background radiation, and quasars.
2. High energy astronomy (X-rays, γ-rays from the cosmos)
3. Particle astronomy (like today Neutrino astrophysics).

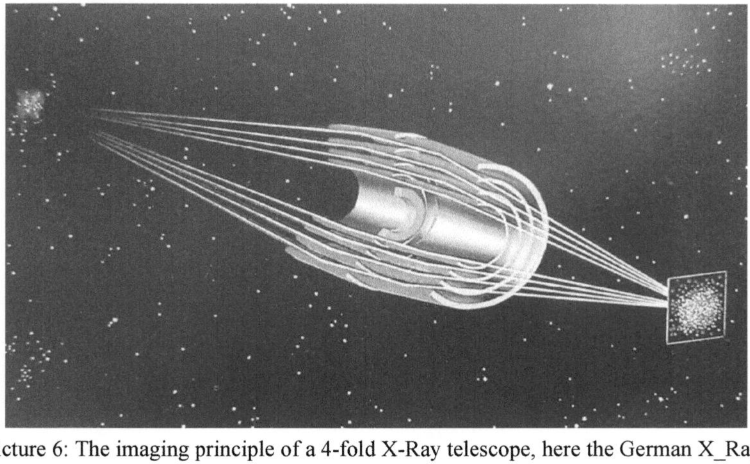

Picture 6: The imaging principle of a 4-fold X-Ray telescope, here the German X_Ray-Satellite ROSAT, 1990-1998 – credit MPG,Germany

Infrared and ultraviolet astronomy also developed quickly, like high energy astronomy, supported by the development of space-satellite technology. Only one "window" of space information is still closed: Einstein's gravitational waves have not yet been detected directly, but in the last two years very powerful and promising new research devices have been built.

The "laboratory for astrophysics" around 1900 changed after World War II to "Big Science". More and more it became necessary to turn to international cooperation, like in the case of ESO (European Southern Observatory), which recently built the four 8-m-telescopes of VLT (Very Large Telescope) in Chile.

More and more, since the nationalistic 19th century, astronomy has become a measure of political power and cultural superiority; also because there was a strong demand for the most advanced technology. In fact, the technology necessary for such a prime science gained more immediate popular reputation than that used for other purposes.

But I think there are other reasons, too, why astronomy/astrophysics in the collective consciousness of our modern world is still more than an important prime science:

1. It tries to give answers about the origin and the end of the whole world – this is an eschatological role this prime science plays and therefore touches on religion.
2. The cosmos occupying unimaginable space still plays the metaphoric role of the "unknown ocean". On earth, all discovery adventures seem to come to an end, but in astronomy there is still a vast amount open, for example the problems connected to black holes, dark matter, dark energy. One could also mention the close connections with space flight and science fiction.
3. Astronomy contains the question about the singularity of mankind or the singularity of life in general: Was life only caused by an accident, for example, by a very unique crash between the earth and a celestial body which created the moon as a stabilizing factor to the earth? Are there pluralities of inhabited worlds? And still more philosophical: Is there a possibility of other existing universes without observers like us?

References

Cohen I.B.:1985, Revolution in Science. Cambridge, Mass. and London. (In German:1994, Revolutionen in der Naturwissenschaft. Frankfurt/M.)

Gingerich, O.(ed.): 1984,Astrophysics and Twentieth-Century Astronomy to 1950. Cambridge, GB (Vol. IV, Part A of the General History of Astronomy).

Hearnshaw, J.B.: 1986, The Analysis of Starlight. One Hundred and Fifty Years of Astronomical Spectroscopy. Cambridge/GB.

Jaki, St. L.:1972, The Milky Way An Elusive Road for Science. New York

North J.: 1994, The Fontana History of Astronomy and Cosmology. London. (In German: Viewegs Geschichte der Astronomie und Kosmologie. Braunschweig/Wiesbaden 1997)

Pannekoek A.: 1961 (Reprint 1989), A History of Astronomy. London.

Solla Price D.de: 1975, Gears from the Greek-The Antikythera Mechanism, a Calendar Computer from ca.80 B.C. New York – The standard work; but see newer interpretations by M.T.Wright in: Antiquarian Horology,Vol.18, 2003, p.270-79 and Vol.20, 2005.p.51-63. See also the Internet, e.g. www.antikythera-mechanism.gr, for actual investigations of the apparatus.

Teichmann J.: 1999 (4th ed.), Wandel des Weltbildes – Astronomie, Physik und Messtechnik in der Kulturgeschichte. Stuttgart, Leipzig

Thanks so much to Arthur Stinner for help to translation.

Do mathematics by hands: two cases from ancient Chinese mathematics

YOUJUN WANG

Department of Philosophy, Shanghai Normal University, Shanghai, 200234, China

Abstract: In modern mathematical teaching, it has become increasingly emphasized that mathematical knowledge should be taught by problem-solving, hands-on activities, and interactive learning experiences. Comparing the ideas of modern mathematical education with the development of ancient Chinese mathematics, we find that the history of mathematics in ancient China is an abundant resource for materials to demonstrate mathematics by hands-on manipulation. In this article I shall present two cases that embody this idea of a hands-on approach in ancient Chinese mathematics, at the same time offering an opportunity to show how to utilize materials from the history of Chinese math in modern mathematical education.

Traditionally, mathematics has been viewed as a discipline where success is limited to a minority of children. During recent years there has been a conscious effort that focuses on making mathematics accessible and enjoyable for all children. Reform efforts all over the world emphasize some topics, such as the need for providing real-life examples, the relationship between natural science and mathematics, collection and organization of data, and problem solving as opposed to rote memorization of procedures. So it has become important that mathematics should be facilitated through increased emphasis on more problem-solving, hands-on activities, and interactive learning experiences to construct mathematical knowledge.

Comparing the ideas of modern mathematical education with the history of Chinese mathematics, we find that ancient Chinese mathematics is an abundant resource for materials to demonstrate mathematics by hands-on manipulation, because many aspects of mathematics in ancient China could be demonstrated by such techniques which could be used to enrich the mathematics curriculum for many grade levels and even some university courses. I shall present two illustrative cases that embody this idea of hands-on experience in ancient Chinese mathematics.

The principle of congruency by addition and subtraction

The most famous Chinese mathematics book of all time is the Jiuzhang suanshu or, as it is more commonly called, the Nine Chapters on the Mathematical Art. The book contains contributions to mathematics that had been made over quite a long period and was compiled during the Han Dynasty (206 B.C. - 220 A.D.). Therefore Nine Chapters is regarded as a major paradigmatic work of ancient Chinese mathematics. This important work dominated mathematical developments and styles in China for 1500 years. Many later developments of mathematics in China came via commentaries on this book (Liu Dun: 1993).

Among all commentaries on the Nine Chapters, Liu Hui's commentary was the most important work. Liu Hui (circa 220 – circa 280) wrote his commentary circa 263. In his work, Liu Hui gave a more mathematical approach than did earlier Chinese texts; in particular, he provided principles on which his calculations are based (Guo Shuchun: 2004). For example he found approximations to inscribe regular polygons with sides in a circle. Liu created also a method to present and prove geometrical problems, which is called 'the principle of congruency by addition and subtraction'. This principle is based upon the idea that an area or volume of a definite figure remains invariant in moving its position. In other words, after dissection of a definite figure, the total areas or volumes of the parts remain equal to the original ones. The principle is one of the chief characteristics of geometry in the Nine Chapters, is also an important theoretical component of ancient Chinese mathematics and indeed represents a general method to show or discover new propositions in mathematics.

Now let's show how Liu Hui used the principle to prove the Pythagorean Theorem and other examples by hands-on manipulation.

In Chapter 9, Problem 3 we find the Pythagorean Theorem: $a^2 + b^2 = c^2$.

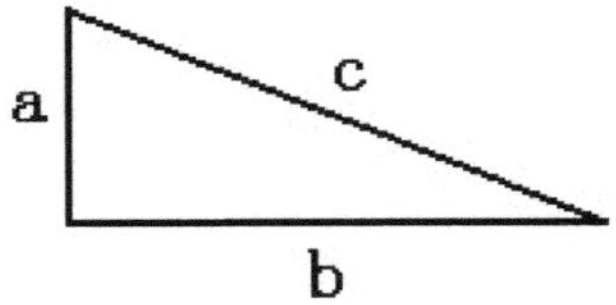

Figure 1. The right triangle

The method of proof given by Liu Hui tells us how to transform one figure to another figure in a very simple language. We can demonstrate this in the following figure:

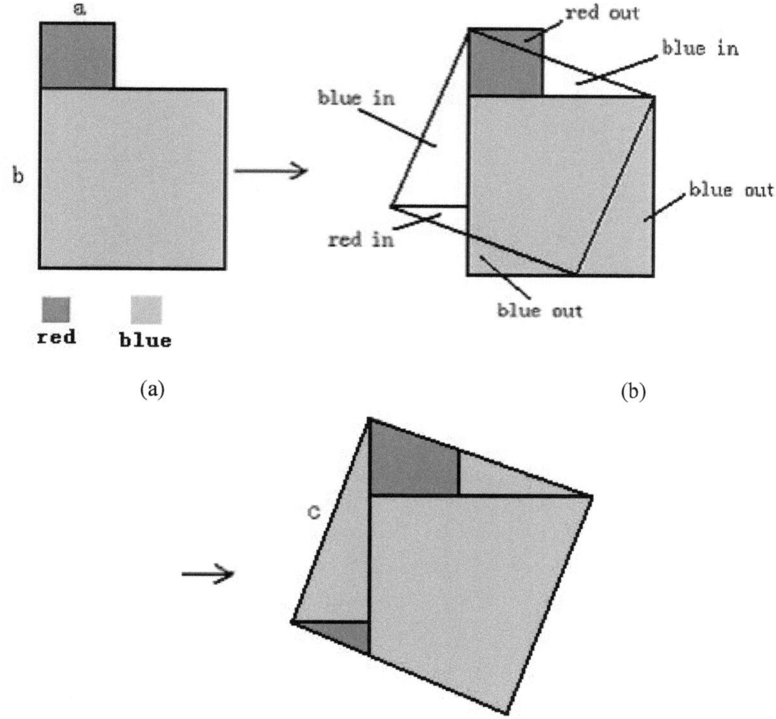

(c)

Figure 2. Dissection proof of Pythagorean Theorem

The area of the two squares in Figure 2a, which is $a + b$ is exactly equal to the area of the tilted square in Figure 2c, which is c^{1}. In this way, the Pythagorean Theorem was proven by Liu Hui's operating process. Furthermore, Liu Hui used this method to solve other problems, such as problems 15 and 16 in Chapter 9.

Problem 15: Let x be the side of a square inscribed in a right triangle. Let a and b be the lengths of the two legs and c the length of the hypotenuse of the triangle. Find the side x of the square. The method of finding x is shown in the following sequence of figures:

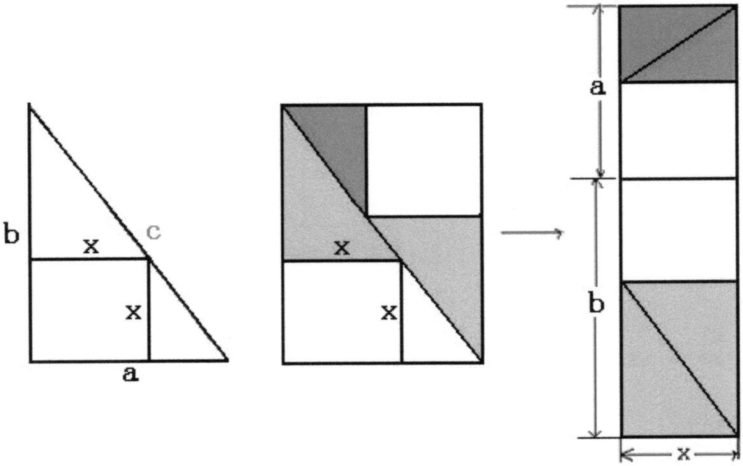

Figure 3. Dissection method to find the side of a square inscribed in a right triangle

Figures 3b is made of two right triangles whose lengths of the two legs are a and b and the length of the hypotenuse is c. Comparison of Figures 3b and 3c indicates that ab = (a + b) x. Thus $x = \frac{ab}{a+b}$

Problem 16: Let 2r be the diameter of the circle inscribed in a right triangle. Let a and b be the lengths of the two legs and c the length of the hypotenuse of the triangle. Find the diameter 2r of the circle.

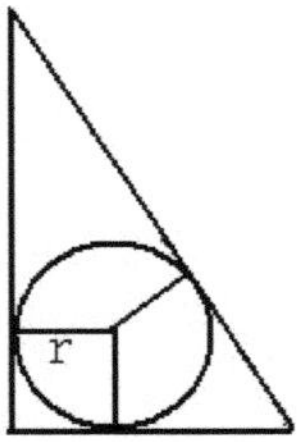

Figure 4. Diameter of the circle inscribed in a right triangle

The method of finding 2r can be shown in the following sequence of figures:

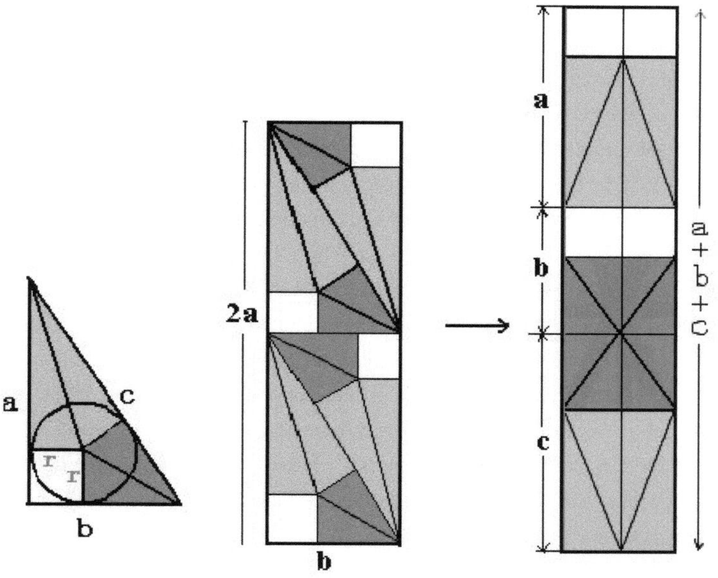

(a) (b) (c)

Figure 5. Dissection method to find the diameter of the circle inscribed in a right triangle

Figures 5b is made of four right triangles whose lengths of the two legs are a and b and the length of the hypotenuse is c. Comparison of Figures 5b and 5c indicates that:

$2a \cdot b = 2r \cdot (a+b+c)$, thus: $r = \dfrac{ab}{a+b+c}$

These are some typical examples of the use of principle of congruency by addition and subtraction. In fact, Liu Hui also used an extension of this principle to explain the theory of limit, which Liu Hui called the method of validation with Chinese chess whose shape is cube (Wu Wenjun: 1982). By using this principle, the propositions pertaining to areas of figures with linear sides and to volumes of simple solids, such as column, cone, and so on could be validated. It was easy to infer the formulas of Chongcha which established the foundation of technology of measurement in ancient China. The principle also played a very important role in the field of solving equations. In general, this unique Chinese mathematical tradition of doing mathematics by hands-on manipulation or by experiments was further developed since the time of Liu Hui. It influenced the future development of Chinese mathematics, such as during the Song dynasty (960—1279)and Yuan dynasty (1279—1368).

Yang Hui's work on the summing series

In the thirteenth century, the tradition of hands-on mathematics was further developed by Yang Hui (circa 1238 - circa 1298), who was a major figure of Chinese mathematics. His most famous work is the Xiang Jie Jiu Zhang Suan Fa (Detailed analysis of the mathematical rules in the Nine Chapters and their reclassifications) written in 1261. In this book Yang wrote that he had obtained a fine edition of the Nine Chapters on the Mathematical Art which contained additional notes by Jia Xian, commented further on by Liu Hui, and later by Li Chunfeng. The notes by Jia Xian have not been preserved, so we know them only by the references from Yang. Yang did not intend to write a further commentary on the ancient classic; instead he selected 80 of the 246 problems for his own discussion. He chose these 80 problems since he felt that they were representative of the different techniques that were presented in the Nine Chapters. Yang's Detailed Analysis contains twelve chapters of which nine correspond to those of the Nine Chapters, but there are three further chapters: one containing geometrical figures, one containing fundamental methods, and one in which Yang presents a new classification of the problems (Li Yan: 1955).

Yang Hui's other works were collected in the Yang Hui suanfa (Yang Hui's methods of computation), which appeared in 1275. He described multiplication, division, root-extraction, quadratic and simultaneous equations, series, and computations of areas for a rectangle, a trapezium, a circle, and other figures. He also gave an account of magic squares and magic circles (Li Di: 1984). One of the most valuable results of Yang Hui was his wonderful method to infer formulas of sum of series by hands-on operations.

Yang's basic intention was, as he wrote in his Riyong Suanfa (Mathematics for everyday use): "... to assist the reader with the numerous matters of daily use and also to instruct the young ones in observation and practice." In Yang's work, the character of doing hands-on mathematics is even more evident than in the work of Lui Hui. By using this idea from geometry in other fields such as number theory, he found a lot of new mathematical formulas. For example, in the chapter named Shanggong, Yang Hui compared some solids that were created by Liu Hui in "Jiuzhang suanshu zhu" to some real things that were called "Duo" by Yang Hui, as in Figures 6:

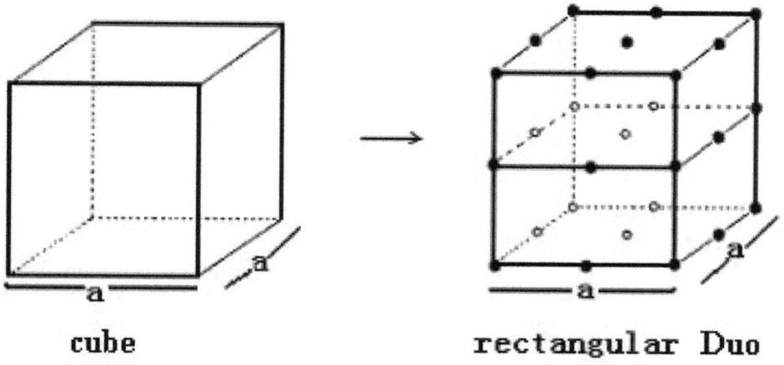

Figure 6. Changing a cube to a rectangular Duo

In fact, the spheres could be replaced by small cubes, or some things cubical which Yang Hui thought to represent units (Liu Dun: 1993).

By calculating the formula of a series by the help of the sum of cubes, Yang Hui extended Lui Hui's principle in geometry to the theory of numbers. He obtained many formulas on the sum of some series. In order to show Yang Hui's methods, some typical examples will be given below. In order to seek the sum of the series 1 + 2 + 3 +... +n (=S), Yang Hui assimilated this series to the shape which has S cubes, as shown in Figures 7a and 7b below:

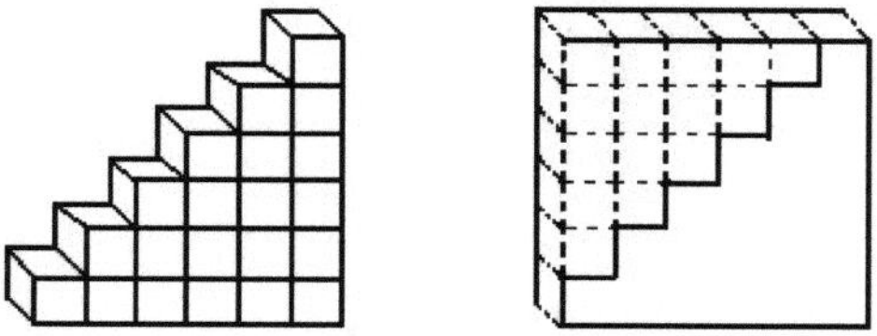

(a) (b)

Figure 7. The sum of cubes

The solid figure in Figure 7b is composed out of two parts: one is the original figure (shown in Figure 7a) which has S cubes; the other part is called the supplementary part and has S - n cubes, therefore (Fu Hai-lun, Chen Huan-fa: 2003):

$$S + (S - n) = n \cdot n \quad \rightarrow \quad 2S = n \cdot n + n$$

$$\rightarrow S = 1 + 2 + 3 + 4 + \ldots\ldots + n = \frac{n(n+1)}{2}$$

Another wonderful result of Yang Hui was the sum of the series $1+3+6+ \ldots\ldots + \frac{n(n+1)}{2}$,

Yang Hui called the corresponding solid (see Figure 8) the triangular Duo.

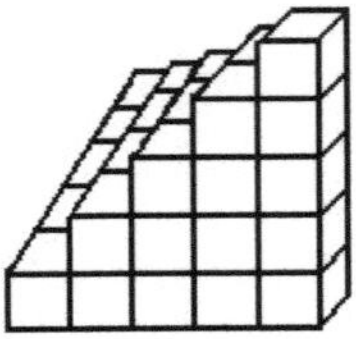

Figure 8: Triangles Duo

From Figure 9 below, we see that six solid triangular Duos will form one hexahedron with sides of length n, n + 1, and n + 2. Therefore, the formula for the sum is (Liu Dun: 1993)

$1+3+6+\ldots+\frac{n(n+1)}{2} = n\,(n+1)\,(n+2)/6$

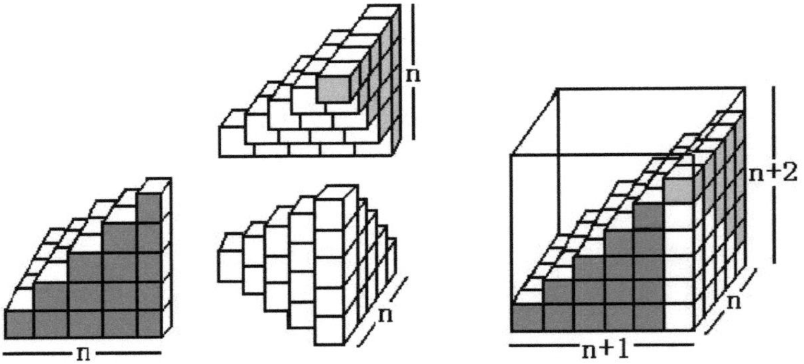

Figure 9. The sum of cubes in six solid triangular Duo

In order to seek the sum of the series $1^2+2^2+3^2+4^2+\ldots\ldots+n^2$ Yang Hui considered the correspondence of this series to the following solid (depicted in Figure 10) which Yang Hui called a quadrangular Duo and which consists of small cubes arranged in square slabs which are piled on top of one another and consist respectively, of $1^2\ 2^2\ 3^2\ 4^2\ \ldots\ldots\ n^2$ small cubes.

Figure 10. Quadrangular Duo

From Figure 11, it is apparent that three quadrangular Duos would form one hexahedron whose top slab's high is a half. So the hexahedron's each side length is n, n+1, n+1/2, thus

$3*(1^2+2^2+3^2+4^2+......+n^2) = n(n+1)(n+1/2)$

Therefore the formula for the sum is

$1^2+2^2+3^2+4^2+......+n^2 = n(n+1)(2n+1)/6$

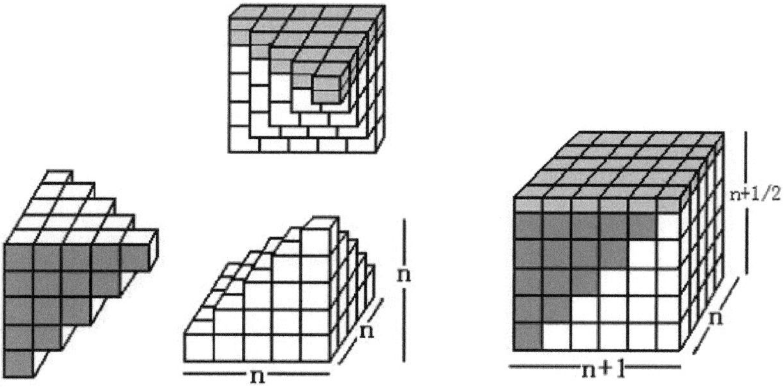

Figure 11. The sum of cubes in three quadrangular Duos

The next works to use hands-on operations were written by the famous mathematician Zhu Shijie the 13th to 14th century in the Yuan dynasty. Zhu Shijie further extended Yang Hui's method in his important book Si Yuan Yu Jian 1303 A.D. Zhu call the method *Duoji Method.* By arranging cubes into varying shapes, he obtained the sums of different series, such as:

$1 + 4 + 10 + 20 + ... + n(n+1)(n+2)/6 = n(n+1)(n+2)(n+3)/24$,

$1 + 5 + 15 + 35 + ... + n(n+1)(n+2)(n+3)/24 = n(n+1)(n+2)(n+3)(n+4)/120$,

$1 + 6 + 21 + 56 + ... + n(n+1)(n+2)(n+3)(n+4)/120 = n(n+1)(n+2)(n+3)(n+4)(n+5)/720$.

$1+2^3+3^3+4^3+5^3+......=n^2(n+1)^2$

Although the methods and shapes used by Zhu Shijie were more complex than those used in earlier works, they represented excellent skill (Li Zhaohua Chen Cheng-Yih 2000).

Traditional Chinese mathematics reached its summit in Song and Yuan dynasty. By then, traditional Chinese mathematics fell into decay and rudimentary knowledge of western mathematics had to be introduced to China by the Jesuits. However, the above method to obtain summation formulae were explained in further detail several centuries later by Du Zhigeng (end of 17th century) and by some other Chinese mathematicians in 19th century (Joseph W. Dauben, Christopher J. Scriba, 2002).

Conclusion

The style and content of Chinese mathematics is very unique, especially the wonderful style of doing mathematics by hands-on manipulation. For mathematical educators, it is clear that we can find a lot of interesting and helpful pedagogic tools in ancient Chinese mathematics. If teachers of mathematics were to implement such approaches to the teaching of mathematics, they may help students get a solid grasp of mathematical concepts and skills, and increase student interest in mathematics and in math-related careers. They also may connect the classroom to the challenges and excitement of the real world, engaging students in higher order thinking skills. Furthermore, any teacher in mathematics can gain useful insights from ancient Chinese mathematics.

Another inspiring idea is to urge our teachers to think about how to integrate the history of ethical mathematics into mathematics teaching. In recent years the importance of the history of mathematics in relation to teaching has been widely recognized and promoted. We can find many articles and books on history of mathematics and some hints on how to integrate western mathematics materials into teaching at school; however, we rarely find some hints on how to use ethnical mathematics materials into mathematical education, especially using Chinese mathematical materials in mathematics education. We should not neglect these mathematical tools from ancient China for promoting our school lessons in mathematics. At least, the history of ancient Chinese mathematics can show us another way to do mathematics which is very different from the western tradition and the system in modern textbooks. This way can enable us to think, to learn and to teach mathematics in a more interesting way, one that is easier to understand and more related to the reality than is now the case.

I express my appreciation to Professor Dr. Jürgen Teichmann for his valuable encouragement and advice.

Notes

[1] Katz (2004) notes that the Indian mathematician Jyesthadeva also presented this method of proving the Pythagorean Theorem in his work the Yuktibhasa, written in the mid-sixteenth century. However the time of Jyesthadeva is about A.D.1300 that is much later than Liu Hui's time. We could also find other similar problems in Yuktibhasa to those in Nine Chapters. It's a very interesting topic to study how the Chinese mathematics and the Indian mathematics influenced each other.

References

Dauben, J.W.; Scriba, C.J.: 2002, *Writing the History of Mathematics: Its Historical Development*, Basel, Boston, Berlin, Birkhauser Verlag,

Di, L.: 1984, *The Brief History of Chinese Mathematics*, Liaoning people publishing company, Shenyang

Dun, L.: 1993, *The Great Mathematics*, Liaoning education publishing company, Shenyang

Hai-lun, F.: 1998, Nine Chapters on the Mathematical Art and the elementary school mathematics teaching, *Hongkong Mathematical Education,* 6(7), 67-71

Hai-lun, F.; Huan-fa, F.: 2003, Modern Educational Significance of Mathematical History in Attaining the Overall Aim of New Curriculum, *Journal of Mathematics Education* 12(2) 70-73

Katz, V.J. (2004). The History of Mathematics: Brief Version, Pearson Addison Wesley, Toronto.

Martzloof, J.C.: 1997, A history of Chinese Mathematics, Springer-Verlag Berlin Heidelberg.

Shang-shu, B.: 1990, *The Modern Commentary on Nine Chapters on the Mathematical Art,* Shandong education publishing company.

Shuchun, G.: 2004, *A proofreading On Nine Chapters on the Mathematical Art,* Liaoning education publishing company, Shenyang.

Straffin, P. D.: 1998, Liu Hui and the First Golden Age of Chinese Mathematics, *Mathematics Magazine,* Vol.71, No.3, pp.163-181.

Wenjun, W.: 1982 '*Nine Chapters' and its commentarist—Liu Hui,* Beijing Normal University publishing company, 58—75

Yan, L.: 1955, *The History of Chinese Mathematics*, Science publishing company, Beijing.

Zhaohua Chen Cheng-Yih, L.: 2000, A Further Investigation on Zhu Shijie's Interpolation Formula, *Studies in the History of Natural Sciences,* 19(1) 30-39

Understanding the Earth and the Cosmos
Magnetism in Cultural History, Geophysics and Astronomy
Three Examples for Contextual Teaching

GUDRUN WOLFSCHMIDT

Institute for History of Science, Hamburg University, Germany

Abstract: I present three examples of instruments used in the field of magnetism. First I discuss the development of the compass in China, in three steps (Sinan, Zhinanyü and Zhinanzhen), followed by its transfer to the western countries in the 13th century. Then I examine pre-1600 experiments with magnets and their applications to navigation, as well as the topic of the Earth as a magnet. Finally I mention the importance of the compass for sundials.

Alexander von Humboldt (1769–1859) initiated a European geomagnetic network in 1829. In 1833 Carl Friedrich Gauß (1777–1855), in collaboration with Wilhelm Weber (1804–1891), founded the "Magnetischen Verein" in Göttingen. I present, in this context, the construction of magnetometers and the further development of geomagnetic observatories.

Finally I discuss the solar-terrestrial relationships discovered in the middle of the 19th century, as well as their effects on the Earth including auroras, geomagnetic storms and radio interference. I show examples of the measurement of solar and cosmic magnetic fields by using radiotelescopes and satellites.

Introduction

I would like to present three examples of contextual teaching in the field of magnetism, first the development of the compass in China and the transfer to the occident, then the "Magnetic Association", founded by Gauß and Weber, and finally the role of magnetic fields in the cosmos.[1]

Transfer of the compass from china to europe

INVENTION OF THE COMPASS IN CHINA

First I discuss the invention of the compass in China – in three steps (Sinan – the south pointing ladle, Zhinanyü – the floating fish compass and Zhinanzhen – the magnetic hanging compass for navigation).[2]

The so-called Sinan (south-reigning, south-guiding), this first magnetic compass came into being in the time of the Zhou dynasty, Warring States Period ("Streitende Reiche") (475–221 B. C.) – (3rd century B. C.). Wang Shong described it in his book LUN HENG in the Eastern Han dynasty.

Figure 1 a,b : a) Sinan – the south pointing, free gliding (suspended) ladle, Zhou dynasty, Warring States Period (475–221 B. C.) b) Floating fish compass (Zhinanyü) in a porcellan compass bowl c) Magnetic hanging compass for navigation (Zhinanzhen, southward pointing needle), Song dynasty (960–1279) (Photographed by Gudrun Wolfschmidt)

The Sinan consists out of two parts:

- a square disk, which represents the Earth; this Earth plate is divided into 28 segments (4 x 7).
- free gliding (suspended) ladle, made from magnetite; the handle pointed to the southern direction.

The Chinese discovered that a magnetite or an iron magnet, which is floating in a bowl filled with water, always points to the south (Zhinanyü); this was described in Wu Jing Zong Yao (Compendium of the most important military technologies). The floating or swimming compass was realized with a wooden fish in a bowl; a magnet was fixed to the fish.

LOADSTONES AND MAGNETIZATION

Already in the ancient time one knew natural magnets, certain iron-ore (Magnetite Fe_3O_4), which attracted iron particles (loadstones).[3] They were found especially near the city of Magnesia in Asia Minor (today Turkish the west coast) – the inhabitants were called Magneter); according to Lukrez DE RERUM NATURA the word magnetic is derived from this city.

Figure 2: A scientist interested in navigation with a floating compass
Apian, Peter: Cosmographia. Antwerpen 1533.

The magnetisation was made according to Shen Kuo (1031–1095) by rubbing with loadstone. The attraction and repulsion between iron and magnetite was compared with Yin und Yang. William Gilbert (1544–1603) describes the magnetisation by hammering. The crusaders reported from dangerous magnetic mountains, which attracted the nails of the ships and smashed the ships.

In the medieval script HÜBSCHE UND LIEBLICHE GESCHICHTE VON HERZOG ERNST VON BAYERN UND ÖSTERREICH, around 1450, this story is reported in the following way in medieval German.[4]

In the 12th century, during the Song dynasty (960–1279), the magnetic compass for navigation was developed (Zhinanzhen, with a needle pointing southward). The needle was hanging on threads. The different kinds of hanging were discribed by Shen Kuo in his book MENGXI BITAN.

The Luopan (azimuth bearing pan) is a compass for navigation with a magnetic needle and a compass rose with directions. This compass, described in the book MENG LIANG LU – South Song dynasty – was used in the Song and Yuan dynasty for navigation of ships. The itineraries were called needle routes.

Compasses were also used in the country (Southward pointer). The compass or magnetic chariot (tschinan-tsche) had a figure fixed on a pivot in front, easy turnable; in the stretched arm the figure held a magnet in form of a bar, which pointed southward. The direction was kept by a complicated gear mechanism. This compass chariot was developed in the Han period (206 B. C.–220 A. D.); it served for traveling over land and served in addition for measuring the distance with the help of counting the rotation of the wheels (hodometer).
Already in the Song dynasty Shen Kuo (1031–1095) discovered the magnetic declination, the deviation of the magnetic north-south-direction from the geographic north-south-direction. In the west it was not discovered until the 15th century.

RECEPTION IN THE OCCIDENT: MIDDLE AGES AND EARLY MODERN TIME

Now I examine pre-1600 experiments with magnets and their applications to navigation, as well as the topic of the Earth as a magnet.
In the Middle Ages, in the 13th century, the compass was transferred through the arabic countries to Europe, to the western countries, and it was used for navigation.
Petrus Peregrinus – (Pierre de Maricourt) (around 1220 – after 1269), a French noble man and crusader, compiled in 1269 with his EPISTOLA DE MAGNETE the first book with experiments in physics.

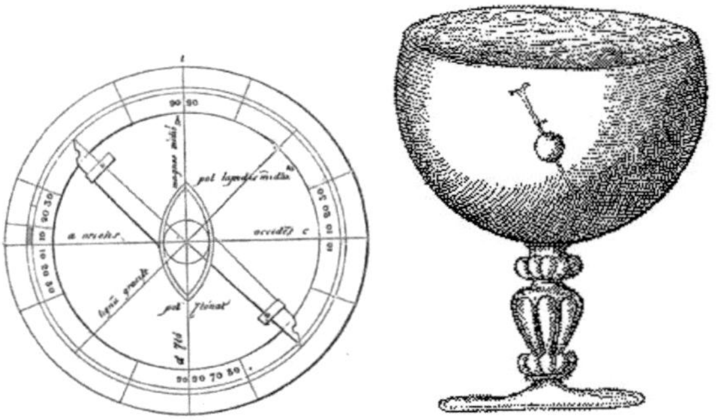

Figure 3 a,b : a) Compass needle in Petrus Peregrinus (around 1220 – after 1269) – (Pierre de Maricourt) EPISTOLA DE MAGNETE 1269 b) Inclination needle, Robert Norman – THE NEWE ATTRACTIVE (London 1581)

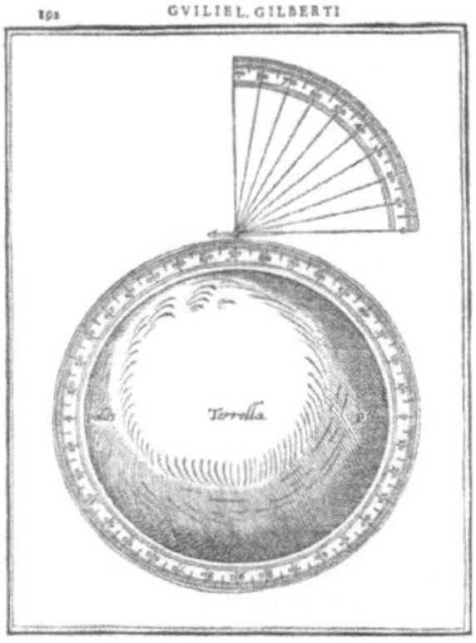

Figure 3 c : “Terrella”, model for the magnetic Earth – William Gilbert (1544–1603): DE MAGNETE. London 1600, S. 192.

The compass also was used for astronomical and geodetical instruments. I mention the importance of the compass for sundials with indication of the magnetic deviation. If one opens the diptych sundial, a thread is stretched, which casts the shadow. The sundial has to be orientated in north-south-direction; for this purpose a compass was needed; the magnetic aberration was marked for the first time by Georg (von) Peurbach (1423–1461) on the compass of his sundial in 1451; the eastern deviation of the magnetic needle is 7°.

After the declination also the inclination was discovered. Georg Hartmann (1489–1564) of Nuremberg, had already recognized in 1544 that the magnet pulls downward.[5] But Hartmann could not explain this observation.

The London compass maker and former sailor Robert Norman (fl. 1581) published his discovery of the inclination in 1581 with the title: THE NEWE ATTRACTIVE. Here he stressed, that he had solved all problems by experiments. He emphasizes, that *“experience, rationality and demonstrative presentation, which are the basis of the arts”*. He examined in addition the question, how the magnetism and especially the aberration of the compass is important for navigation.

William Gilbert (1544–1603), a scientist and physician from Colchester, was not only experimenter and admired the work of his predecessors Peregrinus and Norman, but he also developed theories of the magnetism of the Earth in his book DE MAGNETE, MAGNETICISQUE CORPORIBUS, ET DE MAGNO MAGNETE TELLURE; PHYSIOLOGIA NOVA, PLURIMIS & ARGUMENTIS, & EXPERIMENTIS DEMONSTRATA, published in the year 1600 in London. With magnetic iron balls Gilbert tried to make the phenomena

understandable and to explore by experiments everything, which is important for navigation. Gilbert's book is an early example for experimental science and research. His book was dedicated to the new tradition of people, who *"are not investigating in books for getting new knowledge, but in the things themselves"*.[6]
The compass was further developed in Europe: It consisted now – how we know it – out of a pointed pin, on which a round paper disk with a wind rose was floating; below the paper was a curved wire. Such a compass with dry disk (Hanluopan), how we know it, was used by the Chinese not before the 16th century. Here one can see the reception of European ideas for the construction of a compass in China.
The LO P'AN is a geomantic compass, which is used in the traditional Chinese FENG SHUI doctrine for judging architecture and landscape. It is also used in the Chinese naturopathy. The Lo P'an is a complex system; it has around 15 rings with much information in detail, for example it contains the ten heavenly genealogy and the twelve Earth genealogies with the YIN and YANG dragon. According to traditional Chinese ideas, the Lo P'an consists out of two parts: below there is the square "Earth plate" with the cardinal points and on the top there is the round turnable "heaven plate". Already a report from the Han period (206 B. C.–220 A. D.) mentions the segmentation of the 360° of the compass in 24 parts, so that you get 15° sections.

PORTULAN MAPS: COMPASS, MAGNETIC VARIATION AND CARTOGRAPHY

Already in the 16th century magnetic variations were discovered and described by Levinus Hulsius in his work DESCRIPTIO ET USUS VIATORII ET HOROLOGII SOLARIS (Nuremberg 1597).
Precondition for navigation is the cartography. In mariner guide books all information like wind directions and other special features were compiled. With the introduction of the compass in the 14th century, the so-called Portulan maps arose, which showed on the Sea the compass directions and distances; but the degrees of latitude and longitude were missing. The eight cardinal directions were marked black, the eight half winds green, the 16 quarter winds red. The Portulan maps had their flourishing time only until the time of Gerhard Mercator (1512–1594), who in 1546 emphasized, that these maps don't show the aberration of the needle. In addition one had recognized that the aberration is changeable in time and in space.
In 1698, Edmond Halley (1656–1743), astronomer and geophysicist, started to undertake three trips with a sailing ship, in order to compile a declination chart of the Earth, published in London in 1701: THE DESCRIPTION AND USES OF A NEW AND CORRECT SEA-CHART OF THE WHOLE WORLD, SHEWING THE VARIATIONS OF THE

COMPASS.[7] Halley hoped that this declination chart with the lines of equal declination would help for navigation (longitude problem). Thus his chart had huge dimensions (122 x 52 cm) and the lines of equal declination were not only shown on the Sea but also on the continents. Alexander von Humboldt admired Halley's chart very much.[8]

"Magnetic Association" and the Development of Geomagnetic Observatories

HUMBOLDT, GAUSS AND WEBER

Alexander von Humboldt (1769–1859), who observed since 1806 variations of the declination, initiated – after his trip to Russia – an European geomagnetic network in 1829. He wanted to study in a global way the terrestrial magnetism. He could measure during his trips the relative intensity of the terrestrial magnetism by the help of a swinging inclination magnetic needle. Humboldt started to draw zones of equal intensity of the geomagnetic field parallel to the magnetic equator. His lines of equal intensity were called isodynames, published as ERSTER ENTWURF EINES ISODYNAMISCHEN SYSTEMS.

In 1833 Carl Friedrich Gauß (1777–1855), in collaboration with Wilhelm Weber (1804–1891), founded the "Magnetischen Verein" (Magnetic Association) in Göttingen. In 1839 an international magnetic congress in Göttingen under the leadership of Gauß took place; participants came from Germany, England, Ireland and Russia.[9] Under the leadership of Humboldt, Gauß and Weber the "Magnetic Association" in Göttingen errected until 1841 fifty stations all over the world: In the German states for example Berlin, Göttingen, Breslau, Freiberg, Leipzig, Marburg, Augsburg and Munich, in the Scandinavien countries Altona, Uppsala, Kopenhagen, in Italy Milan and Palermo and further stations in America, Asia, Australia and Africa.

Figure 4: Magnetic Laboratory of Gauß and Weber Becquerel, A.C.: Magnétisme terrestre. In: Traité de l'électricité et du magnétisme. 7 vols. Paris 1834–1840, here vol. 7, planche V.

The magnetic observatory in Göttingen was supplied with instruments (costs of 797 Thaler) for measuring the following geomagnetic values:

- the declination = the deviation of the compass needle of the geographic north-south direction,
- the intensity (strength) of the horizontal component of the vector of the magnetic field,
- the inclination = the inclination angle of the magnetic needle from the horizontal direction in the magnetic meridian.

A magnetometer is a horizontal magnetic bar, hanging on a thread; a mirror on the thread serves for reading the deviations and torsions. The deviation angle is measured and when the magnet is swinging then the oscillation period is measured.

FIRST ELECTROMAGNETIC TELEGRAPH 1833

Inspired by the principle of the magnetometer, Gauß and Weber developed the first electromagnetic telegraph in 1833.

Figure 5: First Electromagnetic Telegraph, Made by Gauß and Weber (1833)
Deutsches Museum Munich

The transmitter is an induction coil, which is moved over a bar magnet or removed from it. A horizontal hanging magnetic bar, surrounded by a large coil, is deviated by the magnetic field; the deviations are indicated by a mirror on the thread; the received signs (a combination of plus and minus signs similar to the Morse code) are observed on a scale with a telescope. Using this idea, Gauß wanted to construct a telegraph connecting the Physical institute were Weber was working with his astronomical observatory at the other end of the town. Gauß thought – as he wrote to Humboldt in 1833 – that he will need for the transmission of one sign less than one minute. In a letter to his friend Wilhelm Olbers (1758–1840) in Bremen Gauß expresses his hope to connect by the telegraph immediately Göttingen with his friends in Hannover and elsewhere. The original electromagnetic telegraph did not survive; it was damaged by lightning in 1845; in 1873 for the World fair in Vienna a new one was re-constructed by Weber himself.

GEOMAGNETIC OBSERVATORIES

Since the 1830s the development of geomagnetic observatories began. The first three important observatories were in Berlin, initiated by Humboldt, in Göttingen founded by Gauß and Weber and in Munich-Bogenhausen founded by Johann von Lamont (1805–1879).

- Berlin, initiated by Alexander von Humboldt
- Göttingen, founded by Carl Friedrich Gauß and Wilhelm Weber
- München-Bogenhausen, founded by Johann von Lamont, transferred to Maisbach and then to Fürstenfeldbruck
- The Imperial Marine Observatory in Wilhelmshaven, founded in the 1870s, was later transferred to Wingst caused by the serious disturbances of the magnetic measurements by the electrification.
- Geomagnetic observatory on the Telegraphenberg in Potsdam, founded in the 1880s, transferred to Niemegk.

Today around 200 geomagnetic observatories exist in the world.

MAGNETIC FIELD OF THE EARTH AND BIRD FLIGHT

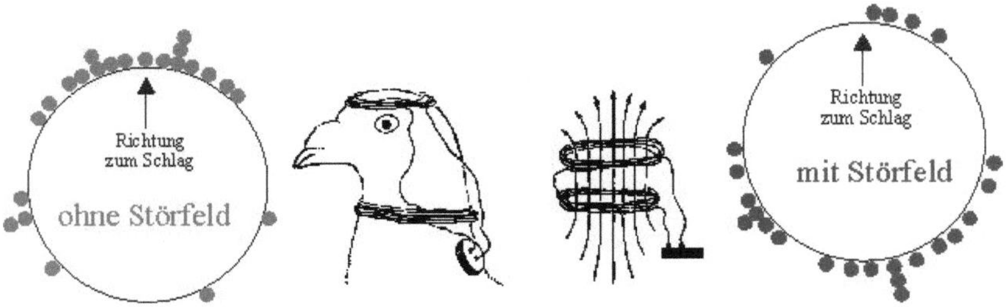

Figure 6: Pigeon's magnetic sensitivity with and without a disturbing field
http://www.bio.uni-frankfurt.de

The magnetic field of the Earth enables the orientation of birds flying to the south in the winter; a kind of a compass was found in the eye and in the pecker by scientists of Frankfurt University. This compass may be a magnetite or a biochemical process: a light receptor should transform magnetic information into visual perception. There are two parameters for orientation:

1. The intensity of the field, which is characteristic for each place and
2. the inclination of the field lines.

Also pigeons have a magnetic sense (sensitivity up to 50 Nano-Tesla). If one switches off the Earth's magnetic field by using a disturbing field, made by induction coils with electric current around the head of the pigeon, then they loose their orientation. Also the Sea turtles have a magnetic sense for orientation.

Solar-terrestrial relationships and magnetic fields in the Cosmos

SOLAR-TERRESTRIAL RELATIONSHIPS

Finally I discuss the solar-terrestrial relationships discovered in the middle of the 19th century, as well as their effects on the Earth including auroras, geomagnetic storms and radio interference.

Already in the 18th century one recognized the coincidence of auroras with disturbances of the magnetic needle – especially by Anders Celsius (1701–1744) and Olof Peter Hjorter (1696–1750) in Uppsala in 1741. Also Alexander von Humboldt reported from a disturbance of the declination during an aurora in 1837.

In 1852 four scientists, Johann von Lamont (1805–1879) in Munich, Edward Sabine (1788–1883) in London, Rudolf Wolf (1816–1893) in Zürich und Alfred Gautier (1793–1881) in Geneva, discovered another relationship between the eleven year solar cycle and magnetic variations. Sabine, president of the Royal Society, founded during his „magnetic crusade" different additional magnetic observatories, for example in St. Helena and in Canada. He was also a participant of the antarctic expedition of James Clark Ross (1800–1862)[10].

In 1934 the famous Göttingen geophysisist Julius Bartels (1899–1964) named places on the Sun, which emit particles, disturbing the terrestrial magnetic field and making auroras, as M-regions, where M should indicate magnetic. Sidney Chapman (1888–1970) announced that particles from the Sun could cause changes in the magnetic field of the Earth. The so-called Van Allen Belt catches and slows down the particles and storages them. So it serves as a protection shield for life on Earth.

In 1958 these charged particles rich of energy could be detected in the high atmosphere by Explorer rockets. These charged particles, the solar wind, consist of electrons and protons. With velocities of several hundred km/sec the solar wind flows out of the so-called holes of the corona, these are cooler regions of lower gas density with magnetic lines open into the cosmos. They were detected by Skylab in 1973 with

X-Ray observations. They are identical with the M-regions mentioned before. The magnetic loops in the corona were observed by several satellites, especially by SOHO (2003). The temperature of the corona is a million Kelvin, discovered by spectroscopy of 89times ionized iron lines in the UV region. Coronal holes can last several solar rotations. They produce recurring magnetic storms at Earth every 27 days.

GEORGE ELLERY HALE AND THE SUN'S MAGNETIC FIELDS

At the end of the 19th century astronomers suspected relationships between sun spots and magnetic fields. In 1908 George Ellery Hale (1868–1938) on Mount Wilson in California could prove this. He used a solar tower with a spectrograph. The magnetic fields in sun spots are 1000 times stronger than the Earth magnetic field (0,5 Gauß). Their strength is up to 3000 Gauß.

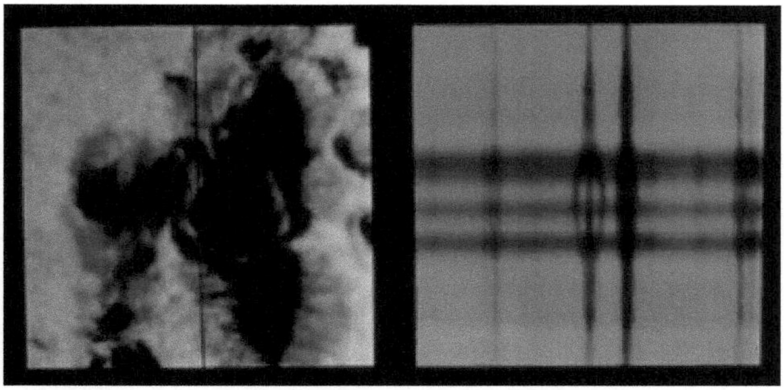

Figure 7: Sun spot and the Zeeman splitting; indicating the intensity of the magnetic field
Kitt Peak National Observatories, USA

George Ellery Hale used the Zeeman-effect, discovered by Pieter Zeeman (1865–1943) in 1896. Hale measured the line splitting in two or three components, which indicates the strength of the magnetic field. For example a line splitting of 0,2 Å (2 nm) indicates a magnetic field of 4000 Gauß (10.000 Gauß = 1 Tesla).
For measuring the magnetic field of the Sun one has developed special solar tower telescopes – very famous is the Einstein tower in Potsdam, erected in 1924. In the NS time in Germany four new solar towers were built, Göttingen, Wendelstein (Munich), Kanzelhöhe (Graz) and Freiburg, paid by the military authorities because the disturbances and radio interfernce caused by the Sun should be studied in detail. Modern solar telescopes use a vacuum tower telescope to prevent heating in the interior. Best examples are 70 cm-Vacuum-tower-telescope of Izana, Teneriffa (1987) of the Kiepenheuer-Institute in Freiburg or Kitt Peak and Sacramento Peak in the USA.

In 1952 Horace Welcome (1912–2003) und Harold Babcock (1882–1968) succeeded to measure the weak general magnetic field of the Sun with a magnetograph, indicated by an oszillograph.
What about magnetic fields of stars? Regarding the relationship between rotation, mass, radius and magnetic field, one can suspect that stars could have large magnetic fields, because many stars are fast rotating and can have large masses and radii. Already in 1946 Horace Welcome Babcock succeeded to measure the first time the magnetic field of a star with the 2,50 m-reflector of Mount Wilson Observatory. Extreme magnetic fields are known nowadays: 53 Camelopardalis has magnetic variations of +37.000 to -5100 Gauß in the range of eight days. Magnetic fields of stars play an important role for stellar structure and development, activity in the convection zone and for the production of flares (high energy explosions on the surface).

COSMIC MAGNETIC FIELDS
Cosmic magnetic fields can be measured by radiotelescopes and satellites.

Figure 8: 100-m-Radiotelescope Effelsberg, MPI for Radioastronomy Bonn and Magnetic field of the spiral nebula NGC 6946, MPI for Radioastronomy Bonn

Today one can even measure the weak magnetic field of the Milky Way or in a spiral nebula using radioastronomy. The largest stearable radiotelescope is the 100-m-Radiotelescope Effelsberg, MPI for Radioastronomy Bonn, built in the 1970s. Very interesting are also the magnetic fields of pulsars and quasars; synchrotron radiation is measured originating from places with large magnetic fields.

Conclusion

I illustrated a variety of information in the field of magnetism from ancient time to nowadays. The historical context is a good method for motivating the students. I did not mention – for example – magnetic prospection in archaeology – but there is an impact of magnetism in many fields of science; it is really an interdisciplinary topic.

- In the first case students should understand that important contributions in science and technology came from outside of Europe. So I showed the transfer of the compass from the Chinese culture to the Western world and in a modified form back to China.

  Students can make experiments with magnets and find out about attraction, repulsion and so on – the laws of magnetic forces – and this could be compared with the description of the experiments in Gilbert's book DE MAGNETE. Then further on students learn how to magnetize wires and to build an own floating compass. These experiments are suitable also for very young children to get them interested in science.
- In the second case one can see an early example of an international cooperation in the beginning of the 19th century. Performing the magnetometer of Gauß and Weber is much too complicated in the classroom. But the great interest in the topic Humboldt and Gauß can be seen in the recent publication „Vermessung der Welt". In addition the measuring principle of the magnetometer with a hanging magnetic bar in an induction coil was transferred to the construction of the electromagnetic telegraph and later to physical instruments.
- In the last case I discussed the solar-terrestrial relationships discovered in the middle of the 19th century, as well as their effects on the Earth including auroras, geomagnetic storms and radio interference. In addition I showed an interesting application of the Zeeman effect and the importance of magnetic fields in order to understand the cosmos.

The last two cases are for older students and are suitable also for project teaching. In Bavaria it was called "Studientag" (day or week of studies) where all teachers of different subjects (e. g. sciences like physics, chemistry, astronomy, geoscience, biology, and humanities like history, art and literature) were working together on an interdisciplinary topic like for example "water" or "energy" – and here you can also choose "magnetism"; there can be experiments and presentations, student lectures and exhibitions.

Finally magnetism is also a good tool for popularization of science; students can have fun by using magnets. There exist even magnetic games, one example from 1800 in the city museum of Cologne or a modern one for building the Platonic bodies, crystallographic structures or chemical molecules.

Notes

[1] More detailed information about this topic can be found in Wolfschmidt 2005

[2] See e. g. Needham 1959, 1962 and Zhao & Zhao 1989.

[3] Thanks to Elizabeth Cavicchi <ecavicch@mit.edu> I can give this reference, which describes an early magnetic object from the American continent: Carlson 1975.

[4] *„wann ferst du zu der gelincken hand*
so bist du verloren:
wann do ligt ein stein
der zeucht alles eysen, das zu jm kommen mag
und hat auch die kiel verderbt
mit leuten und mit gut.
der selb stein ist genannt Magnet ... "
Quoted after Fraunberger 1985, p. 13.

[5] *„... so finde ich auch dies an dem Magneten, daß er sich nicht allein wendet von der Mitternacht und lenket sich gegen den Aufgang, um 9 Grad mehr oder minder, [...], sondern er zeucht auch unter sich [...] um 9 Grad mehr oder minder."* Hartmann, Georg: Letter to Albrecht Duke of Brandenburg-Prussia. Nuremberg March 4, 1544. Transscription in: Hellmann 1898, p. 65–66.

[6] *„nach dem Wissen nicht in Büchern forschen, sondern in den Dingen selbst".*

[7] The original chart is in the British Museum; in 1895 existed further copies in Hamburg and in Paris, according to Hellmann 1895, p. 10.

[8] *„Dies Unternehmen hat Epoche in der Geschichte des tellurischen Magnetismus gemacht. [...] Nie vorher, glaube ich, hat ein Gouvernement eine See-Expedition zu einem Zwecke angeordnet, von dessen Erreichung die praktische Nautik sich zwar viel versprechen durfte, der aber doch recht eigentlich ein wissenschaftlicher, physiko-mathematischer genannt zu werden verdiente."* Quoted after Balmer 1956, p. 180.

[9] Participants were: Carl August Steinheil (1801–1870), Edward Sabine (1788–1883), Humphrey Lloyd (1800–1881) and Adolf Theodor Kupffer (1799–1865), since 1843 director of the Magnetic-meteorological Central Institute of Russia.

[10] Wiederkehr 1983/84.

References

Balmer, Heinz: 1956, 'Beiträge zur Geschichte der Erkenntnis des Erdmagnetismus.' Aarau: H.R. Sauerländer (Schweizerische Gesellschaft für Geschichte der Medizin und der Naturwissenschaften).

Carlson, John: 1975, 'Lodestone Compass: Chinese or Otmec Primacy? ' In: *Science* 189 (1975), 753–760.

Fraunberger, Fritz: 1985, 'Illustrierte Geschichte der Elektrizität.' Köln: Aulis Verlag Deubner & Co.

Gilbert, William: 1600, 'De magnete, magneticisque corporibus, et de magno magnete tellure; Physiologia nova, plurimis & argumentis, & experimentis demonstrata.' London: Petrvs Short 1600. Faksimile Brüssel 1967.

Hellmann, Gustav: 1895, 'Neudrucke von Schriften und Karten über Meteorologie und Erdmagnetismus. No. 4. E. Halley, W. Whiston, J.C. Wilcke, A. von Humboldt, C. Hansteen: Die ältesten Karten der Isogonen, Isoklinen, Isodynamen 1701, 1721, 1768, 1804, 1825, 1826.' Berlin: A. Asher 1895.

Hellmann, Gustav: 1898, 'Neudrucke von Schriften und Karten über Meteorologie und Erdmagnetismus. No. 10 Rara magnetica 1269–1599.' Berlin 1898. Reprint Nendeln, Liechtenstein 1969.

Needham, Joseph: 1959, 1962, 'Science and Civilisation in China.' Cambridge: Cambridge University Press, Vol. 3: Mathematics and the Sciences of the Heavens and the Earth, 1959, Vol. 4, Part I, 1962.

Wiederkehr, Karl-Heinrich: 1983/84, 'Über die Auffindung des nördlichen und südlichen Magnetpols der Erde, die Antarktisexpedition von James Clark Ross (1839–1843) und die Verbindung zu Göttingen.' In: Mitteilungen der Gauß-Gesellschaft 20/21 (1983/84), S. 7–38.

Wolfschmidt, Gudrun (ed.): 2005, 'Vom Magnetismus zur Elektrodynamik.' Herausgegeben anläßlich des 200. Geburtstags von Wilhelm Weber (1804–1891) und des 150. Todestages von Carl Friedrich Gauß (1777–1855). Hamburg: Schwerpunkt Geschichte der Naturwissenschaften, Mathematik und Technik 2005.

Zhao, Käthe & Hsi-lin Zhao: 1989, 'Wissenschaft und Technik im alten China.' Hrsg. vom Institut für Geschichte der Naturwissenschaften der Chinesischen Akademie der Wissenschaften. Basel, Boston, Berlin: Birkhäuser.

The nature of science in the classroom – sociology to the rescue?

GÁBOR Á. ZEMPLÉN
TU Budapest, (HAS, HPS Research Group)
Karinthy Frigyes Dual Language High School, Budapest
Max Planck Institute for the History of Science, Berlin

Abtract: Contrasting two examples from 2005, a creationism-trial and a recent textbook, the article shows how differently social considerations are used to demarcate science from non-science. Drawing lessons from the comparison, the paper describes a 6 week module that tackles major „nature of science" issues from a perspective that grows out of sociological and anthropological considerations of expertise and views on legitimation. The broad sociological starting point can help students understand the presence of values in science, the often criticised bias that scientists at times show, and still appreciate science as a privileged form of knowledge-production.

Prelude: Lessons from creationism

If one reads carefully the 139 page long argument supporting John E. Jones III, District Judge' decision in the Kitzmiller/Dover case (from now on referred to as KD)[1], the last of a series of trials on the teaching of creationism in the USA, one can draw some important lessons from the document concerning the nature of science.

As with creationism-trials generally, the question again was: is creationism, or intelligent design (ID) in its new disguise, a) a religious movement and b) a scientific enterprise. For the present paper only the second question is of importance. What positions should or could the expert witnesses testifying about the nature of science take? Earlier court rulings have also caused some stir, and the Dover trial was not different.

Just to recall, in the 1981 Arkansas Creationism trial,[2] as expert witness Michael Ruse testified, he basically listed a number of epistemological desiderata that distinguish science from non-science. These included the use of law to effect explanation, the ability to predict hitherto unknown phenomena, that a theory should be testable, and must be falsifiable, and has to be of necessity tentative (Ruse 1982). Going through the criteria Ruse confidently stated that creationism is not science, as it fails to live up to the criteria he listed. But this testimony was not greeted unanimously by other experts. Larry Laudan strongly criticised Ruse's position, claiming that creationism made many testable statements, the theory changed at several points, and after listing several criticisable moves, he concluded that creationism is not non-science, but simply bad science. As he summarised the case:

> The victory in the Arkansas case was hollow, for it was achieved only at the expense of perpetuating and canonizing a false stereotype of what science is and how it works. If it goes unchalleneed bv the scientific community, it will raise grave doubts about that community's intellectual integrity. ... No one familiar with the issues can really believe that anything important was settled through anachronistic effort to revive a variety of discredited criteria for distinguishing between the scientific and the nonscientific. ... this time, the pro-science forces are defending a philosophy of science which is, in its way, every bit as outmoded as the "science" of the creationists. (Laudan 1982, p. 19).

So epistemic criteria to decide what science is and what it is not have already been criticised in the aftermath of the 1981 trial. Similar issues have been raised in the debate over Steve Fuller's expert testimony[3]. What I want to focus on in the next paragraphs, is how Judge Jones argued for the position that creationism is not science, and what role social criteria played in his decision. Obviously arguments that can be considered epistemological or even ontological have not been dropped completely. Thus the unscientific notion of supernatural causation and other religious aspects of the movement were highlighted, like Behe's statement that "the plausibility of the argument for ID depends upon the extent to which one believes in the existence of God" (KD, p. 28). But these were mostly used to argue that ID is religion.

Where Judge Jones argued that ID is not science (KD, pp. 64-89), the sociological considerations played a surprisingly significant role. The judge cited statements made by NAS and AAAS, major scientific organisations in the USA. Without investigating the arguments these organisations made in support of their position concerning ID, the judge relied basically on the status of these organisations in society (the NAS being the "most prestigious scientific association in this country, (KD, p. 69), the AAAS the "largest organization of scientists in this country", KD, p. 70.). Not only *ad verecundiam*, but also *ad numeram* arguments have been used: "an overwhelming number of scientists, as reflected by every scientific association that has spoken on the matter, have rejected the ID proponents' challenge to evolution." (KD, p. 83.)

And a significant part of the reasoning is even more closely connected to non-epistemic factors. As the judge stated: "Although ID's failure to meet the ground rules of science is sufficient for the Court to conclude that it is not science, out of an abundance of caution and in the exercise of completeness, we will analyze additional arguments advanced regarding the concepts of ID and science." (KD, p. 71.) Investigating the nitty-gritty details of one of the most important books on ID, *Darwin's Black Box* by Michale J. Behe, the judge again and again, whether concerning the bacterial flagellum or the blood-clotting system, referred to articles that 1) contradict Behe's argument, and 2) are peer-reviewed.

This issue reappears in the final section of the argument: "A final indicator of how ID has failed to demonstrate scientific warrant is the complete absence of peer-reviewed publications supporting the theory." (KD, p. 87) And to cite a crucial passage in extenso:

> "Expert testimony revealed that the peer review process is "exquisitely important" in the scientific process. It is a way for scientists to write up their empirical research and to share the work with fellow experts in the field, opening up the hypotheses to study, testing, and criticism. In fact, defense expert Professor Behe recognizes the importance of the peer review process and has written that science must "publish or perish." Peer review helps to ensure that research papers are scientifically accurate, meet the standards of the scientific method, and are relevant to other scientists in the field. Moreover, peer review involves scientists submitting a manuscript to a scientific journal in the field, journal editors soliciting critical reviews from other experts in the field and deciding whether the scientist has followed proper research procedures, employed up-to-date methods, considered and cited relevant literature and generally, whether the researcher has employed sound science. The evidence presented in this case demonstrates that ID is not supported by any peer-reviewed research, data or publications." (KD, p. 87, leaving out the citations in the original).

What we see here is that what is science or not is not decided upon demarcation criteria that directly utilise epistemological or methodological considerations, but rather sociological ones. The historically developing norms of the scientific establishment are "exquisitely important". What appears in peer-reviewed journals, what is endorsed by leading scientific organisations is science, movements and ideas that fail to achieve these are not scientific. This stress on the sociological aspects of science and on norms is even more pronounced at the concluding part of the argumentation in this section:

> "After this searching and careful review of ID as espoused by its proponents, as elaborated upon in submissions to the Court, and as scrutinized over a six week trial, we find that ID is not science and cannot be adjudged a valid, accepted scientific theory as it has failed to publish in peer-reviewed journals, engage in research and testing, and gain acceptance in the scientific community." (KD, pp. 88-9).

As no satisfactory epistemic criteria of demarcation have been found and embraced by the majority of experts, the judge – wisely in my view – resorts to much safer and non-epistemic criteria to define what is science and what is not. If one publishes in peer-reviewed journals, does research and testing that is accepted by the scientific community, then one does science. The lesson might be hard to swallow, but clear: if today one wants to defend the boundaries of science, one's strongest ally is not epistemology, outdated notions of falsification, vague definitions of simplicity, explanatory power, or testability, but sociology.

Sociology in "Nature of Science"

How is this great potential of sociological insight utilised in courses that teach students about the nature of science[4]? Not much. To contrast the 2005 trial, I take just one example, a book published the same year, by no means untypical of the genre (Alchin 2003a, 2003b; Woolman 2000).

The book on "Theory of Knowledge" by van de Lagemaat has a separate chapter on the natural sciences, preparing 17-19 year old students for the International Baccalaureate examination, a worldwide network of schools comprising more than 1000 high-schools with an examination centre in Geneva (van de Lagemaat 2005).

The chapter discusses what is generally considered as suitable and adequate material on the nature of science including philosophical, epistemological, and sociological considerations. The sequence of subheadings, however, shows the order and the emphasis. After a short introduction the first tackled topic is science and pseudoscience, followed by a few pages each on the scientific method, problems with observation, testing hypotheses, the problem of induction, falsification, finishing with science and society, and science and truth. From these I will only discuss the part on pseudoscience, and the one on society.

The book clearly stresses that demarcation criteria are needed for distinguishing science from pseudo-science. (van de Lagemaat 2005, p. 223). The original demarcation that "distinguishes a pseudo-science is that it claims the status of science while lacking its substance", however, is clearly not operationalised. On the next pages we learn that actually testability is the main criterion[5], and pseudo-science protects its statements from testability by being vague and/or using ad hoc explanations.

Needless to say these criteria are inadequate – one simply has to look at the history and philosophy of science to see that while these categories have often been used in arguments for or against specific movements, it is unclear how they could be used as either sufficient or necessary criteria. Darwin's theory of evolution failed to provide anything but utterly vague mechanisms for the emergence of variations, and the Bohr model of the atom used ad hoc assumptions concerning the shape and positions of the molecular orbits. Even if these examples are found wanting, it has clearly and forcefully been argued that no sufficient or necessary epistemic criteria of demarcation can be given (Laudan 1983). But in this part of the chapter none of the sociological criteria that play such an eminent role in Judge Jones' argument surface.

And what does the book say about science and society? After introducing students to inductivism and falsification, and showing that they cannot "give us an adequate account of the nature of science" a third perspective is introduced, that of Thomas Kuhn[6]. The problem of rationality in the Kuhnian framework is investigated, and here social and individual (i.e. psychological) factors are listed. These include "ambition, vanity, envy, … public recognition, … the military's desire for power and big business's desire for profit" (van de Lagemaat 2005, p. 242). Furthermore "Ambitious scientists may be attracted to areas in which there is plentiful supply of money to fund research", etc. Individual and social factors are not seen as epistemically important resources and as assets, but as at the least only directing research interest, and at the most hindering progress (the Lysenko affair and cold fusion is mentioned when talking about the self-correcting mechanisms of science – by the way the only "positive" role for a broadly viewed social mechanism).

When assessing Kuhn's position, the writer falls back to a normativist philosophy of science. A long quote from Popper, stressing that we ought to be sorry for the 'normal' scientist and that he "has been taught badly" is cited on the discussion of normal science. In his evaluation of scientific revolutions he reasserts continuity in scientific development, by claiming that Newtonian mechanics is a special case of Einstein's theory of relativity, And as for the rationality of paradigm-choice he states that "we should distinguish between the *origin* of a belief and its *justification*. For the origin of a belief is not of any great relevance to science. All that matters is that the belief should be *testable*. If it is confirmed by experiment, then we provisionally accept it; if it fails then we reject it." (van de Lagemaat 2005, p. 244). In these cases the writer has recourse to the very view of science that has been found wanting and necessitated the introduction of a "third perspective", i.e. a sociological one. I do not want to evaluate these answers here, the sad fact remains: the insights from the last forty years of sociology of science, of social studies of science, and of the sociology of scientific knowledge are not present and utilised[7].

Is upside down the right way up?

The two examples I contrasted indicate that in today's courtroom sociological considerations play an important, and, if compared to earlier trials on creationism, an increasingly important, role. These insights, however, are generally not utilised in the context of the demarcation problem and fail to surface in courses or modules on the nature of science. Today the trend in these courses is to start with "philosophical" or epistemic criteria, stress some simplified "scientific method", and the social considerations only emerge at the end of the chapters or courses, where they are usually used to *weaken* claims on the applicability of the method: "let us not forget that scientists are humans" (Alchin 2003b, p. 22)". Thus, naturally, seeing science as a primarily social phenomenon and subscribing to a social constructivist view is generally seen as endorsing relativism, giving up the hope of demarcating between science and non-science.

But is this necessarily the fate of a sociological approach? In the following I will argue that simply by shifting the position of the social in these modules (without a major change in the contents of the course) a radically more constructive role can be given to sociological considerations, which, by the way, is also closer to what historians, sociologists and philosophers of science today think about science.

A module on Nature of Science – putting sociology first

In the following I describe a six week (weekly 90 minutes) module on the nature of science carried out in 2006 in a Theory of Knowledge (TOK) class, a compulsory course of the International Baccalaureate Organization (IBO) Diploma Programme. A few words about the course. TOK was launched a in the diploma programme in 1999[8]. It is an interdisciplinary requirement with at least 100 hours of teaching time during the programme's two years. The aim is to bring students to appreciate different perspectives and "to stimulate critical reflection on the knowledge and experience gained inside and outside the classroom. The course challenges students to question the bases of knowledge, to be aware of subjective and ideological biases and to develop the ability to analyse evidence that is expressed in rational argument"[9].

The subject is structured around the so called TOK diagram with strong focus on the knower(s), surrounded by four ways of knowing (Reason, Emotion, Language, Perception), and the six areas of knowledge (Natural Sciences, Human Sciences, History, The Arts, Ethics, Mathematics). In case all "areas" and "ways" are treated with equal emphasis, appr. 10 contact hours are left for discussing and studying the natural sciences. Earlier I have introduced the programme in more detail and showed potential conflicts between the general approach taken by the critical-thinking skills oriented TOK and traditional science subjects, and have also highlighted some possible solutions (Zemplén 2007 (accepted)). In the following, I will give only the objectives of the classes and the rationale for this structure, with a quick summary of the classes themselves including at times the reaction from students.

Objectives

The objective of the course is to introduce science as a social institution, and scientists as experts in a system of expertise. For this students are introduced to different systems of expertise – some institutionalised and some not. Once this social starting point has been established, norms of science are discussed – some ethical, some epistemic. The necessity of these norms are investigated and their role in maintaining the proper functioning of the expert system in a) producing knowledge (i.e. epistemic norms) and b) functioning properly within society. Different communities claiming to do science are investigated, and the changing boundaries of science, pseudoscience and fringe science are discussed and the notion of paradigms is introduced. Some norms are discussed in detail, like the historically developing and changing "scientific method" with special emphasis on the logical structure of inductivism, deductivism, and falsification. These norms are connected to science's constant strive to eliminate or at least control different sources of errors. Finally, connecting science to the students' lifeworlds, the objectives of the educational system and science education are investigated.

Rationale

Why is this sequence introduced? A major problem with current treatments of the nature of science in the classroom is the separation of "philosophical" from historical cum sociological insights. The scientific method is generally introduced in a framework that implicitly endorses the statements view of science, where the job of the scientist is to come up with statements that have certain characteristics (can be verified, falsified, corroborated, reduced to other types of statements, etc.). This statements view does not allow for easy introduction of social factors – and already requires non-obvious notions of truth, truth value, etc. Also, the complexities of laboratory work and the scientific established are rendered invisible in this view, and the method of science often boils down to simplistic logical insights focusing on white swans and the like.

When putting sociology first, these problems are sidestepped. The students realise the difficulty of organising systems of expertise in society to achieve certain goals (in the case of science, a major goal is knowledge). They are therefore more open to pluralistic approaches, more open to appreciate the historically changing and contingent development of scientific method. At the same time the social framework allows for the discussion of changing societal needs, the power-struggle between different social groups, where scientists have their own agenda and aims, and socioscientific issues can easily be connected to NOS insights.

Social insights are also more obvious in the students' life world than epistemic considerations, and a course aiming to stimulate critical thinking needs to tackle these issues. Furthermore they are easier to grasp, as from early childhood on the students have experienced different social relations, but not the complex processes of knowledge-production. Also, students can be expected to make reasoned choices in their picking of experts, but can hardly be expected to pick reasoned positions in complex fields like science. And this is especially so in the increasing number of cases, where expert opinions also differ.

This last point is not trivial, but a short survey of the social aspects of epistemology will make it clear. Let us take as an example the forefronts of science. As has been argued by Hardwig and generally accepted in most research areas, epistemic dependence is present and a necessary phenomenon of present-day science (Goldman 2001; Hardwig 1985). And this does not only mean reliance on the literature and on the expertise of other laboratories and the trustworthiness of data by these, but dependence also exists within a laboratory and co-authors of an article. Thus even within science a crucial element enabling the joint production of reliable knowledge is trust in the expertise of the co-workers. Previous research, academic position, institutional background can all help the picking of experts, but these are social categories. The product of this cooperation is then submitted to journals, where again the editors pick the reviewers based on their position, role in the scientific community, track-record in science, and ask them to review the article in the light of what they consider sound science. The results appear in journals that are ranked according to how other experts cite these for their own work.

The social aspects of this procedure are quite transparent and easy to grasp. To refer back to the beginning of the paper, it is easy to understand this way why creationism is not science. But the epistemological aspects are much more difficult to grasp. Especially as recent strategies by movements like creationism include the mimicking of epistemic norms (like testability, the notion of crucial experiments). These can easily lead students to consider such movements as real alternatives to scientific views, as they are unfamiliar with the intricacies of the specific scientific field. But to mimic social desiderata, like the general acceptance by other scientists or publishing in peer-reviewed publications, is much more difficult.

The module

LESSON 0

As there are 10 different modules in the teaching of TOK at the school, the modules are connected with single 90 minute lessons to lead students from one topic to another. The modules have no necessary or even preferred ordering: many alternatives are open to the teacher(s) teaching it. In this case, the previous module was perception. During the class examples of conceptual art have been discussed, leading over to perceptual art, and art-works that directly utilise specific perceptual mechanisms. A catalogue of the recent and opening exhibition of the Kunsthaus Graz was used (Pakesch and Bucher 2003), that directly connected issues in cognitive science research and contemporary art.

Once attention has been directed to perceptual mechanisms, several illusions were shown and discussed in class. These included well-known illusions like the Müller-Lyell illusion and Ponzo illusion, but also Mach-band-effect, colour-afterimages, motion-artefacts, etc. like Purkinje, etc. The artistic work of Nicholas Wade has been used and some of his historical writings have been alluded to (Wade 1998). The class-discussion was directed toward the problem of illusions and the questions of perceptual errors. Does this category make sense? If yes, in what use?, etc. Finally the problem of observation and systematisation of phenomena has been raised and the students received their homework assignment:

HW 0 – Give 5 scenarios (as different as possible) that can hinder "scientific understanding", and where science/the scientist can go wrong!

LESSON 1

Expertise and expert-systems were introduced during the class. My aim was "to get to know the class better". At the beginning of the class students are asked to take a paper and pencil, put everything else away, number lines on the paper from 1 through 10 (a setting resembling a quick in-class test, common to the Hungarian school system), but are then asked to write names of their comrades who

1. knows best parties
2. can help with physics HW
3. can trust which movies to see
4. tell which medicament to take in case I feel unwell in school
5. help in an emotional crisis

The students could pick the last five questions that they were interested in (if it was accepted as appropriate by all members).

The "expert-game" showed, and it was shortly discussed, that students have expertise in a number of areas, and they treat each other as experts in social, educational, and various other matters. The form of the game was used to contrast "knowledge" that school-tests usually test and other forms of knowledge important for the everyday lives of the students. After this, a short introduction to the development of science was given, and the gradual institutionalisation of science was introduced through examples. These included astronomy and other divinatory techniques in ancient Babylon, where dynasties of rulers were served by dynasties of scribes and other experts, and where handing over knowledge and expertise was still tied to familiar relations. Other examples included the early Modern period with the birth of academies, like the Royal Society, contemporary Chinese medicine, where centralised educational system is still mixed with traditional discipleship, and a discussion of the students' own experiences of gaining knowledge via different structures and modes of knowledge-transmission. After this, the class gave a rough classification of how experts gain acceptance, which was similar to Weber's tripartite grouping of legitimating into 'charismatic', 'legal', and 'traditional'.

In the second half of the lesson the test-results were added up (by the student who has not done his homework), and in the last minutes of the class the results, showing at times nearly unanimous agreement in choice of experts, were discussed. Before that the class in groups of 4-5 compared their homework and the groups devised classifications of error sources for all the examples of the group. Three of the four classifications showed surprising similarity:

Group 1	Group 2	Group 3	Group 4
Human misinterpretation/ fallacy	scientist	human	Language/physical limitation
	sample	specimen/environment	Mental limitation
lack of technology	method/equipment	equipment	Placebo effect

Once the error source-typologies were discussed, scientific examples were given, where the predictive power is so small, that the layman is not much worse in predicting the outcome than the expert using expensive research tools and a complex methodology (i.e. 10 day weather forecasts, or cases where there is no expert consensus). This was introduced before the second homework assignment, an argumentative essay on the following question (the class is regularly asked to write short argumentative essays, and such an essay contributes to 60 % of their diploma grade in TOK):

HW 1: Why do we follow extremely complex scientific methodologies, even in cases where much simpler methods could yield similarly (im)precise answers? Is this good practice or not?

LESSON 2

After rehearsing concepts of the previous lesson (notion of error, expertise) the class was split into two groups of approximately equal size, depending on their position taken in HW 1. The groups were asked to consult each other's essays and together make the "best possible" 3 minute speech in defence of their standpoint.

The following notable arguments were made by the "pro" group (in square brackets the key issues raised):

- We need to tolerate science in areas where they are not too successful to allow for the development of these fields [chance for developing expertise].
- It is a society's choice to empower a group of people with authority for an area, and there is no reason trusting the laymen more than the experts in these cases, plus we have to trust the people who are best at something to do as they wish [epistemic dependence].
- Expert-systems offer more jobs [science as institution connected to socioeconomic problems].
- We need expensive experiments, as we want to be sure or as sure as possible [error-avoiding].

The second speech, in general less adequate and containing no epistemic considerations, in contrast highlighted:

- Funding of science as a highly problematic area that takes money from important issues [relevance, public opinion].
- The problem of unidirectional trust: society puts trust in scientists but cannot control them or check if they manipulate citizens [highlighting problematic nature of maintaining expert-systems].
- The recognition that the better the science is, the better it is for the society, and raising the question of how can one ensure that science is becoming better? [The question and significance of norms.]

The preparation, the talks, and a discussion of the talks took up most of the lesson. The exercise was also used to rehearse concepts that the students have met earlier in a module on reasoning and argumentation (for this class, this was the first module of their TOK classes). Students were reminded of the notion of strategic manoeuvring and balancing of rhetorical and dialectical aims of the argumentation. The arguments were analysed both as to how "strong" they were (i.e. were there any fallacies, can the arguments be used to make reasoned choices, etc.) and also as to how "effective" they were (the rhetorical aspects of persuasiveness, the question of target-group and audience were discussed).

HW 2: The homework was to read a one-page subversive text on "Economics meets science" (McErlean 2000, p. 205)[10], where induction and deduction are displayed as simply different strategies to get grants.

LESSON 3

The lesson started with a discussion of HW 2. Most students found the text funny, and some had vague ideas about induction and deduction, so a very simplified answer was given as to what the two terms "really" mean. (These were later discussed in lesson 5, and only one student in the final essay-test referred to the highly misleading notion as given in the text for HW 2, all others clearly realised that it is meant to be funny but not true.)

After this, however, students were asked why they found HW 2 funny. Soon they pointed to the discrepancy of what scientists say they do and what they really do, how methodology is used to create an image of the scientist that legitimates her in the eye of the public. So issues of ideology, self-image, and public image were raised, and students were asked to tell what they thought the communicated image of science was, and what the actual aims of the scientists. After collecting ideas, the students were asked to figure out how society (including the group of scientists) can ensure that the individual scientist does what is expected of him by society or by the scientific establishment. This discussion introduced the notion of norms (here the teacher suggested the terminology), as bridging the two tasks, something like this (reduced version of blackboard-image)

Science as a social institution

Communicated task	to guarantee that actual practice	actual task of scientists
("ideally")	is as close to expectation	utility, fame, truth
"truth" production	NORMS	new questions, failure
Privileged source of	methodological	money, conspiracy
knowledge in society	ethical (Merton,	
	anti-Merton)	

Mertonian norms were introduced and some general methodological considerations discussed.

LESSON 4

Once science has been located as a social institution and a system of experts and expertise, the historically changing borders of science have been shown. In a frontally coordinated discussion the class was to list branches and modes of knowledge production, and were asked to classify them. Accepted and rejected science, pseudoscience, and fringe-science were suggested as categories. Well-known examples like astrology and other divinatory techniques were mentioned, and the historically and culturally changing position of these modes of knowledge-production in the given scheme of classification was pointed out. Other examples included the works of Däniken, pi-water, acupuncture, etc. Radical shifts (later introduced as paradigms) were mentioned, and their effects considered. In the case of acupuncture the possibility of radical methodological differences were highlighted, but the possibility of joint research and interaction of the different traditions were shown (Kim 2006; Scheid 2002).

Rather simplistic differences between pseudoscientists (as mimicking the norms of real science) and fringe scientists (as striving to confirm to the norms of science) were proposed. This is in line with recent interest in the pseudoscientist as opposed to pseudoscience (Dersken 1993), but the problems of the approach were also hinted at.

Following this, the students were introduced to two, for them unknown, historical examples: phrenology and the practice of frontal lobotomy, more specifically Dr. Freeman's transorbital lobotomy, based on the work of Egas Moniz, who received a Nobel Prize for frontal lobotomy in 1949, but Freemen's overzealous use discredited the practice. Reading short texts they were asked to draw lessons from them, and to come up with similar examples they have heard of or knew about.

The general impression was that the students realised that historical contingency does not lead to relativism, as the cultural background at any given time period will have more or less clear norms according to which decisions are made in a tradition, even though these norms themselves are subject to change.

LESSON 5

As one of the most demanding lessons, this 90 minutes was used to rehearse some of the logic material students have already learnt and to introduce them to different views on te scientific method. First the well known Wason card selection task was used. The task asks students to test propositions that are in a conditional form. The first example runs like this (with drawings on the blackboard):

Here is a set of four cards each of which has a letter on one side and a number on the other side. The visible faces of the cards show E, K, 4, and 7. Which cards should you

turn over in order to test the truth of the proposition that if a card shows a vowel, then its opposite face shows an even number?
As expected from the literature and my earlier experience, the students did rather poorly on the task, but the solution was not yet given.
Following this, the task was slightly changed
You are in a bar where four costumers are consuming beverage. You either know the type of drink they drink, or their age. In this case you know that the customer drinks beer, carrot juice, is 20 years old and 4 years old. Which customers should you ask for the other piece of information, in order to test the truth of the proposition that if someone drinks an alcoholic beverage in the bar, she has to be over 16?
Here, again as expected, nearly all students gave the right answer. After this, both examples were formalised and the formal identity was pointed out. Both have a structure p ε q, but one is much easier to solve then the other. The four cases have been reconstructed as follows

Form:	p ε q	p ε q	p ε q	p ε q
	p	~p	q	~q
	q	~q	p	~p
	modus ponens	negating antecedent	affirming consequent	modus tollens
1st ex.	E	K	4	7
2nd ex.	Beer	Carrot juice	20	4

The students were reminded that the right solutions (to check the other side of the first and fourth card) coincided with the logically valid forms. With this the possibility of errors in reasoning was highlighted. After this the truth table was given, and students were asked to recall similar truthtables from their module on argumentation.
Truth table

p	q	p ε q
1	1	1
1	0	0
0	1	1
0	0	1

Once the students, through a number of examples, had found their orientation in the truth-table easily, they were asked to practice the Wason test with partner. (A short break was introduced, as students were unused to concentrate long on abstract problems.) They had to devise as contrived examples as they could, and the partner had to be as quick as possible in supplying the correct answer. At first, many of the students giving the example failed to solve the task correctly, but after approximately 15 minutes and lots of heated debates, apart from one student they all were confident and got used to using the logical structure in devising and answering the problem. (The one student who had difficulty still received extra homework and by the coming week was confident to share her examples with the class.)

Finally the class was introduced to "the scientific method", referring back frequently to the formalisation of the Wason-test and the truth table of the conditional. Deduction was defined as moving from hypothesis (H) to observation (O), while induction as moving from observation to hypothesis. The earlier formalisations were used to show the "problem of induction", and the deductively valid falsification (as a *modus tollens*). Still the same forms were used to introduce the problem of underdetermination, and students were asked to give specific scientific test-situations and explicate what is to be tested. Thus the conditional Hε O was extended: $(H\&A_1\&A_2\&A_3\& \ldots\& A_n)\ \varepsilon\ O$.

At home the students were asked to think through these examples, and give an example of complex experimental situation and the problems of testing.

LESSON 6

The lesson was used partly to wrap up the course, to repeat and deepen the insights from the previous lesson, and to investigate science as it appears from the "inside" and from the "outside".

As underdetermination is often perplexing for students, they are introduced to some of the simplified ideas of Duhem, Neurath, and Quine. Importantly, none of these writers considered underdetermination as an insurmountable problem of science, and Duhem's *bon sens* (Duhem 1954), Neurath's antifoundationalism (Neurath 1913), and Quine's pragmatism (Quine 1951) all show how choice is not arbitrary even though not logically determined.

The largest part of the lesson was used to collect the impressions of the previous lessons, according to whether they belong to the "inside" or the "outside" of science. So problems of methodology and controlling error sources would belong to the "inside", while the picture of science in the media, complex interactions of science, society, policy-makers, etc. to the outside. The students were asked to compare their primary and secondary school education to what they knew about university and post-graduate education. They clearly saw the gradual move towards the "inside" of science from the "outside", even though the names of subjects have stayed the same during their education. In primary school, science had for them an unquestionable authority, but during their IB Diploma Course they meet more and more of the intricacies of scientific activity. As their knowledge of science increased, the extreme complexity of the issues became clearer and clearer. The two categories helped students to place their experiences concerning the popularisation of science, as well arguments they were already familiar with either strongly supporting the scientific establishment (scientism) or strongly critical of it and seeking for alternatives ("romantic" attitudes). Their common impression was that, just as in the case of demarcating science, these general attitudes help little when a decision is to be made or action is to be taken in case of a specific complex problem. It also helped them to appreciate in a new light the personal trajectories they are taking – some moving more and more "inside" science, and planning to continue their studies at university science faculties, others remaining "outside", but still facing science-related issues every day[11].

The next double lesson included the composition of an in-class essay and the introduction of the next module (in this case history).

Discussion

The module described above is certainly simplified, but attempts to provide an alternative to modules on the nature of science where social considerations play little or no role. It is structured in a way as to embed the epistemic considerations (Lesson 4 and 5) in a historically and socially informed view of science (Lesson 1-3 and 6). Key concepts employed included experts and expertise, as well as the legitimating systems of expertise, as crucial for our understanding of contemporary science and its role in society.

With this, the approach outlined here stands close to the "third wave" of science studies (Collins and Evans 2002). Even though sociological approaches have been equated with an attempt to *replace* epistemological questions with social question, the "third wave" is thoroughly sociological yet brings back epistemological considerations and is an attempt to lay the basis of a *normative* theory of expertise. This is partly employed by Collins and Evans to find a new place for sociology of science, and define it neither as serving the scientific establishment (as was characteristic of what they call the first wave), nor as focusing exclusively on the "extra-scientific factors", as "scientific method, experiments, observations, and theories are not enough" (ibid. p. 239), typical of the second wave aiming to dissolve categories that differentiated scientific knowledge and other forms of knowledge. In this envisioned third wave scientists are experts that interact with other groups of experts and the public. At the same time sociologists and other analysts of science are also experts in areas that at times can help in public and technical decision making, and can provide prescriptive statements about the nature of expertise[12].

Another novel element of the approach is to help students place different claims about science and a wide array of rhetorical strategies they are already familiar with concerning the status of scientific knowledge, the usefulness of science and its place in today's society as well as the historical and cultural dimensions that lead to this specific role and function. For this the somewhat unconventional notions of science from the "inside" and "outside" have been utilised. That this distinction is highly significant is dramatically shown in an article by Shapin (Shapin 2001). Shapin collected a number of "antiscientific" statements, including "There is no such thing as the Scientific Method", "Scientists do not find order in nature, they put it there", "New knowledge is not science until it is made social", and other statements generally associated with constructivist and sociological approaches (Shapin 2001, pp. 99-100). Shapin after the provoking list on the nature of science discloses his informants: they include Nobel-prize winning physicists, biologists; the list includes Peter B. Medawar, Albert Einstein, Niels Bohr and others, but none of the feared social constructivist sociologists, or postmodern theorists. Shapin used the examples to show and argue that

> "it cannot be the claims themselves that are at issue, or the claims themselves that must proceed from ignorance or hostility. Rather, it is *who has made* such claims, and what motives can be attributed – plausibly, if often inaccurately and unfairly – to the *kinds of people* making the claims." (ibid., p. 101).

Without repeating the arguments, the playful but meticulously documented hoax of Shapin points to an underestimated problem of science education, the general neglect of social considerations when nature of science is discussed in the classroom or in publications in science education[13].

As science develops and its complexity increases, the laymen and students are less and less likely to make informed decisions as to what is sound science and what is not. The Royal Society's motto "Nullius in verba" (On the words of noone) is not a fruitful approach in a period where scientists themselves have to rely more and more heavily on the words of others. Current emphasis is still on "Robinson Crusoe" epistemologies, but by giving the rightly deserved attention to social considerations in NOS, we can better prepare students to find the right expertise and thus the reliable knowledge in our society.

Conclusions

The paper described a 6 week module (12 x 45 minutes) that tackles major „nature of science" issues from a perspective that grows out of sociological and anthropological considerations of expertise and views on legitimation. The broad sociological starting point can help students understand the presence of values in science, the often criticised bias that scientists at times show, and still appreciate science as a privileged form of knowledge-production. This approach runs counter to general trends, where science is first introduced as characterisable with a distinct and specific method, reliance on empirical data and logical reasoning (as opposed to other social institutions), and then this view is gradually "softened" once it is recognised that there is no distinct scientific method, there is no guarantee to cut nature at its joints, or a guarantee that the observations will favour one theory as opposed to another. The traditional approach has to make concessions to account for science in practice. The road the pilot module takes is the opposite: starting from "soft", social considerations one can appreciate the renewing attempts of scientists to tackle sources of errors and acquire as reliable knowledge as possible about the natural world.

Notes

[1] *Tammy Kitzmiller, et al. v. Dover Area School District, et al.* 2005 WL 578974 (MD Pa. 2005). For the memorandum opinion see:
http://www.pamd.uscourts.gov/kitzmiller/kitzmiller_342.pdf .

[2] *McLean v. Arkansas Board of* Education, 529 F. Supp. 1255 (E.D. Ark. 1982).

[3] See for example the debate on the testimony on the HOPOS list server: http://listserv.nd.edu/archives/hopos-l.html, that was the major focus of the November 2005 discussions.

[4] Here and in the following I will use the term "sociological" in a rather loose sense. For a more detailed discussion in the journal see (Allchin 2004). The reason is that in the rather limited form these insights can appear in the classroom does not necessitate the strict separation of the very different streams in sociology. But it will be quite evident from the context which movements are implied in the discussion. Another reason why the traditions are not separated is to avoid "gut-feelings" and the general attitude that approves of structural functionalist or Mertonian sociology and automatically rejects e.g. SSK approaches.

[5] The original definition recalls attempts like Bunge's: "pseudoscience is a cognitive field advertised as a science though failing to be one" (Bunge 1982, p. 378). But instead of listing the rich and detailed set of criteria that characterise Bunge's work or other attempts from recent decades, the writer falls back on testability as a criterion.

[6] The division of science into normal and revolutionary is fairly correctly described, but a figure (8-10. on van de Lagemaat 2005, p. 241) is used to show "reality according to Kuhn", with time on the X and progress on the Y axis, where the smooth slope of progress during normal science is interrupted with vertical lines (no time, huge progress) standing for

revolutions. This is certainly not what Kuhn thought reality is like, as exactly between paradigm-shits the meaning of progress itself becomes questionable.

[7] The suggested reading section only mentions Chalmers's *What is this thing called science* and an essay by Feynman.

[8] The IBO was founded in 1968, and has around 1,595 schools (August 2005) in 121 countries (more than 1,300 teach the Diploma Programme to approximately 200,000 students).

[9] http://www.ibo.org/ibo/index.cfm/en/ibo/programmes/prg_dip/prg_dip_cv.

[10] Originally from (Weller 1985), see also http://www.besse.at/sms/smsintro.html.

[11] As students progress in their studies their position is constantly changing, and this change needs to be addressed and explicated, especially in courses that focus on the gradual development of reflective thinking and responsible citizenship.

[12] For a critique see (Jasanoff 2002). Her critique, however, does not hinder the use of the approach in science education. Even if simplified (and here Jasanoff's critique rightly points to a weak spot) the model is one of the few approaches that take into account political factors and the role they play in decision-making, yet appreciate scientific consensus-building as an epistemically different activity.

[13] A result of overestimating the significant aspects of the social is the – at times near desperate – attempt to be able to say "something true" about science. Recall the expert consensus statements that include "truths" like: "Scientific knowledge while durable has a tentative character." (see e.g. McComas et al. 1998). The ambiguity of these statements have been pointed out in science education literature (Good and Shymansky 2001), and called "irresponsibly vague, if not self-contradictory" (Allchin 2004).

References

Alchin, N.: 2003a, *Theory of Knowledge - Teacher's Book*. London: John Murray.

Alchin, N.: 2003b, *Theory of Knowledge*. London: John Murray.

Allchin, D.: 2004, 'Should the Sociology of Science Be Rated X?' *Science & Education* 88:934-946.

Bunge, M.A.: 1982, 'Demarcating Science from Pseudo-Science'. *Fundamenta Scientiae* 3 (3/4):369-88.

Collins, H.M., and R. Evans: 2002, 'The Third Wave of Science Studies: Studies of Expertise and Experience'. *Social Studies of Science* 32 (2):235-296.

van de Lagemaat: 2005, *Theory of Knowledge for the IB diploma,* Cambridge: Cambridge University Press.

Dersken, A.A.: 1993, 'The Seven Sins of Pseudo-Science'. *Journal for General Philosophy of Science* 24 (1):17-42.

Duhem, P.: 1954, *The Aim and Structure of Physical Theory.* Princeton, N.J.: Princeton University Press.

Goldman, A.: 2001, 'Experts: Which Ones Should You Trust?' *Philosophy and Phenomenological Research* 63:85-109.

Good, R., and J. Shymansky: 2001, 'Nature-of-Science Literacy in Benchmarks and Standards: Post-Modern/Relativist or Modern/Realist?' *Science & Education* 10 (1 - 2):173-185.

Hardwig, J.: 1985, 'Epistemic Dependence'. *The Journal of Philosophy* 82:335-349.

Jasanoff, S.: 2002, 'Breaking the Waves in Science Studies'. *Social Studies of Science* 33 (3):389-400.

Kim, J.: 2006, 'Beyond Paradigm: Making Transcultural Connections in a Scientific Translation of Acupuncture'. *Social Science & Medicine* 62:2960-2972.

Laudan, L.: 1982, 'Commentary: Science at the Bar - Causes for Concern'. *Science, Technology, and Human Values* 7 (41):16-19.

Laudan, L.: 1983, 'The Demise of the Demarcation Problem'. In *Physics, Philosophy and Psychoanalysis*, edited by R. Cohen and L. Laudan. Dordrecht: Reidel. pp. 111-128.

McComas, W.F., H. Almazroa, and M.P. Clough: 1998, 'The Nature of Science in Science Education: An Introduction'. *Science & Education* 7 (6):511-532.

McErlean, J.: 2000, *Philosophies of Science: From Foundations to Contemporary Issues*. Belmont, CA: Wadsworth.

Neurath, O.: 1913, 'Die Verirrten des Cartesius und das Auxiliarmotiv (Zur Psychologie des Entschlusses)'. *Jahrbuch der Philosophischen Gesellschaft an der Universität zu Wien* 1913:45-59.

Pakesch, P., and K. Bucher: 2003, *Einbildung: Das Wahrnehmen in Der Kunst*. Köln: Walther König.

Quine, W.v.O: 1951, 'Two Dogmas of Empiricism'. *Philosophical Review* 60:20-43.

Ruse, M.: 1982, 'Creation-Science Is Not Science'. *Science, Technology, and Human Values* 7 (40):72-78.

Scheid, V.: 2002, *Chinese Medicine in Contemporary China: Plurality and Synthesis*. Durham, NC: Duke University Press.

Shapin, S.: 2001, 'How to Be Antiscientific'. In *The One Culture? A Conversation About Science*, edited by J. A. Labinger and H. M. Collins. Chicago: University of Chicago Press. pp. 99-115.

Wade, N.J.: 1998, *A Natural History of Vision*: MIT Press Cambridge, Mass.

Weller, T.: 1985, *Science Made Stupid: How to Discomprehend the World around Us*: Houghton Mifflin Company.

Woolman, M.: 2000, *Ways of Knowing*. Victoria, Australia: IBID Press.

Zemplén, G.Á.: 2007, 'Conflicting Agendas: Critical Thinking Versus Science Education in the International Baccalaureate Theory of Knowledge Course'. *Science and Education* accepted for Volume 16 (2): 167-196.

About the Authors

Douglas Allchin taught high school biology for several years before returning to the University of Chicago, where he received an M.S. in Evolutionary Biology (1991) and a Ph.D. in the Conceptual Foundations of Science (1991). His research now focuses on historical case studies of scientific error and the philosophical lessons they provide. His educational publications include Doing Biology, the "Sacred Bovines" column for American Biology Teacher, and (co-edited) An Introduction to the History of Science in Non-Western Traditions and Proceedings of the Third International Conference on History, Philosophy and Science Teaching. He also edits the SHiPS Resource Center website for teachers (URL: ships.umn.edu). allch001@umn.edu

Jeff Babb is Chair of the Department of Mathematics & Statistics at the University of Winnipeg, Winnipeg, Canada. Earlier, he was a consulting statistician with the University of Manitoba and the Canadian Grain Commission. He has collaborated with researchers and practitioners in many fields and has publications in several journals, including Biometrika, Canadian Journal of Plant Science, Early Childhood Research Quarterly, Journal of the AOAC International, Journal of Cereal Science, Journal of Early Childhood Research, Science & Education, and Theoretical and Applied Climatology. Jeff has taught the History of Calculus course at the University of Winnipeg numerous times. His research interests include history of mathematics, statistical climatology and applications of statistics to agriculture, anthropology, biology chemistry, and childcare. j.babb@uwinnipeg.ca. **James Currie** is Professor of Mathematics at the University of Winnipeg, Winnipeg, Canada. He introduced the History of Calculus course at the University of Winnipeg in 2000. Currently he is designing a companion course, Mathematics Prior to 1640, for delivery in 2007. His major research interest is in combinatorics on words, and his work has appeared in many journals, including American Mathematical Monthly, Discrete Mathematics, Combinatorica, European Journal of Combinatorics, Order and others. j.currie@uwinnipeg.ca

Elizabeth Cavicchi teaches at MIT and in the Honors Program at the University of Massachusetts Boston. She has been a postdoctoral fellow at the Dibner Institute for the History of Science and Technology at MIT. She received the EdD in 1999 from Harvard University, following her prior masters' degrees earned at Harvard, Boston University, and MIT, and her two bachelors' degrees in physics and humanities at MIT. Her publications include "Nineteenth century developments in coiled instruments and experiences with electromagnetic induction", Annals of Science, 2006; "Faraday and Piaget: Experimenting in Relation with the World", Perspectives on Science, 2006; "Exploring Water: Art and Physics in Teaching and Learning with Water", in Facilitating Watershed Management: Fostering Awareness and Stewardship, R. France, ed., Lanham MD: Rowman & Littlefield, 2005, 173-194. Her research of nineteenth century electromagnetic induction devices involves her in recreating historical apparatus and experiments, and in sharing these experiences with students. elizabeth_cavicchi@post.harvard.edu

Peter Heering is currently professorial substitute of the chair for physics didactics at the University Augsburg. He studied physics and chemistry to become a secondary school teacher. After completing his PhD (Das Grundgesetz der Elektrostatik: Experimentelle Replikation, wissenschaftshistorische Analyse und didaktische Konsequenzen) and two years training in secondary school he became lecturer in the physics department of the Carl-von-Ossietzky Universitaet Oldenburg. In 2000, he was Research Fellow at the Bakken Library and Museum of Electricity in Life (Minneapolis), in 2004, he was Scholar in Residence at the Deutsches Museum Munich. In 2006, he completed his habilitation in the field of history of science at the Department Mathematik der Fakultät Mathematik, Informatik und Technik der Universität Hamburg (Experimentierstile in der zweiten Hälfte des 18. Jahrhunderts). His major research fields are the experimental practice of 18th and 19th century physics, which is analysed with the replication method, and the implementation of historical aspects in physics teaching both on university and school level. peter.heering@uni-oldenburg.de

Stephen Klassen teaches in the Department of Physics at the University of Winnipeg. His BSc and MSc were earned in Physics, and his PhD in Science Education. He has presented research at many conferences with special emphasis on including history of science in classroom activities at the college and university level. His present research interests focus on the basis of narrative and learning theories for contextual learning and for pedagogical thought experiments. s.klassen@uwinnipeg.ca

Pannagiotis Kokkotas is an emeritus professor at the Pedagogical Department of the University of Athens and was the coordinator of the Comenius 2.1 project entitled "The MAP project" and now is also coordinator of a new Comenius 2.1 project entitled "STeT project" . Both projects are in History of Science and Science Education.

He teaches "Science Education", "Multimedia (audio, visual etc.) teaching tools" and "Museum Education" to both initial and in-service teachers. His recent books include Science Education I (Athens, 2000), Science Education II - The constructivist approach to teaching and learning science (Athens, 2002). He is also writer of many Greek science textbooks for Primary and Seconday students. He is the Foundation president of the "The Hellenic Union for Science Education (EDIFE)" and the Editor of the Greek journal: Science Education: Research & Practice. kokkotas@primedu.uoa.gr **Katerina Plakitsi** is a lecturer in the Faculty of Education at the University of Ioannina in Greece. She teaches Natural Sciences and also Science Education in Early Childhood. She teaches Museum Education related to Science Education to initial and in-service teachers as well. **Ioannis Vlachos** is a Science teaching Advisor, working at the Secondary Education. He has a PhD in teaching science methodology. He has contributed, as co-author, in the creation of innovative physics textbooks that are used in Greek secondary education since 1983. **Panagiotis Piliouras** is working as a teacher, is a mathematician and he has a PhD in science education. **Katerina Malamitsa** is working as a teacher, and she has a PhD in science education. **Euthymios Stamoulis** and **Mironas Maurogiannakis** are teachers and PhD students in the Pedagogical Department of Primary Education at the University of Athens.

Pierre Lauginie has been a Physics lecturer in Orsay, the scientific centre of Paris-Sud University (France), and a researcher in condensed matter Physics. During the 1990's years, he introduced an experimental approach of Science and History of Science teaching through the use of historical experiments within a "self-directed learning" attempt, involving a close cooperation between physics lecturers and historians of science. Finally, he joined the "Groupe d'Histoire et de Diffusion des Sciences d'Orsay" (GHDSO), in which he continues working from his official retirement three years ago. He is also involved in local popularization of science events.
P. Lauginie has recently contributed to several French and International conferences (see published Proceedings), in the fields: gravitational and electrostatic torsion balances, the "weighing of the Earth" challenge, Foucault pendulum and Earth's motion, velocity of light in its relationship to light models, namely:
— IHPST Conferences: Winnipeg (2003) and Leeds (2005);
— ICHSSE Conferences and workshops: Pognana sul Lario (2003); Kesthely (2004); Oldenburg (2006);
— "Vitesse de la lumière ?" in: Actes du Congrès d'Histoire des Sciences et des Techniques (Société Française d'Histoire des Sciences et des Techniques, Poitiers, 2004),

Cahiers d'Histoire et de Philosophie des Sciences, numéro hors série, Paris, 2006, p. 109-117.
— "De la Mécanique à l'Histoire des Sciences. Expérimenter en Histoire?", in : La pluridisciplinarité dans les enseignements scientifiques: tome 2, Place de l'expérience, CDRP de Basse-Normandie, Caen (France), 2003, p. 15-30. pierre.lauginie@u-psud.fr

Shu-Chiu Liu is currently a post-doctoral research fellow in the research group of Physics Education and History and Philosophy of Science at University of Oldenburg, Germany. She also takes part in the interdisciplinary graduate program "Educational Reconstruction- Teaching and Learning Research" in the Education Center. Her research interests include students' understanding of scientific knowledge, and knowledge restructuring and innovative approaches in science teaching and learning, in particular with a cross-cultural perspective. shu.c.liu@mail.uni-oldenburg.de

Barbara A. McMillan is an assistant professor in the Department of Curriculum, Teaching, and Learning, Faculty of Education, University of Manitoba. She received her B.Sc. (Art Education) at the University of Wisconsin-Madison, and a B.Sc. Honours in Botany, a M.Sc. in Mycology, and Ph.D. in Science Education from the University of Manitoba. Her research interests include Early Years science education, teacher education, and the application of science knowledge, both historical and contemporary, to science teaching. bmcmill@cc.umanitoba.ca

Don Metz, PhD, Associate Professor, Faculty of Education, University of Winnipeg, Winnipeg, MB. Don Metz received his B.Sc. (Physics) from the University of Winnipeg and his M.Ed. and PhD from the University of Manitoba. Don taught physics, math, and computer science in the Collegiate Division of the University of Winnipeg for over twenty years before joining the Faculty of Education program at the University of Winnipeg in 2000. He is the founding president of the Manitoba physics teachers association and the current editor of The Manitoba Science Teacher. Don has been active in curriculum development in Manitoba as the principal writer for the Physics 30S and 40S program. Don's interests include the history and philosophy of science and science teaching. d.metz@uwinnipeg.ca

Daniel Osewold graduated in physics, mathematics and education at the Carl von Ossietzky Universität Oldenburg. After the examination he continued to work in the research group Physics Education/ History and Philosophy of Science in the Physics

Department and was also a member of the graduate school titled "Educational Reconstruction". He completed his PhD thesis on students' conceptions and historical ideas on the wave conception. His main research interests include the investigation and analysis of students' conceptions and the implementation of historical units in secondary physics teaching on the basis of the investigated students' conceptions. Currently he teaches physics and mathematics at a secondary school (Gymnasium) in Germany. daniel.osewold@uni-oldenburg.de

Falk Riess, b. 1944, studied physics, philosophy and pedagogy, professor for physics education and history of physics at Oldenburg University, research interests: relations between science concepts of children and students and the historical development of science; history of experimentation in physics; reworking historical experiments. falk.riess@uni-oldenburg.de

David Wÿss Rudge (B.S. 1986 Duke University; M.S. 1990, M.A. 1992, Ph.D. 1996 University of Pittsburgh) researches how the history and philosophy of science (particularly evolutionary biology) can be used to inform the teaching and learning of science. He holds appointments in both The Department of Biological Sciences and The Mallinson Institute for Science Education at Western Michigan University. Dr. Rudge is a published scientist and philosopher, and is currently writing a book on H.B.D. Kettlewell's famous experiments on industrial melanism, with special attention to how the episode has been portrayed in science textbooks and the popular media. david.rudge@wmich.edu

Christian Sichau is curator for physics, geophysics and surveying at the Deutsches Museum in Munich. He has studied physics at the University of Oldenburg and at King's College London; between 1997 and 1999 he worked at the Technisches Museum Wien. His major research interests are the history of physics in the 19th century with a special focus on Britain (J.P. Joule, W. Thomson, J.C. Maxwell), history of experimenting, history of scientific instruments, historiography, museology, and history of science in science education. c.sichau@deutsches-museum.de

Arthur Stinner is a professor of science education at the Faculty of Education, University of Manitoba. He received his early education in Hungary and Germany, and his secondary and university education in Canada. He holds undergraduate degrees in physics, modern languages and education from the University of Alberta, an MSc in physics from York University, and a doctorate in science education from the University of Toronto. He has taught high school and college physics for over 20 years before moving to the University. His research interests are focused on relating the history and philosophy of science to science teaching, and the development of large context problems in physics. See his website for detail: www.ArthurStinner.com, stinner@ms.umanitoba.ca

Roger H. Stuewer is Professor Emeritus in the Program in the History of Science, Technology, and Medicine at the University of Minnesota. He has written extensively on the history of quantum and nuclear physics, as well as on the use of history of physics in physics teaching. rstuewer@physics.umn.edu

Jürgen Teichmann, Professor at Ludwig-Maximilians-University, Munich, Germany. Retired Director for education of Deutsches Museum, Munich. Studied physics and history. Ph.D. in history of electricity, habilitation in history of solid state physics. Research fields: history of physics and astronomy, the pedagocical use of history of science. j.teichmann@lrz.uni-muenchen.de

Youjun Wang is an associate professor in the department of philosophy in Shanghai Normal University. She graduated from Department for the History & Philosophy of Science, Shanghai Jiaotong University in 2003 and received the Ph.D. for a thesis entitled: "The History of Laplace's Theory of Probability and its Transmission in China" which has been published. She visited Berlin Technology University and Deutsches Museum as visiting scholar. Her research interests are the History of Mathematics and the Education of Science. wangyj@shnu.edu.cn

Gudrun Wolfschmidt, born in Nürnberg, Dissertation "Analyse enger Doppelsternsysteme", Dr. Remeis-Sternwarte Bamberg, Astronomisches Institut der Universität Erlangen-Nürnberg, 1st and 2nd state examination (physics and mathematics), teaching in "Gymnasium" in Bavaria. Since 1987 research in history of science in the Deutsches Museum Munich; conceptual design and realisation of the permanent gallery "Astronomy" in the Deutsches Museum (opening in 1992, catalogue 1993). 1992-1995 scientific assistent at the research institute for history of science and technology of the Deutsches Museum, several exhibitions (e.g. Copernicus in 1994), teaching and habilitation "Genese der Astrophysik" (1997) at the Ludwig-Maximilians-Universität in Munich; since 1997 Professor at the Institute for History of Science of Hamburg

University. Field of research: especially history of astronomy (Early modern time and 19th/20th century) as well as scientific instruments. Books covering the following topics: "Copernicus – Revolutionär wider Willen", "Milchstraße – Nebel – Galaxien" and "Popularisierung der Naturwissenschaften". Schwerpunkt Geschichte der Naturwissenschaften der Universität Hamburg, Bundesstr. 55, D-20146 Hamburg http://www.math.uni-hamburg.de/home/wolfschmidt
wolfschmidt@math.uni-hamburg.de

Gábor Á. Zemplén, after studying Biology, Chemistry, and English Literature, received his PhD at the Department of Philosophy and History of Science (An Eye for Optical Theory - Newton's Rejection of the Modificationist Tradition and Goethe's Modificationist Critique of Newton) summa cum laude from the Technical University, Budapest (BUTE). In 2001-2002 he was invited Assistant Professor at the University of Bern, Switzerland, Institute for Philosophy, Theory of Science and History of Science, teaching a course on classical theories of light and colour. A book on the course appeared recently (The History of Vision, Colour, & Light Theories - Introductions, Texts, Problems. In: Bern Studies in the History and Philosophy of Science. Bern, 2005). He recently (2005-6) received a postdoctoral fellowship from the Max Planck Institute for History of Science, Berlin, and is currently working as a Békésy György postdoctoral Fellow in Budapest, on a book discussing 17th-18th c. optical controversies. His recent articles include: Zemplén, Gábor Á. 2006. The development of the Neurath-principle: unearthing the Romantic link. Studies in History and Philosophy of Science A 37 (4):585-609. Zemplén, Gábor Á. 2007. Conflicting Agendas: Critical Thinking versus Science Education in the International Baccalaureate Theory of Knowledge Course. Science and Education 16:167-196. zemplen@filozofia.bme.hu